Anders Bengtsson

Higher Spin Field Theory

Texts and Monographs in Theoretical Physics

Edited by
Michael Efroimsky, Bethesda, Maryland, USA
Leonard Gamberg, Reading, Pennsylvania, USA

Anders Bengtsson

Higher Spin Field Theory

Volume 1: Free Theory

DE GRUYTER

Physics and Astronomy Classification 2010
52.30.-q, 52.30.Cv, 52.80.-s, 02.70.-c

Author
Dr. Anders Bengtsson
Akademin för textil, teknik och ekonomi
Högskolan i Boras
SE-501 90 Boras
Sweden
Anders.Bengtsson@hb.se

ISBN 978-3-11-045053-8
e-ISBN (PDF) 978-3-11-045177-1
e-ISBN (EPUB) 978-3-11-045059-0
ISSN 2627-3934

Library of Congress Control Number: 2020934839

Bibliographic information published by the Deutsche Nationalbibliothek
The Deutsche Nationalbibliothek lists this publication in the Deutsche Nationalbibliografie; detailed bibliographic data are available on the Internet at http://dnb.dnb.de.

Cover image: Raycat / iStock / Getty Images Plus
Typesetting: VTeX UAB, Lithuania
Printing and binding: CPI books GmbH, Leck

www.degruyter.com

To Marianne, Olof, Erik and Monika

Preface

No one knows at the present time if higher spin gauge fields have anything to do with reality. But the subject is interesting in itself, and – as its founder P. A. M. Dirac wrote in 1936 – it might be good to have a theory at hand if they do turn out relevant for fundamental physics [1].[1] Dirac studied massive fields, and it took another 40 years for massless higher spin gauge fields to come into focus with the work of C. Fronsdal in 1978 [3]. The subject is now in its ninth decade. This is a first book in a planned two volume project aimed at trying to cover some parts of this fascinating subject.

Origin of the subject

The theory of higher spin fields dates back to the beginnings of quantum field theory in the 1930s. For a long time, the focus was on massive fields, as such could describe massive matter particles with spin. From Wigner's classification of the representations of the Poincaré group in 1939, it was known that there are massless and massive representations of integer and half-integer spin [4]. But experimentally there was no need to go beyond spin 1/2 until the end of 1950s when massive higher spin resonances were found in strong interaction physics.[2] Not until the 1970s were massless fields studied theoretically (except being mentioned in passing) – with one notable exception – when the gauge theory of free massless fields was constructed by C. Fronsdal [3] and J. Fang and Fronsdal [6]. The exception was S. Weinberg's S-matrix argument [7] from 1963, showing that massless fields of spin greater than 2 cannot generate long-range forces. And experimentally such fields were not – and still are not – seen. The history of this fascinating subject will have its own chapter in the present book.

Free fields are not so interesting in themselves. The standard recipe for interactions followed in many early references was to try to couple massive higher spin fields to electromagnetism and gravity, or later, to couple conserved matter currents to massless higher spin fields.[3] It did not work out well, as we will have occasion to study in detail. The investigation of self-interactions came even later with the work of Fang and Fronsdal, who explicitly formulated the research program of finding higher spin interactions as deformations of free field theories [8]. This program – designated as *the generalized Gupta program* (after S. Gupta, who was one of the pioneers of this approach to gravity) – was a generalization of earlier work to construct gravitational interactions in that manner.

If a starting time for positive results in the study of massless higher spin self-interactions can be found, it is presumably in the early 1980s with the work of

1 There is actually what can be considered as a co-founder of the subject: E. Majorana with the paper [2] from 1932, pre-dating Dirac.

2 See, for instance, [5], Chapter 21.

3 It can be argued that coupling to gravity is an absolute requirement so that having free higher spin fields would be impossible in principle.

https://doi.org/10.1515/9783110451771-201

I. Bengtsson, L. Brink and myself, of F. A. Berends, G. J. H. Burgers and H. van Dam and of E. S. Fradkin and M. Vasiliev. But I think it is safe to say that during most of its history, the theory of higher spin fields, in particular massless fields, has been considered to be very difficult, fraught with consistency problems, experimentally not relevant and perhaps even a totally misguided endeavor. Still a small number of researchers pursued the subject up until the first years into the new millennium when new and young researchers were attracted to it. In the mid 2010s, it was a thriving subject with an established community of researchers working on it. A slight cooling off may be observed as we move into the 2020s. Much has been done during the last 35 years, but the subject is still far from closed. What will happen in the future, we have to wait and see. The basic higher spin problem – researching consistent interactions and investigating their role in nature – is one of those problems that are relatively easy to state, yet so hard to solve, and therefore allures the brave, or foolhardy.

The present state of the subject

The present state of the subject offers a strange picture. The Vasiliev theory is a background independent formulation of higher spin theory that can be expanded around an anti-de Sitter background (AdS), but apparently not around a Minkowski background.[4] It partly solves the interaction problem by the so-called "gauging" approach which works very well for spin 1, can be made to work for spin 2, but for higher spin forces a number "workarounds" that threatens to remove – in my opinion – the resulting theoretical construction from the basic intuition of the original problem. This is of course a rather common fate to fall upon fundamental questions in theoretical physics, but here it is aggravated by the lack of experimental input. Barring that there may be data, perhaps cosmological, that we do not yet interpret as pertaining to higher spin, there is no phenomenological guidance, except the non-occurrence of higher spin gauge fields at presently attainable energy scales.

For quite a long time, the interaction problem could be considered to be essentially solved in AdS. The problems with expanding the theory around Minkowski space was taken as an indication – sometimes even as a proof – that interacting higher spin gauge theories do not exist in flat space-time. The dominating approach in Minkowski space, the "deformation" approach, also called the *Noether procedure* or the *Fronsdal program,*[5] has been very slow in producing positive results, instead leading to severe difficulties at the quartic level of interaction. It should be noted, though, that the gauging approach does not escape the need for deformation.[6]

4 "AdS" is often used as a shorthand for 'Anti-de Sitter space-time'. "dS" is used for "de Sitter".

5 What Fang and Fronsdal called the "generalized Gupta program".

6 Gauging is, in a sense that will be clarified, basically a kinematical procedure of making a global symmetry local. It remains to find the self-interactions of the introduced gauge fields. Here, one is helped by the gauge algebra, but it does not provide the full dynamics.

From another point of view, particle physics is done in flat background, and coming from this environment, it would seem natural to look for higher spin theory in flat space. But to get to higher spin, one must first pass spin 2 – that is gravity – which is naturally interpreted as a curved space-time theory. So it can be argued that flat space higher spin theory is not very natural to consider. On the other hand, gravity can be viewed as a highly nonlinear theory of spin 2 fields, and that theory is not likely to go unaffected by an eventual higher spin theory. Phenomenologically, it seems that we are living in a de Sitter universe, rather than an anti-de Sitter universe. This fact confounds the question even more, if one has hopes for finding a role for higher spin gauge fields in nature. We just do not know enough.

Many readers may have heard that there are quite a few no-go results that rule out either the existence, or at least the relevance, of higher spin theories. Of lately, new difficulties having to do with locality issues has been discovered, both in the Minkowski and the AdS approaches to the theory. These issues are still researched at frontiers of the subject. These problems, and the no-go results, concern the theory of interacting higher spin fields, and belong to a planned second volume of the present work. In the present work, we will mainly treat the free, noninteracting theory, which is of course a prerequisite for any attempt at studying interactions.

A personal note

On a personal note, in the very late 1980s, after having worked on higher spin theory since my graduate studies, I was discouraged by the difficulties, and this coupled to a lack of new ideas and a wish to spend my time with my children as well as pursuing other intellectual interests, made me temporarily withdraw from the subject. This explains why I did not write a single paper for 14 years. I was not completely idle though, doing calculations for the drawer.[7] Then in 2003, I saw papers "rediscovering" what I had done in 1986–1988, apparently unaware of my work. Of this, I was irritated, but also glad to see the new interest and I decided to take up the subject again. I was very happy to be able to make a come-back.[8] When some years later, I started to meet the new researchers in the field, I felt very welcomed back, in particular by Mikhail Vasiliev, who I had actually met only once before, at Ingemar Bengtsson's place in Göteborg some time in the late 1980s.

There were a few exceptions to the relative lack of interest during the 1990s, apart, of course, from M. A. Vasiliev's own work on the AdS formulation of higher spin theory. One is the work of E. Sezgin and P. Sundell [10] starting in 1998. Higher spin excitations also did turn up, and were discussed, in the context of membrane theory and infinite

7 I tried to do BRST theory for singletons in AdS space-time, in that way hoping to find another approach to higher spin fields. This was inspired by the Christian Fronsdal paper [9].

8 This was to some extent helped by the presence of the arXive and TeX-ing, that made "home research" feasible. The engineering school I was working at had no tradition at all in theoretical physics.

dimensional algebras, but they were not the main subject of focused interest. This was the high tide of superstring theory, and higher spin gauge field theory was definitely at the fringe of theoretical physics. But strangely enough, massless higher spin fields were to turn up in string theory in connection to the AdS/CFT conjectures. That gave a boost to the interest in the Vasiliev theory.[9]

Now, with the subject fairly well established as an interesting part of theoretical physics to pursue, it may be a good time to try and condense and explicate parts of it in book form. The present book came about by a very unlikely – and funny – coincidence. In January 2015, I was at the AMS/MMA Joint Mathematics Meeting in San Antonio, Texas, and gave a presentation on a calculus textbook I was writing with a colleague. When browsing the exhibition hall where the publishers were showing their books, I happened upon the De Gruyter stand. I was leafing through a book when I was approached by Konrad Kieling who was reading my name tag. Now "Bengtsson" is a quite common name in Sweden but not so internationally. Konrad asked be if I knew Ingemar Bengtsson, which I of course did since we were working on higher spins in the 1980s as graduate students and post-docs. This lead to talk about how Konrad knew about Ingemar and about what I was doing and eventually to the question if I had ever thought about writing a book about higher spin. The idea had actually crossed my mind – at least in the form of a review article – but it was very far from being realized. My guarded answer was "yes". After this encounter, the idea started to take on concrete form in the course of an exchange of emails during the spring of 2015. I must take the opportunity to thank my (book writing) colleague and professor in mathematics, Dragu Atanasiu, for prompting me to go to San Antonio.

Audience

I have tried to make this book accessible to graduate students who have had standard courses in classical mechanics, special and general relativity and quantum mechanics and some quantum field theory.[10] Although massless higher spin fields and particles are yet to find their proper place in a fundamental description of nature, I have tried nevertheless to connect the subject to basic physical intuition.

When planning the book, I played the game of imagining how some other generic author might write a book on higher spin, and then tried to deviate from that imagined book. In this way, my text would not compete with other eventual texts, but rather be a complement to them. This is my basic philosophy of doing research and writing: I'm not that interested in redoing what others have already done or written. I guess in the end, the book I'm presenting here simply reflects my understanding and point of

9 The subjects mentioned in this paragraph: membrane theory, infinite dimensional algebras, superstring theory, AdS/CFT conjectures and the Vasiliev theory, are all very extensive. I refrain from providing any fair subset of references at this point.

10 Roughly where I was myself when getting started on higher spins in 1983.

view (and to be honest, lack of knowledge) of this very fascinating subject. However, inevitably much of the material is of course standard for the subject – as is appropriate – but hopefully the approach and emphasis is a bit different. I hope the book contains much of what you need to know to get started on the subject. Apart from technical skills, you will need ideas.

Contents

The book was initially planned as one volume, but due to the size of the subject, and the time it takes to write about it, the book is divided into two volumes: the free theory in the present Volume 1, and the interacting theory in the planned Volume 2. Higher spin theory has indeed become a very large subject with many aspects and approaches. Even if I had the knowledge and understanding to write about it all, even two volumes would hardly cover it. Rather than trying to write a full review of the whole subject, I have tried to focus on a number of parts of the subject, treating them quite thoroughly. The guiding principle has been to write a text that is concrete and computational, physical and intuitive and conceptual and abstract. The present Volume 1 has the following contents.

Chapter 1 is an introduction. I try to motivate the subject and put it in context, as well as introducing the basic conventions and notation used.

Chapter 2 is historical. I think it is a bit more detailed than has been written on the subject before. I often find historical comments on an area of research both interesting in themselves and helpful for understanding. I hope it will be useful to the reader, too.

Chapter 3 is collection of background knowledge that is essential for working in higher spin theory. Much of the material in this chapter will be used first in Volume 2 on interactions.

Chapter 4 reviews knowledge about the well established theories of spin 1 and spin 2. However, it does so from the perspective of trying to take advantage of such knowledge when approaching the higher spin problem.

In Chapter 5, we come to higher spin theory itself. Here, I treat the free field theory in its various formulations. The focus is on the Minkowski space-time theory.

In Chapter 6, we develop the basics of the light-front approach to massless higher spin fields. Again, as a preparation for a review of what is known about higher spin interactions on the light-front. With this content, the problems of interacting higher spin fields can be discussed in Volume 2.

Referencing

The number of papers in some way pertaining to higher spin theory may soon run into the thousands, if they have not already done so. I may have seen many of them, but of those I've seen I must admit I have not read them all, and certainly not worked through all the details.

I've tried to reference original work as fairly as possible. The position of the subject at the outskirts of theoretical physics has, however, had the effect that ideas and methods have been rediscovered over and over again from different perspectives and in different guises – ostensibly unaware of earlier work,[11] making back-tracking particularly frustrating. Of topics outside higher spin proper, my referencing has been guided by the wish to help the reader to find relevant information in textbooks and reviews, mostly the ones I have found useful myself.

How the book is written

As many authors have said before: it may be that I wrote the book that I would myself have liked to read when I started out in the subject. Another philosophy that has guided me is something Ingemar Bengtsson said when we were graduate students together: "A good book must have a point of view.". This I agree with. In my opinion, the readable books are often the ones which are based on a few basic unifying ideas that are pursued – not single-mindedly – but pragmatically. This is not an easy ideal to live up to, specially not in a subject like the present one that is far from mature and that has recently undergone rapid development in various directions by many researchers employing different methods, techniques and formalisms. To review all this material, I find out of my reach. There are many different models depending on space-time dimension, back-ground geometry and symmetry groups. This generality – which is of course interesting in itself since it corresponds to a desire to map out the terrain of possible higher spin theories – easily becomes bewildering. The subject tends to look like botany.[12] It is implicit in my way of looking at the subject, that there are other points of view that one can stress. These you will certainly find in other texts.

To write is therefore as much about what not to include as to what to include. It is only natural if an author wants to express his/her knowledge of the subject (while in the preface perhaps humbly professing to limited knowledge, as I have already done). One often feels that by leaving things out, one does harm to the subject. However, every student of theoretical physics understands that there are more examples of exact solutions to Newton's equation than a particle in a constant force field or in a harmonic potential, but perhaps one does not want to read about them all, at least not just for now.

I have employed another strategy. Almost everything is done in four space-time dimensions and for small symmetry groups and algebras. Thus rather than using my allotted space to listing and classifying various models, I will use the space to try to explain things thoroughly, following a few lines of thought. I will take a walk through the terrain rather than surveying it. I also adhere to the philosophy that formulas do

11 A fault, which I have myself been guilty of: having seen, even read, but forgot.

12 My apologies to the real world botanists.

not speak for themselves, there must be a story surrounding them.[13] I have tried to provide such a story for higher spin theory.[14]

Higher spin field theory exploits quite a few tools from mathematics. Although I had from the outset planned a chapter on such things, it eventually grew to one of the longest. I anyway had to work things through to get signs and factors right, and fix notation. For the expert, the contents of this chapter may seem well known or even trivial. I am not quite sure it is so for every nonexpert reader that picks up the book, or for that matter, to newcomers and graduate students. Explaining well-known things in a separate chapter, allows for a shorter and more succinct treatment of the higher spin theory itself. I also found that writing down some of these things was useful in itself, in that hidden assumptions and glossed over details, came into focus.[15]

Another aspect of the large set of mathematical tools used in the subject, is the occurrence of different notational systems, often confusing, sometimes even conflicting. I have tried to streamline notation, at least to harmonize it, but also kept parallel notational systems (the ones that seem most commonly employed). I will explicate this in Section 1.4, in the historical chapter and in Chapter 3.

As an aside, I would actually say that devising formalism is a neglected part of higher spin theory. Having studied computer science for some years, I came to think in terms of *objects* and *processes*. In theoretical physics, objects can be fields, processes can be transformations. A symbol for an object needs to carry enough information for it to be able to – in the circumstance – convey, perhaps not all, but its salient properties. This may vary with the context. A process must be symbolized with enough information for the reader to be able to carry through the intended computations reliably. As we all know, this is no easy task. What works well in the personal notebook, may not work so well in print.

An example of a very well devised formalism is Leibniz's formalism for the differential and integral calculus. This is a formalism that computes almost by itself. It

13 In the hilariously funny book on mathematics teaching [11], M. Kline writes: "Many authors seem to believe that symbols express ideas that words cannot. But the symbolism is invented by human beings to express their thoughts. The symbols cannot transcend the thoughts. Hence, the thoughts should first be stated and then the symbolic version might be introduced where symbols are really expeditious. Instead, one finds masses of symbols and little verbal expression of the underlying thought."

14 Regarding the question of the number of dimensions of space-time and size of groups, let me be honest. Limiting myself to $D = 4$ and small groups is not just a matter of limiting the scope of the subject. At a deeper level, it is, and has always been, my contention that the world is actually four-dimensional. Fascinating as the subjects of higher dimensions, supersymmetry and large groups undoubtedly are, they have never really spoken to me. My fascination is focused on four dimensions and small groups where I feel there might be depths still not investigated.

15 Let me also risk a quite personal opinion. In Yang–Mills theory, and even gravity, it seems that one can get away with rather sloppy concepts, since one is corrected by – if not reality – by accepted canon. This will not do in higher spin theory. Clear understanding of fundamental concepts of theoretical physics is needed. That is not the same thing as mathematical rigor.

does so at the price of not completely hiding, but certainly circumventing, the subtleties of analysis. This may bother a mathematician, but is generally no great issue with a theoretical physicist. That is instead what we like about good formalisms. Leibniz's construction of his formalism for the calculus may have been a stroke of genius or inspiration, but certainly not of good luck. As the story goes, Leibniz spent a lot of time on thinking about effective and transparent formalisms in mathematics and philosophy [12].[16]

A physics example of a very well devised notational system is Dirac's bra and ket notation for quantum mechanics, introduced in 1939 in [13]. A quote from the first paragraph of the paper says it all.

> In mathematical theories, the question of notation, while not of primary importance, is yet worthy of careful consideration, since a good notation can be of great value in helping the development of a theory, by making it easy to write down those quantities or combinations of quantities that are important, and difficult or impossible to write down those that are unimportant.

In research articles, and even in review articles, it is only natural to hurry toward what is new, and focus on recent developments. In a book, even if the subject is extensive, one should have the freedom to dwell on the basics of the subject. If not there, where else?

Many books in theoretical physics make a point about developing the subject logically rather than historically. That is a good point, but it is one-sided. Science is a human endeavor, and history is interesting in itself, and it often – or at least sometimes – can shed light on the logic of the subject. Perhaps striking a balance at 25 % history to 75 % logic, is more productive than a balance 5 % to 95 %.

For all these reasons, the book is therefore to a large extent, a book in the "re-thinking" tradition. And I must admit, as the author, a "relearning" experience. For the reader who finds the treatment unsystematic: think of it as an exploration of unknown territory, rather than as designing a garden.

? Some people like exercises and they may be a crucial part of learning a subject. Myself, I don't, always having preferred to choose for myself what to work through and what to trust. The reader of this book will have to do that, too. There are no regular exercises. But there are questions, often of a conceptual rather that technical nature, in certain places in the text. Sometimes tentative answers are given. Sometimes I may not even know – at the time of writing – the answer myself. Other questions may relate to obscure, but interesting, passages in papers. Still others to reproducing results in historical papers. There are cases where I felt a section risked to develop into tedious detail, deflecting from the main goal, so I relegated some material to a question. Questions of these kinds are marked as shown here. I hope readers who enjoy exercises will appreciate these questions. Sometimes the reader has to figure out for herself what the question is.

---

16 Lars Brink used to say: "The formalism is smarter than we are.". In higher spin theory, a smart formalism would certainly be of help.

Acknowledgments
First of all, I would like to thank my good friends, Gunnar Orrskog and Bobo Ohlsson, who encouraged me to go for early retirement. Upon getting the book writing contract with De Gruyter, I seriously considered that option. Their example made the decision an easy one.

The book is of course a result of years after years of thinking about, and working on higher spin theory. No doubt conference talks I've listened to, conversations – both real and digital – with researchers in the field, and of course paper reading, have contributed to the contents. In particular, I would like to thank Bo Sundborg and Per Sundell for discussions throughout the years and not dismissing my sometimes somewhat orthogonal views on the subject. I have also benefited from discussions with D. Ponomarev and E. Skvortsov when we have met at various conferences during the writing. Thanks also to N. Boulanger and R. Bonezzi during an informal workshop in Mons in January 2018. The same to E. Skvortsov and K. Mkrtchyan for the invitation to the very good Potsdam workshop in December 2018.

During the writing, I have been helped by e-mail conversations with S. Deser, S. Ouvry, M. Tsulaia and G. Barnich. Christian Fronsdal has kindly answered questions and read a draft of the book. So has Lars Brink done, saving me from some embarrassing errors. Ingemar Bengtsson read a draft, found quite a few errors, but otherwise wrote encouraging comments! Remaining errors are of course my responsibility.

Furthermore, I would like to thank the librarians at the University College of Borås for the efficient help with acquiring references, even after having retired from my teaching position at the college. My former co-author and colleague Mats Desaix picked them up and mailed them to me. I found several old books on theoretical physics in the antiquarian bookshop Faust in Göteborg. One of them was the Corson book which turned out to be a gold mine when writing parts of the historical chapter.

The financial support from Stiftelsen Längmanska kulturfonden has been very helpful.

Contents

1 Introduction and motivation

The description and explanation of physical reality puts very heavy burdens on mathematics. Physics is not, not even theoretical physics, a deductive mathematical science. Experiment or observation must eventually, in the last analysis, decide what is true or not, no matter how beautiful principles may be involved. Of course, physics is a pragmatic science, and fundamentally discarded theories – such as Newtonian mechanics – are still used wherever they are applicable, not the least in technology, and they may even be axiomatized or be given a neat mathematical formulation.

Now and then, new powerful principles subsume and explain large tracts of physics, and this is part of the beauty of physics. But grand principles tend to clash with each other. The "quantum principle" and the "relativity principle" are not easily united. Where they meet, there is dissonance, and quantum mechanics and special relativity are almost incompatible [14], but have been reconciled in the *theory of quantum fields*, of which the observationally very successful *Standard Model* is a particular example. "Quantum gravity" – an umbrella denotation for quite a few attempts at reconciling "the quantum principle" with the "general relativity principle" – is still an unsolved theoretical problem.

Attempts to axiomatize, or even to put large parts of theoretical physics on firmer ground, often run into problems, as the example of "axiomatic field theory" has shown. This is not to say that one should not try, if one finds such a line of research interesting. But is seems that we often have to make do with robust, and practically working mathematical formulations, such as renormalized quantum Yang–Mills theory.

Higher spin theory – the very subject of this book – seems to be in several awkward conflicts with the grand principles of theoretical physics, even though the theory is superficially easy to couch in the language and formalisms of relativity and quantum mechanics. Most physicists take these conflicts as clear indications that higher spin gauge theories are completely out of the picture as regards fundamental understanding of physical reality. A few hope that higher spin gauge fields will eventually help reconcile the grand principles. As the author of this book, I shall remain neutral on this very question.

Having stated this piece of basic philosophy behind the book, let us continue towards our particular subject. In this first chapter, I will introduce the problem and motivate why it is interesting to study. Let us begin by playing the game how we could explain our field of study to an educated person who is not a physicist.

1.1 Spin

Physics is about describing systems and explaining the changes they undergo. This is *motion* or *dynamics*. But in order to change – in order to move – there must be something that has a freedom to move. What is moving? We call it *degrees of freedom*. These are variables that describe the *states* of the *system*.

https://doi.org/10.1515/9783110451771-001

The simplest thing that can move is a structureless particle, what we call a *point particle*. We describe it mathematically by a zero-dimensional point. Such a point particle is thought of as moving in a space of higher dimension, in reality in three dimensions. It thus has at least 3 degrees of freedom since it must be somewhere in *space*. This somewhere is its *position*. But it moves and in order to describe the motion we introduce the concept of *velocity*. Since motion can be in three independent directions of space we get another 3 degrees of freedom.

However, the introduction of velocity requires the concept of *time*. Or rather, the notion of a particle moving requires the notion of a *function* describing the position of the particle depending on some *parameter*. This is the *trajectory* or *history* of the particle. Is this new parameter – time – a new degree of freedom or what is it?

And what about the velocity itself? Is it constant or is it changing? Do we need still more degrees of freedom corresponding to what is normally designated by *acceleration*? Since Newton, the answer is almost exclusively, a no. The acceleration is determined by the *force* acting on the particle. This gives us the *equation of motion* that allows us in simple cases to compute the motion of the particle. We are doing *classical mechanics*, as we say. In the process, velocity is refined into two concepts of quantity of motion, *momentum* and *kinetic energy*. And the question arises: what are forces?

Thinking in another direction, what happens when the particle is not without structure? A system of two similar particles would have $2 \times 6 = 12$ degrees of freedom. But if the particles are connected to each other in some way, perhaps by a thin rigid rod, the number of degrees of freedom (d. o. f.) may be less than 12. We can think of the particles as forming a dumbbell shaped system and the rigid rod we can abstract to a constraint on their motion so that their relative distance is constant. If the particles are similar, the motion can be described by the motion of the midpoint on the distance between them – *the center of motion* – and a rotation around the midpoint. Even if the relative distance between the particles can change, rotation of the two-particle system still seems to be an interesting concept. The change of relative position could be a *vibration*. It becomes interesting to figure out the number of degrees of freedom and their nature.

As soon as we consider more than one particle, *rotation* becomes an inevitably physical concept to understand. We analyze it using the related concepts of *angular velocity* and *angular momentum*. But so far, everything said has been within Newtonian physics. Relativity forces modifications regarding time, for instance about whether we think of time as a parameter or a degree of freedom, or if it is both depending on context and point of view. Relativity also forces modifications to the notion of space itself.

When we move beyond classical physics, on to *quantum physics*, there seems to be a need for a dramatic shift of ontology[1] and epistemology.[2] At least, new concepts

1 The nature of reality.

2 The nature of our knowledge about reality.

and theoretical structures arise. In particular, a new type of degree of freedom, sharing properties with rotation, emerges: the concept of *spin*. It is independent of space-time angular momentum and cannot be reduced to it, yet it emerges – theoretically – from a union of quantum mechanics with special relativity. Experimentally, its presence was first seen in the spectrum of the simplest of atoms, the hydrogen atom. It is a truly quantum phenomenon and concept. Mathematically, it is inherent in the *Poincaré group*, the group of symmetries of special relativistic space-time. Quantum relativity predicts that spin is a degree of freedom that can take any one of the values $0, 1/2, 1, 3/2, 2, 5/2, 3, \ldots$ continuing up to infinity.[3]

At this point, we may have lost our nonphysicist audience. Let us cut short by saying that electromagnetic and nuclear forces are connected to spin 1 while gravitational forces are connected to spin 2. What this means is that these forces are mediated by fields or particles that are themselves carrying the spin degree of freedom; spin 1 and spin 2, respectively. The question may arise: what is the role of spin $3, 4, 5, \ldots$? This is indeed a very natural theoretical question to ask. Fundamental matter particles, such as the electron, are described by fields of spin 1/2. Spin 3/2 fields occur in theoretical extensions to gravity called supergravity theories. The Poincaré group of space-time symmetry have representations of all integer and half-integer spin. What is the role, if any, played by this infinite spectrum of particles and fields?

This book is therefore about one set of aspects of this intriguing concept of spin, namely the theory of *higher spin gauge fields*, where the word "gauge" – to be explained in the following – denotes a concept that is central to the whole subject.[4] Experimentally mapped-out nature, as it is captured by the fundamental equations of theoretical physics, however, makes do with low values of spin. If this situation persists – and there are no indications that it will change in the near future – the theory of higher spin may have little to do with fundamental matter and forces. Then it may just be a piece of interesting mathematics. If it does play a physical role, then it might be a signal from the deep structure of reality. We just do not know.

Degrees of freedom in general, and the electromagnetic field in particular

i

In the context of field theory (in dimension $D = 4$), the six degrees of freedom (d. o. f.)[5] of a structureless point particle: the position $\mathbf{x}$ and momentum $\mathbf{p}$, count as just one scalar field ϕ degree of freedom with $\phi(t, \mathbf{x})|_{t=t_0}$ and $\partial_t\phi(t, \mathbf{x})|_{t=t_0}$ at some initial time t_0. When confusion might arise, we may write *particle* d. o. f. and *field* d. o. f., respectively. But in most cases, context will show what kinds of d. o. f. are implied. The description of a field with $\phi(t, \mathbf{x})$ is called *configuration space* description.

3 For the reader who remembers from school physics that a physical quantity must be measured in some "unit", spin is measured in the unit "energy·time", that is, Js in SI-units.

4 Model railroaders will recognize the word "gauge" as the same word as in the measure of railroad track gauges – the inner distance between rails – both in reality and in the model.

5 The abbreviation d. o. f. will be used in both the singular and the plural depending on context.

This, however, shows that a field d. o. f. actually corresponds to two initial values: one for the field itself, and one for its time derivative. Therefore, counting in phase space where the time derivative of the field is traded for the field momentum, one could also say that a scalar field carries two phase space d. o. f., namely the field itself ϕ and its conjugate momentum $\pi \sim \partial_t \phi$. The spin degree of freedom is almost always discrete – running over a finite set of values – and it is often represented with a label on the field, for instance as ϕ_σ.

The electromagnetic field is a particularly interesting example. It can be described by an electric field $\mathbf{E}$ and a magnetic field $\mathbf{B}$, both being three-dimensional vectors. They therefore together constitute six field degrees of freedom. Are these configuration space d. o. f. or phase space d. o. f.? This innocent looking question immediately throws us into the depths of our subject.

The electromagnetic field can also be described by the vector potential $\mathbf{A}$ and the Coulomb potential $\Phi = A^0$. The relation between the two descriptions is given by $\mathbf{E} = \partial_t \mathbf{A} + \nabla\Phi$ and $\mathbf{B} = \nabla \times \mathbf{A}$. The first of these equations gives a hint that the electric field $\mathbf{E}$ should be counted as a field momentum d. o. f., while the second is consistent with taking the magnetic field $\mathbf{B}$ to be a field d. o. f., and we have six phase space degrees of freedom.

However, there are four partial differential equations connecting the electric and magnetic fields – the Maxwell equations – and these relations reduce the number of degrees of freedom. Two of the equations, namely $\nabla \cdot \mathbf{E} = \rho$ and $\nabla \times \mathbf{B} - \partial_t \mathbf{E} = \mathbf{j}$ (where ρ and $\mathbf{j}$ are the electric charge and current densities), reduce the number of phase space d. o. f. from six to four. The other two Maxwell equations, namely $\nabla \cdot \mathbf{B} = 0$ and $\nabla \times \mathbf{E} - \partial_t \mathbf{B} = 0$, just reduce to the mathematical identities $\nabla \cdot (\nabla \times \mathbf{A}) = 0$ and $\nabla \times (\nabla\Phi) = 0$. Alternatively, these two equations can be thought of as allowing us to express the electric and magnetic fields in terms of the vector potential and the Coulomb potential as we have done above. Identities of this type are called *Bianchi identities*.

Counting in configuration space, the electromagnetic field thus carry two field degrees of freedom. These d. o. f. correspond to the two polarizations of light. Alternatively, one says that the *photon* – the massless spin 1 particle of light – carry two field degrees of freedom. It will turn out that all massless particles, regardless of their spin, will carry two field degrees of freedom in three-dimensional space.

1.2 Quantum mechanics and relativity

The last 5 years of the nineteenth century saw dramatic new discoveries in physics: X-rays, the electron, the Zeeman effect and radioactivity, to name a few of the most prominent. Together with unsolved puzzles like the ultraviolet catastrophe of black-body radiation, the photoelectric effect and the absence of an electromagnetic ether for light to propagate in, they eventually lead to a fundamentally new understanding of the laws of physics in terms of quantum mechanics and special relativity. General relativity, however, was not prepared by any widely felt crisis in physics.[6]

Quantum mechanics is often thought of, not as a physical theory per se, but rather as a scheme that any theory of fundamental physics must conform to. The scientific understanding, and indeed everyday experience – if contemplated – has led to a thinking about reality as being about states of systems and processes transforming such

6 This history is told in many places, a few of which are reference [5] which tells the “physics” story, reference [15] which tells the “physicist” story and reference [16] tells the “conceptual” story.

systems. The states are in physical sciences described in mathematical language in terms of variables and functions, and the processes as transformations of the states and equations relating them.[7] This is very clear from Paul Dirac's formulation of quantum mechanics [17]. As such, the description of states and processes in quantum mechanics is different from that in classical mechanics.

Special relativity, on the other hand, is a classically formulated theory of particles and fields in space-time and of space-time transformations. One can say that when special relativity is formulated quantum mechanically, what results is quantum field theory. Aspects of this topic will be reviewed in the following Chapters 2 and 3.

1.3 The standard model and general relativity

The conceptual difference between special and general relativity is much greater than the names of the theories – or the mathematical formalisms – suggest. Whereas special relativity is a static theory of a background geometry of space-time, general relativity is a dynamical theory of the gravitational field where the field and the space-time geometry is intrinsically related. Perhaps then it is not such a surprise that a quantum theory of gravity still does not exist. As a classical theory, gravity is well understood, and although there are conceptual problems, there are no mathematical problems with the formulation of the theory.

The *Standard Model* of particle physics is the result of experimental investigations and theoretical developments from the 1930s up to the end of the 1970s. In retrospect, this is a very short time for such a successful theoretical understanding of fundamental matter and forces to be worked out. Verification of crucial predictions of the model, such as the existence of the vector bosons of the weak interactions, properties of the strong interactions and the discovery of the Higgs particle, was achieved in the period from the 1980s to the near present.

But the standard model neglects the effects of gravity, for the simple reason that these play no measurable role for elementary particles at the energies that can be reached in the laboratory, now and in the foreseeable future. The general consensus among theoreticians is however that this state of affairs is unsatisfactory.[8]

---

7 This thinking is also inherent in modern computer science. One way to confront fundamental puzzles in physics in a new way could be to challenge this thinking, which has its expression in language with the noun-adjective-verb structure.

8 As an aside, one could ask the question: what does the Standard Model actually achieve? It explains the new phenomena discovered and investigated around the turn of the nineteenth and twentieth centuries: atomic spectra, cathode rays, radioactivity, etc.. In order to do that, it was needed to probe nature at smaller scales, in the process discovering even more phenomena that were also explained as the theory developed. In short, there was a substructure to atomic and nuclear physics, now explained, or at least described, by the Standard Model. It is an intriguing question whether there are more new

1.4 Basic notation and conventions

It is impossible to devise a system of notation and conventions that is convenient in all circumstances. Choices as to what alphabets and what parts of alphabets to use for different objects and indices on objects, may seem trivial (and often seem to be treated as trivial) but are not so if one cares about aesthetics, typography and readability, while of course also being a matter of taste. The problem is furthermore aggravated by the fact that the degree to which the choice of factors of $1/2$, $\sqrt{2}$, π, i, signature of the metric, etc., varies from the trivial to the deep. Here, we will record some choices that will be adhered to throughout. They should be sufficient to read the historical chapter. Complete conventions will be developed in Chapter 3.

Minkowski space-time coordinates are denoted by x^μ with $t = x^0$ and $\mathbf{x} = (x^1, x^2, x^3)$ and the corresponding momenta by p_μ with $E = p_0$ and $\mathbf{p} = (p_1, p_2, p_3)$. We will choose a mostly plus metric $\eta = (- + ++)$. Unless otherwise stated, the dimension of space-time will be $D = 4$. When explicit four-vectors are written, the order of indices is thus $(0, 1, 2, 3)$. Somewhere, as for instance in Section 3.5, for practical computational reasons, the order of indices will be $(1, 2, 3, 0)$.

I find it inconvenient to adhere to consistent index conventions throughout, therefore, when we come to general relativity and curved space-times, the "curved" or "world" indices will be denoted by x^μ, etc., while the Minkowski local tangent and cotangent spaces will be indexed according to x^a.[9] For two-component spinor indices, I will choose dotted and undotted Greek letters $\alpha, \beta, \gamma, \ldots$.

The transition from classical Poisson brackets $\{\cdot, \cdot\}$ to quantum commutators $[\cdot, \cdot]$ is done through the convention:

$$\text{If classically: } \{A, B\} = C, \quad \text{then quantum mechanically: } [\hat{A}, \hat{B}] = i\hbar\hat{C} \tag{1.1}$$

where the *Poisson bracket* is defined by

$$\{A, B\} = \frac{\partial A}{\partial x}\frac{\partial B}{\partial p} - \frac{\partial A}{\partial p}\frac{\partial B}{\partial x} \tag{1.2}$$

for a mechanical system of one degree of freedom (x, p).[10] In quantum mechanics, we have for the operators $\hat{x}_\mu$ and $\hat{p}_\mu$

$$[\hat{x}_\mu, \hat{p}_\nu] = i\hbar\eta_{\mu\nu} \quad \text{with } \hat{p}_\nu = -i\hbar\frac{\partial}{\partial x^\nu} = -i\hbar\partial_\nu \quad \text{and} \quad \hat{x}_\mu = x_\mu \tag{1.3}$$

---

phenomena to discover, or not. This is not the same questions as to whether the Standard Model can be improved on, or not.

9 The point of this choice is that special relativistic physics and classical and quantum field theory, will look "normal". Using indexing such as, for instance, $x^{\underline{m}}$ or x^M, in order to be consistent throughout, in my opinion, make large parts of physics look unnecessarily awkward, if not baroque.

10 The generalization to several degrees of freedom and continuous systems follow naturally.

From now on, $\hbar = 1$ (unless introduced for a specific reason) and where no confusion can arise, hats on operators are dropped. Related to this, we choose the following *Fourier transform* pair

$$\phi(x) = \frac{1}{(2\pi)^2}\int d^4p\phi(p)e^{ip\cdot x} \tag{1.4}$$

$$\phi(p) = \frac{1}{(2\pi)^2}\int d^4x\phi(x)e^{-ip\cdot x} \tag{1.5}$$

The integral $\int$ is over all of momentum space, or configuration space, respectively (i. e., $\int_{-\infty}^{\infty}$ in all directions with due care to poles and cuts). We see that a derivative ∂_μ in configuration space is represented by ip_μ in momentum space, consistent with quantum mechanics (1.3), with p_μ the eigenvalue of the momentum operator $\hat{p}_\mu$. We also remember that the Dirac delta function is represented as

$$\delta(x) = \frac{1}{(2\pi)^4}\int d^4pe^{ip\cdot x} \tag{1.6}$$

The Lagrangian density for a scalar field is

$$\mathcal{L} = -\frac{1}{2}\partial_\mu\varphi\,\partial^\mu\varphi - \frac{m^2}{2}\varphi^2 = \frac{1}{2}\varphi\Box\varphi - \frac{m^2}{2}\varphi^2 \tag{1.7}$$

where $\Box = \eta^{\mu\nu}\partial_\mu\partial_\nu = \nabla^2 - \partial^2/\partial t^2$. The signs in $\mathcal{L}$ ensure that the Hamiltonian $\mathcal{H}$ is positive definitive. Likewise, for a spin half-field, we have

$$\mathcal{L} = -\bar{\psi}\gamma\cdot\partial\psi - m\bar{\psi}\psi \tag{1.8}$$

where $\bar{\psi} = i\psi^\dagger\gamma^0$ and where $\dagger$ denotes Hermitian conjugation (complex conjugation will be denoted by *). Our conventions for gamma matrices are those of Weinberg [18], in particular,

$$\{\gamma_\mu, \gamma_\nu\} = 2\eta_{\mu\nu} \tag{1.9}$$

The energy-momentum dispersion relation is, in these conventions, $p^2 + m^2 = 0$.

Regarding units, remember: setting $c = 1$, as we have done above, equalizes the units for time and space. It also equalizes the units for energy and mass. Furthermore, setting $\hbar = 1$ equalizes the units for energy and frequency (inverse time). This also subsumes the unit for force and, therefore, also all the electromagnetic units. For dimensional analysis, we count in mass units so that, for instance, the dimension of p is +1 and for x the dimension is −1.

Gamma matrices and spinors

A convenient representation of the four-dimensional Dirac matrices in terms of Pauli matrices, that we will use, is the following (also from Weinberg):

$$\gamma^0 = -i\begin{pmatrix} 0 & \sigma^0 \\ \sigma^0 & 0 \end{pmatrix} \quad \text{and} \quad \gamma^k = -i\begin{pmatrix} 0 & \sigma^k \\ -\sigma^k & 0 \end{pmatrix} \tag{1.10}$$

where the 2×2 σ^0 unit and σ^k ($k = 1, 2, 3$) Pauli matrices are given by[11]

$$\sigma^0 = \begin{pmatrix} 1 & 0 \\ 0 & 1 \end{pmatrix} \quad \sigma^1 = \begin{pmatrix} 0 & 1 \\ 1 & 0 \end{pmatrix} \quad \sigma^2 = \begin{pmatrix} 0 & -i \\ i & 0 \end{pmatrix} \quad \sigma^3 = \begin{pmatrix} 1 & 0 \\ 0 & -1 \end{pmatrix} \tag{1.11}$$

By writing $\sigma^\mu = (\sigma^0, \sigma^k)$ and $\bar{\sigma}^\mu = (\sigma^0, -\sigma^k)$, the equations (1.10) can be collected into

$$\gamma^\mu = -i\begin{pmatrix} 0 & \sigma^\mu \\ \bar{\sigma}^\mu & 0 \end{pmatrix} \tag{1.12}$$

For group theoretical reasons, that will be explained further on, the four-dimensional linear (spinor) space upon which the 4×4 gamma matrices (1.12) act, is indexed by *undotted* and *dotted* indices $1, 2, \dot{1}, \dot{2}$. The indices on the σ matrices then appear as $\sigma_{\alpha\dot{\beta}}$ and $\bar{\sigma}^{\dot{\alpha}\beta}$. Correspondingly, the four-dimensional Dirac spinor ψ is written as

$$\psi = \begin{pmatrix} \chi_\alpha \\ \bar{\lambda}^{\dot{\alpha}} \end{pmatrix} \tag{1.13}$$

where the bar over λ has no operational meaning for now. The algebra of two-component spinors will be further developed in section 3.6.4. For instance, the Pauli matrices can be used to set up a $1 \leftrightarrow 1$ correspondence between vectors and spinors through the formula

$$V_{\alpha\dot{\beta}} = \sigma_{\alpha\dot{\beta}}{}^{\mu} V_\mu \tag{1.14}$$

This formula generalize naturally to tensors (see formula (3.300)). The Pauli matrices satisfy a Lie algebra

$$\sigma^1\sigma^2 - \sigma^2\sigma^1 = 2i\sigma^3 \quad \text{and cyclic premutations of } 1, 2, 3 \tag{1.15}$$

In two-dimensional dotted or undotted spinor spaces, antisymmetry is essentially trivial, being captured by the matrix

$$\epsilon_{\alpha\beta} = \epsilon^{\alpha\beta} = \begin{pmatrix} 0 & 1 \\ -1 & 0 \end{pmatrix} \tag{1.16}$$

and likewise for dotted indices. These matrices can be used for raising and lowering indices

$$\psi^\alpha = \epsilon^{\alpha\beta}\psi_\beta \quad \text{and} \quad \psi_\alpha = \psi^\beta \epsilon_{\beta\alpha} \tag{1.17}$$

---

11 This is probably the only convention in theoretical physics that everybody agrees on. Strictly speaking, Pauli matrices refer – historically – only to the matrices σ^i with $i = 1, 2, 3$.

Symmetrization and antisymmetrization of indices are symbolized with (...) and [...] respectively, and is done with unit weight. For instance, $\phi_{(\mu\nu)} = \phi_{\mu\nu} + \phi_{\nu\mu}$.[12]

The scope of derivatives, operators and transformation symbols δ, are always restricted to the object standing immediately to the right of it. Other cases are indicated by parentheses and arrows.

As for summation convention, repeated indices – upper-lower – are (unless otherwise stated) summed over. In cases where the space or internal metric is a Kronecker delta, we sometimes (for notational convenience) do not distinguish between upper and lower indices. For unit matrices in general, we use the symbol I.

Furthermore, I will adopt a convention – often implicit, but seldom acknowledged – regarding operations and the result of operations. Taking complex conjugation as an example, I will use the notation z^* for the "operation" of computing the complex conjugate of the number z. The "result" of the operation is the complex conjugated number $\bar{z}$. Of course, writing $z^* = \bar{z}$ gains nothing, but keeping a distinction between *an operation* and *the result of the operation* is sometimes helpful to clarify formulas as the notation gets more intricate.[13]

Finally, a few words on terminology. The *Lorentz group* always refers to the homogeneous Lorentz group, never to the inhomogeneous Lorentz group, which will be denoted just so, or as the *Poincaré group*. By *field*, we refer either to a classical field (that is a function) or a quantum field (that is an operator), with qualifiers when it makes a difference.[14] By *wave function*, we refer to a quantum state (with the probabilistic interpretation if desired) which is a complex function.

The reader may already have noticed that I adhere to the not-so-common practice of **not** using textual punctuation marks in formulas set on separate lines. Instead I may use words such as "and" to separate expressions, or most often just an extended space. End of sentence is **not** marked by a dot after the formula but by a uppercase letter in the first word of the sentence to follow. A comma after a formula is marked by lowercase letter in the first word of the continuation of the sentence.

12 It would perhaps be better to denote symmetrization with {...} – as some authors indeed do – in harmony with anticommutators. However, round brackets are more common.

13 In computer science, such a distinction is often crucial: an operation is a process, and a result is an object. Conflating them may lead to syntax errors. In physics, it may not always be practical to keep separate symbols, but the distinction is helpful conceptually anyway.

14 For the mathematical concept of a "field", we will use the term "number system".

2 Notes on the history of the subject

While experimental knowledge of matter spin is not more than a 100 years old, investigations into polarization phenomena for light is older, going back to the seventeenth century. The history is intertwined with the scientific investigation of the very nature of light itself, whether it was a corpuscle or a wave, and what kind of wave in that case.[1]

Wolfgang Pauli developed the theory of the spinning electron in nonrelativistic quantum mechanics after the existence of spin had been proposed by S. Goudsmit and G. E. Uhlenbeck. Then spin came out as a direct consequence of P. A. M. Dirac's relativistic equation for the electron. Later on, the concept was understood, by E. Wigner, as a characteristic of the representations of the space-time symmetry group of special relativity. In the meantime, many authors had developed the subject of relativistic wave equations for arbitrary spin.

At this time – the 1930s – only a handful of elementary particles were known to exist and they all were of low spin. Still, for theorists like Dirac, it was natural to consider the question of wave equations for higher spin particles. As Dirac wrote

> [...] it is desirable to have the equations ready for a possible future discovery of an elementary particle with spin greater than a half [...].[1]

The logic in the early investigations was, and continued to be so for a long time, the following: find the free wave equations, or preferably the Lagrangian, and try to couple minimally to the to the existing force fields: electromagnetism and gravity.

Almost from the beginning, difficulties with interactions were discovered. These were first thought to be possible to overcome, but starting around 1960 it began to be understood that the problems were serious – so serious that the subject of higher spin fields and particles got a rather bad reputation. Only with the advent of supersymmetry and supergravity in the late 1970s, was there a revival of interest in the subject. Still there is a large number of scientific papers on higher spin wave equations and interactions, dating from the 1930s up to Christian Fronsdal's 1978 paper. Only a few of them are regularly referred to in current higher spin research. To chart out this literature in detail, remains to be done. However, one very useful survey by S. Esposito is [20].

In revisiting some of these papers, we can try to reconstruct the thinking of the times as it were expressed in the published record. This chapter will tell the story of the old papers. Some of the detailed calculations will be reviewed in Chapters 3, 4 and 5. Revisiting the history of the subject also allows us to formulate fundamental concepts of theoretical physics and record some of the basic equations.

---

1 See, for instance, Chapter 3 of [19].

https://doi.org/10.1515/9783110451771-002

A few words of warning

The early history of the subject is tangled up with the history of quantum theory itself to the extent that it is the history of wave equations for particles interacting with electromagnetic fields. This history of quantum mechanics is very convoluted, as the main characters of the story struggled with deep conceptual problems while accounting for experimental data and constructing the theoretical tools needed for doing the calculations.

Furthermore, when we write *higher spin* in the present chapter we should be a little wary of dressing up these words with twenty-first century connotations. Remember that AdS/CFT is still far in the future. Furthermore, I am not a historian of science, and since I entered the subject of higher spin only in 1983, I missed the first 50 years of it. There is a danger when writing the history of a scientific subject to interpret it in terms of what came later – to write "Whiggish history" – or what Stephen Jay Gould call "textbook cardboard" history [21]: history where the development points to the present as if by inherent logic. This is one reason why I have deliberately chosen not to write the history as if the only outcome, or solution to the difficulties encountered, is, or will be, the Vasiliev theory. Still, a certain amount of anachronisms, can be helpful in understanding the development of a subject. No doubt, I am guilty of that. The particular temptations of Whiggish history in physics is also discussed by S. S. Schweber in [22].[2]

Finally, this is a history of the published record. It remains for a professional historian of science to tell the history of recollections, letters and unpublished material.

In reviewing, brief as it will be, contents of historical papers, I have decided to change the notation of the originals, to a certain degree, into a notation coherent with the one used throughout the present book. The notation of the originals has been kept when no confusion is likely to occur. Again, this is not scholarly work of science history, but an attempt to understand the history of the subject as it is recorded in the published papers. Still, direct comparison of equations between papers, might involve issues with signs and normalizations going back to differing basic conventions. One reference that has turned out to be very useful during the writing of this chapter is the E. M. Corson book from 1953 [23] on relativistic wave equations. Apart from a clear exposition, it also contains a bibliography of the subject up to its year of publication. Furthermore, being finished in the early 1950s, it reflects the status of the subject at that time. Finally, the narrative is not linear, like a chain, since the history itself is rather like a net. As in other familiar contexts, there is no global time, not even in the world of ideas.

2.1 The Majorana, Dirac and Fierz–Pauli era

The early history of higher spin is the history of wave equations, and therefore, it is linked with the history of quantum theory itself. The pioneers of quantum mechanics were thinking in terms of relativity theory from the beginning. Since nonrelativistic mechanics could not explain the quantum phenomena, it was natural to incorporate

2 Where one can also find a reference to H. Butterfield who coined the term in 1931.

relativity in the new mechanics and Erwin Schrödinger tried to do that.[3] Already Louis de Broglie, when he proposed his theory of matter waves, reasoned within the theory of relativity [25].[4] Let us remind ourselves of the basic ideas.

One can start from Einstein's relation between energy and frequency $E = h\nu = \hbar\omega$ and the de Broglie hypothesis embodied in the equation relating momenta to wavelength $p = h/\lambda$ and write it as a three-vector equation relating wave number $\mathbf{k}$ to momenta $\mathbf{p}$ through $\mathbf{p} = \hbar\mathbf{k}$. Then consider a plane wave $\Psi = \exp i(\mathbf{k}\cdot\mathbf{x} - \omega t)$. Differentiating with respect to time and space, we get

$$i\hbar\frac{\partial}{\partial t}\Psi = E\Psi \quad \text{and} \quad -i\hbar\nabla\Psi = \mathbf{p}\Psi \tag{2.1}$$

From the relativistic relation between energy and momenta,

$$E^2 = \mathbf{p}^2c^2 + m^2c^4 \tag{2.2}$$

then follows the Klein–Gordon wave equation (first considered by Schrödinger)

$$\left(\frac{\partial^2}{\partial t^2} - c^2\nabla^2 + \frac{m^2c^4}{\hbar^2}\right)\Psi = 0 \tag{2.3}$$

Note that $\hbar$ appear in this equation only in the mass term and, therefore, the massless equation can be treated as a classical wave equation. The same is true of the massless Dirac equation. The interaction with the electromagnetic fields $\Phi = A^0$ and $\mathbf{A} = (A^1, A^2, A^3)$, is introduced through the *minimal coupling* prescription

$$E \to E - e\Phi \quad \text{and} \quad \mathbf{p} \to \mathbf{p} - ec\mathbf{A} \tag{2.4}$$

with e the elementary charge. In terms of space-time derivatives, we have

$$\frac{\partial}{\partial t} \to \frac{\partial}{\partial t} - i\frac{e}{\hbar}\Phi \quad \text{and} \quad \nabla \to \nabla - i\frac{e}{\hbar}c\mathbf{A} \tag{2.5}$$

We see that $\hbar$ appears through the quantum minimal coupling prescription.

The relativistic equation was rejected by Schrödinger since it gave results in conflict with spectroscopic data for the hydrogen atom.[5] From the nonrelativistic equation $E = \mathbf{p}^2/2m$, follows the Schrödinger equation itself

$$i\hbar\frac{\partial}{\partial t}\Psi = -\frac{\hbar^2}{2m}\nabla^2\Psi \tag{2.6}$$

3 As told by Dirac in [24].

4 L. de Broglie writes – after first telling the story of the turn of the century expectations of a imminent unification of all physics – that Lord Kelvin had brought attention to two clouds at the horizon: the Michelson–Morley experiment and the Rayleigh–Jeans law. L. de Broglie continues with poetically writing that in the beginning of the twentieth century, Lord Kelvin's clouds yielded precipitation. One leading to relativity, the other to quantum mechanics.

5 See [26], Chapter 13, for a textbook calculation.

The general Schrödinger equation

For general quantum systems, the right-hand side of the Schrödinger equation is replaced by a Hamiltonian operator $\mathcal{H}$ acting on a quantum state Ψ in a Hilbert space, while the left-hand side is retained as in equation (2.6).

The relativistic wave equations of Klein–Gordon, Dirac etc., are physically of a "transitional" character of belonging to "relativistic quantum mechanics", a theory which was to be replaced with quantum field theory.[6] From this perspective, one can, at least mathematically, consider these wave equations as classical. The conceptual status of relativistic wave equations is something that we will have occasion to return to in several places. It should also be noted that the pioneers of quantum mechanics were thinking in terms of quantization of fields from the beginning of the mathematical development of the theory [28].

Chronologically, we now know that Ettore Majorana was very early in his thinking [29] and writing on relativistic wave equations [2]. But the story is best started with Paul A. M. Dirac.

2.1.1 P. A. M. Dirac

In the new nonrelativistic quantum mechanics of the mid-1920s, a particle was described a wave function $\Psi(x,t)$ governed by the Schrödinger equation and interpreted as a *probability amplitude*. The wave function, although being complex, describes just one degree of freedom. But spectroscopic data (splitting of lines) indicated that this was insufficient to explain what was seen in the laboratories. This led Goudsmit and Uhlenbeck to assume the existence of an electron intrinsic magnetic moment and spin – not related to orbital angular momentum [30], [31], [32].[7] Such a hypothesis could also explain other phenomena such as the Zeeman effect and the Stern–Gerlach experiment where a beam of electrons was split up in two distinct beams by an inhomogeneous magnetic field.[8] The concept of spin was from the very beginning motivated by the understanding of experimental results.

Mathematically, this property of electrons could be described by introducing a new discrete variable σ with two values and writing the wave function as $\Psi(x,t;\sigma)$. Due to the discreteness of σ, the wave can just as well be described by a two-component

---

6 Relativistic quantum mechanics is developed in the textbook by Bjorken and Drell [27].

7 For Goudsmit's own account, see a talk "The discovery of the electron spin" given at the Golden Jubilee of the Dutch Physical Society 1971.

8 See, for instance, Chapter 16 in [33].

object

$$\Psi(x, t; \sigma) = \Psi_\sigma(x, t) = \begin{pmatrix} \Psi_{+\frac{1}{2}} \\ \Psi_{-\frac{1}{2}} \end{pmatrix} \tag{2.7}$$

where we have anticipated that the values of σ naturally become $\pm 1/2$. The theory of orbital angular momentum was already worked out [34], and could be adapted to the theory of intrinsic spin by W. Pauli [35].[9] Essentially, the spin operator **S** is written in terms of – what became known as – the 2×2 complex *Pauli matrices* $\boldsymbol{\sigma}$ as

$$\mathbf{S} = \frac{\hbar}{2}\boldsymbol{\sigma} \tag{2.8}$$

and the $\pm 1/2$ components of the wave-function (2.7) are the eigenvectors of $\mathbf{S}_z$.

Several physicists contemplated the idea of a spinning electron [36] and A. H. Compton wrote as early as 1921:

> I then conclude that the electron itself, spinning like a tiny gyroscope, is probably the ultimate magnetic particle and is responsible for ferromagnetism.

But it was Pauli who gave the nonrelativistic theory of spin the form it still retains today.[10] At this time, the concept of a spinor had not yet become sharp, and C. G. Darwin tried to describe the electron spin by a vector model, which in retrospect it is easy to see will run into problems (as noted by Pauli). Furthermore, the spin is an intrinsic property, not related to any space-time rotation of the particle (no "tiny gyroscope"). It was realized that such a rotation would run into problems with relativity.

As the story goes, Dirac was searching for a relativistic generalization of the Schrödinger equation. Such a generalization was already known as the Klein–Gordon equation (2.3). This is partial differential equation second-order in both time and space derivatives. It was naturally real, but by taking two fields and combining them, complex solutions could be contemplated. Such a complex scalar field can describe electrically charged spin zero particles and can be coupled to an electromagnetic field through minimal coupling.

The Klein–Gordon equation has negative energy solutions and it is sometimes said that trying to avoid this was Dirac's motivation for looking for an alternative equation. However, Dirac was motivated by several deficiencies – as perceived at that time – of the Klein–Gordon equation. Any relativistic wave equation will have negative energy solutions due to Einstein's quadratic energy momentum relation $E^2 = \boldsymbol{p}^2 + m^2$. This is true also for the Dirac equation, a fact Dirac was aware of when writing his paper [37]. More importantly, Dirac explicitly wanted to have a wave equation linear in the time derivative. This is needed so that

> [...] the wave function at any time determines the wave function at any later time.

9 The mathematics of angular momentum and spin will be reviewed in Chapter 3.

10 Upon reading Pauli's paper, it is clear that the theoretical physics of the late 1920s had taken on a form recognizable today.

This was crucial for the Dirac–Jordan *transformation theory* of quantum mechanics to work. It is also connected to the probability interpretation. The invariant density that can be formed for the Klein–Gordon equation is not positive definite (since it involves time derivatives) and this precludes a probability interpretation.

The Schrödinger equation is first-order in the time derivative and second-order in space derivatives. What was called for was some kind of linear square root of $\boldsymbol{p}^2 + m^2$. Today – in retrospect – we have no problem envisaging such a square root using matrices, but in Dirac's time it was a very clever idea come up with.[11] The procedure was however known by mathematicians [39].

The transformation theory

The *transformation theory* was developed by Dirac and Jordan independently [40, 41] in 1927. It concerned the question of equivalence between the different formulations of quantum mechanics: the Heisenberg matrix mechanics and Schrödinger wave mechanics theories. Both theories were first-order in time derivatives and worked with noncommuting variables. Dirac realized that the equations of motion of Hamiltonian classical mechanics – which are linear in time derivatives – could be reinterpreted in quantum theory. The canonical transformations of classical mechanics become the picture changing transformations of quantum mechanics. It is essential that the evolution equations are linear in time derivatives.

The theory was later elaborated by Dirac in his book on quantum mechanics [17]. In this abstract approach to quantum mechanics, states are represented by bra and ket vectors $\langle\psi|$ and $|\psi\rangle$ and linear operators $\mathcal{O}$ acting on them. The bra and ket notation was invented by Dirac in 1939 in [13]. It did not appear in the first editions of the book. The transformation theory and the equivalence problem is also treated in the first chapter of J. von Neumanns "Matematische Grundlagen der Quantenmechanik" [42].

It should perhaps be remarked that there was nothing wrong, after all, with the Klein–Gordon equation. It just describes spin zero particles. The problem with negative energies was eventually overcome by the method of *second quantization* by M. Fierz, and the direct probability interpretation is not relevant in the quantum field context. There also seem to have been a perception at the time, following the success of the Dirac equation, that relativity forced the spin to be 1/2. This was of course wrong, as was eventually made clear by the classification of the representations of the Poincaré group by Wigner and Bargmann–Wigner.[12]

---

11 Pauli, in his spin paper, admitting the provisional nature of his work, had doubted the possibility of a relativistic theory of electron "[...] as long as one retains the idealization of the electron by an infinitely small magnetic dipole [...] or whether a more precise model of the electron is required [...].". In Dirac's theory, the electron is indeed an infinitely small point particle. No "model of the structure of the electron" was needed. How Dirac may have got the idea is discussed in [38, p. 59].

12 For this misconception, see [20].

The Dirac equation does lead to a positive density, but as we now see it, it is anyway not an equation for a relativistic probability amplitude, but rather an equation for a quantum field operator. This we will discuss at length in Section 3.4.

Second quantization

The deficiencies of the Klein–Gordon equation: negative energies, negative probabilities and being quadratic in time derivatives, are mathematically related. Historically, they are related to the transition from relativistic quantum mechanics to quantum field theory. The transition was surrounded by confusion at the time, and still may lead to confusion. The conceptual bridge was "second quantization": the wave functions of the Klein–Gordon and Dirac equations, were thought of as being quantized once more – introducing creation and annihilation operators – yielding a multi-particle theory. The language of second quantization is in itself confusing since the first field to be quantized was the electromagnetic field which is surely a classical field. The bottom line – as seen from a modern point of view – is to abandon the idea of second quantization and accept that the wave functions of Klein–Gordon and Dirac are not probability amplitudes at all, but quantum operators acting in a Hilbert space of states. A good reference, both on the history and for clarifying the issues, is the historical chapter in [18]. See also Chapter 15 of [5] and [16].

A much more thorough text on the history behind the Dirac equation can be read in H. Kragh's paper [39]. It is interesting to read: "As the understanding of the general quantum mechanical formalism advanced during 1926, the problem of including spin and relativity in quantum mechanics remained essential. It was widely accepted, not only that spin and relativity were intimately related, but also that spin should find its explanation in relativity, either by the special or the general theory.". Kragh makes it clear that relativistic effects and spin were introduced together in many papers written at the time on the spectrum of the "Wasserstoffatom" – in particular, in the Pauli paper – although not in a unified way.

2.1.2 The Dirac equation

Dirac sought a wave equation linear in both time and space derivatives

$$(p_0 + \alpha_1 p_1 + \alpha_2 p_2 + \alpha_3 p_3 + \beta)\psi = 0 \tag{2.9}$$

where (in a slightly modernized notation with $c = 1$ and $\hbar = 1$)

$$p_0 = -i\frac{\partial}{\partial t} \quad \text{and} \quad p_r = -i\frac{\partial}{\partial x_r} \quad \text{for } r = 1, 2, 3 \tag{2.10}$$

He argued that the new dynamical variables α_r and β must be independent of the momenta and of the coordinates (in empty space and time). Therefore, the wave function must indeed depend on more variables than merely x_1, x_2, x_3 and t.

Next, Dirac multiplied the wave equation (2.9) with the conjugated operator $-p_0 + \alpha_1 p_1 + \alpha_2 p_2 + \alpha_3 p_3 + \beta$ and by comparing to the Klein–Gordon equation

$$(-p_0^2 + \mathbf{p}^2 + m^2)\psi = 0 \tag{2.11}$$

he derived quadratic equations for the new operators α_r and β. These equations could be solved by building up 4×4-matrices out of the Pauli-matrices. In the end, the Dirac equation [37] appeared in the nowadays familiar form

$$(i\gamma^\mu p_\mu + m)\psi = 0 \tag{2.12}$$

Dirac also showed relativistic invariance of the new wave equation and introduced electromagnetic coupling by the, then already familiar, minimal coupling scheme, replacing the four-momenta according to

$$p_\mu \to p_\mu + eA_\mu \tag{2.13}$$

Here, A_μ denote the electromagnetic potentials and e the charge of the electron.

By using the two-component spinor formalism of B. L. van der Waerden [43],[13] that we introduced in Section 1.4, the four-component Dirac equation can be written as a pair of coupled two-component equations

$$p_\mu \sigma^\mu_{\alpha\dot{\beta}} \bar{\lambda}^{\dot{\beta}} + m\chi_\alpha = 0 \tag{2.14}$$

$$p_\mu \bar{\sigma}^{\mu\dot{\alpha}\beta} \chi_\beta + m\bar{\lambda}^{\dot{\alpha}} = 0 \tag{2.15}$$

making explicit the occurrence of the Pauli matrices.

This way of writing the Dirac equation soon became common, as theoreticians were exploring the properties of the known low spin wave equations: the Dirac and Maxwell equations [44]. Inserting any of the equations (2.14) or (2.15) into the other, and using two-component algebra, the Klein–Gordon operator $p^2 + m^2$ is recovered acting on either of the spinors χ_α or $\bar{\lambda}^{\dot{\alpha}}$.

The spin 1/2 property of the electron was of course a novel and welcome theoretical discovery at the time. Introduced phenomenologically, but ad hoc, by Pauli, the spin was "explained" by the Dirac theory. But at this time, only vector and tensor representations of the Lorentz group were known among physicists. The half-angle $\theta/2$ that occurred in the Lorentz transformation of a Dirac field (when the coordinates were rotated by the infinitesimal angle θ) was noted [45]. The deeper understanding came with the van der Waerden paper, prompted by a question of P. Ehrenfest, who introduced the name "spinor", as reported in the introduction to [43]. The spinor representations were however already known in mathematics since the work of E. Cartan in 1913.[14] It is in fact possible to build the Dirac theory from such a starting point, as is done in [18].

---

13 To be developed in detail in Section 3.6.4.

14 See Cartan's book, [46], on the subject.

Recognizing spin when you see it

!

It might be handy to have a heuristic way of spotting spin in wave equations. By choosing a convenient representations of the gamma matrices, the Dirac equation (2.12) can be written in "Hamiltonian" form, reminiscent of the nonrelativistic Schrödinger equation

$$i\hbar\frac{\partial\psi}{\partial t} = \frac{\hbar c}{i}\alpha^k\frac{\partial\psi}{\partial x^k} - mc^2\beta\psi \equiv H\psi \tag{2.16}$$

where we have also reintroduced $\hbar$ and c for a physical flavor. The 4×4 matrices α^k are built from Pauli spin matrices σ^k and β from the σ^0 matrix. The message is that when one sees a three-vector object contracted into a the three-gradient or the three-momentum, then we know that spin is involved. In covariant language, expressions such as $\zeta \cdot p$ betray the presence of spin.

2.1.3 Dirac's arbitrary spin wave equations

About 8 years later, in 1936, Dirac returned to the problem of relativistic wave equations. The paper [1] starts with the classical relativistic connection between energy and momentum of free particles – quadratic in all the variables – and the quantization procedure as in equations (2.10). He then briefly reviews the road to the spin-$\frac{1}{2}$ equation.

Quantum mechanics requires a wave equation of the form $(p_0 - H)\psi = 0$, but the obvious equation following from (2.11), namely $(p_0 - \sqrt{\mathbf{p}^2 + m^2}\,)\psi = 0$ is "[...] unsatisfactory on account of the square root, which makes the application of Lorentz transformations very complicated.". Allowing the particles to have spin, one can get an equation linear in all the momenta. An example of such an equation is (2.9) with "anticommuting matrices whose squares are unity" which is of course what we now call the Dirac equation. The concern of the new paper was to generalize this to spin greater than a half. The motivation that Dirac gives for the paper is interesting to quote.

> The elementary particles known to present-day physics, the electron, positron, neutron and proton, each have a spin of a half, and thus the work of the present paper will have no immediate physical application. All the same, it is desirable to have equations ready for a possible future discovery of an elementary particle with a spin greater than a half, or for approximate application to composite particles. Further, the underlying theory is of considerable mathematical interest.

The first part of the paper is a concrete and detailed investigation into representations of spin angular momentum $S_{\mu\nu}$. Dirac returns to the factorization of the Einstein dispersion relation $E^2 = \mathbf{p}^2c^2 + m^2c^4$ in terms of certain matrices, but now does it in general, arriving at arbitrary spin equations, not just spin 1/2. What Dirac does, in modern terms, is to rewrite the Lorentz Lie algebra into two SU(2) algebras that are interchanged under Hermitian conjugation. He does it utilizing the, at that time quite new, now very well-known, two-component spinor notation of van der Waerden and

Laporte–Uhlenbeck. The paper is however rather taciturn when it comes to motivations and the logic is not spelled out very clearly.

The end results of the investigation are coupled higher spin wave equations, linear in derivatives, written in terms of two multispinors A and B

$$A^{\dot\beta_1\dot\beta_2\dots}_{\alpha\gamma_1\gamma_2\dots} \quad 2k \text{ undotted indices down, } 2l-1 \text{ dotted indices up} \tag{2.17}$$

$$B^{\dot\alpha\dot\beta_1\dot\beta_2\dots}_{\gamma_1\gamma_2\dots} \quad 2k-1 \text{ undotted indices down, } 2l \text{ dotted indices up} \tag{2.18}$$

The spinors are symmetric in dotted and undotted indices separately. This means that all traces with the antisymmetric metric spinors $\epsilon_{\alpha\beta}$ and $\epsilon_{\dot\alpha\dot\beta}$ vanish. Group theoretically, the A-spinor corresponds to the $D(k, l-\frac{1}{2})$ representation of the Lorentz group, while the B-spinor corresponds to the $D(k-\frac{1}{2}, l)$ representation.

The wave equations are

$$p^{\dot\alpha\beta}A^{\dot\beta_1\dots\dot\beta_n}_{\beta\gamma_1\dots\gamma_n} = -mB^{\dot\alpha\dot\beta_1\dots\dot\beta_n}_{\gamma_1\dots\gamma_n} \tag{2.19}$$

$$p_{\alpha\dot\beta}B^{\dot\beta\dot\beta_1\dots\dot\beta_n}_{\gamma_1\dots\gamma_n} = -mA^{\dot\beta_1\dots\dot\beta_n}_{\alpha\gamma_1\dots\gamma_n} \tag{2.20}$$

It is clear from comparing the equations to the Dirac equation (corresponding to $k = l = \frac{1}{2}$) that all but one index on each spinor just "tag along", so to speak. This fact is a bit remarkable, because referring back to the Dirac equation in the form (2.9), one would perhaps have expected more complicated generalized gamma "operators" to occur in the higher spin case. As it turns out, it is possible to do with just the Pauli matrices, all gamma matrix structure moved into the simple operators $p_{\alpha\dot\beta}$ and $p^{\dot\alpha\beta}$. We will see further on how this comes about.[15]

Inserting one of the equations (2.19) or (2.20) into the other, and vice versa, yields

$$(p^2 + m^2)A^{\dot\beta_1\dots\dot\beta_n}_{\alpha\gamma_1\dots\gamma_n} = 0 \tag{2.21}$$

$$(p^2 + m^2)B^{\dot\alpha\dot\beta_1\dots\dot\beta_n}_{\gamma_1\dots\gamma_n} = 0 \tag{2.22}$$

As we will see, not requiring the dispersion relation $p^2 + m^2 = 0$ (as Majorana did not do), leads to interesting consequences that has been investigated by many authors.

By contracting equation (2.19) by $\epsilon_{\dot\alpha\dot\beta_1}$ and equation (2.20) by $\epsilon^{\alpha\gamma_1}$, it follows that the divergence of the spinors vanish. These are subsidiary conditions that are needed in order to get the correct number of degrees of freedom. Although Dirac writes of "supplementary" conditions, he does not compute them or present them in this systematic fashion.

---

15 It can be explained from the group theory of the Lorentz group. The same phenomenon occurs in the spinor-tensor formulation of higher spin fermions, where it is enough to use the ordinary 4×4 γ-matrices. However, not requiring the second-order Klein–Gordon equation allows for "more complicated matrices" as we will see.

Dirac did not consider any actions leading to the field equations. Nor did he discuss gauge invariance in the massless case. Although the paper is to a large extent very concrete and calculational, there are not much of motivations in it. It was instead Fierz in 1939 who clarified and developed the theory considerably. But let us first try to understand the logic in what Dirac did.

Trying to understand Dirac

Let us focus on the logic, the specially interested reader will figure out the details for herself.

The spin 1/2 Dirac equation can be written in two-component notation as the two coupled equations $p^{\dot{\mu}\nu}A_\nu = -mB^{\dot{\mu}}$ and $p_{\mu\dot{\nu}}B^{\dot{\nu}} = -mA_\mu$ (we are reverting here to Dirac's convention to index two-component spinors by $\mu, \nu, \ldots$). These equations are special cases of the higher spin equations. Now we want to generalize them.

Dirac starts with the spin 1/2 Dirac equation in the form $(p_t + \alpha_x p_x + \alpha_y p_y + \alpha_z p_z + \alpha_m m)\psi = 0$. Here, we understand the α matrices as 4×4 matrices and ψ as a 4-component spinor. Next, he turns to the six spin angular momentum generators s_{jk} and linearly recombines them into

$$\begin{aligned} \alpha_x &= \frac{1}{2}(s_{yz} - is_{xt}) & \beta_x &= \frac{1}{2}(s_{yz} + is_{xt}) \\ \alpha_y &= \frac{1}{2}(s_{zx} - is_{yt}) & \beta_y &= \frac{1}{2}(s_{zx} + is_{yt}) \\ \alpha_z &= \frac{1}{2}(s_{xy} - is_{zt}) & \beta_z &= \frac{1}{2}(s_{xy} + is_{zt}) \end{aligned} \tag{2.23}$$

These matrices separately satisfy SU(2) Lie algebras, and commute with each other. They are Hermitian conjugates of each other. What we see here is a split of the Lorentz algebra that we will study in more detail in Sections 3.4.2 and 3.5.6. The matrices being angular momenta generators, we also have $\alpha_x^2 + \alpha_y^2 + \alpha_z^2 = k(k+1)$ and $\beta_x^2 + \beta_y^2 + \beta_z^2 = l(l+1)$ with k and l positive integers or half-integers. Here, one should understand that the α are $(2k+1)\times(2k+1)$ matrices and β are $(2l+1)\times(2l+1)$ matrices. We have a generalization of spin 1/2 in which case the α and β together build up the γ matrices (compare also Section 1.4).

Dirac then introduces 2-component notation through the identifications

$$s^1{}_2 = \alpha_x - i\alpha_y \qquad s^2{}_1 = \alpha_x + i\alpha_y \qquad s^1{}_1 = -s^2{}_2 = \alpha_z \tag{2.24}$$

$$s_{\dot{1}}{}^{\dot{2}} = \beta_x - i\beta_y \qquad s_{\dot{2}}{}^{\dot{1}} = \beta_x + i\beta_y \qquad s_{\dot{1}}{}^{\dot{1}} = -s_{\dot{2}}{}^{\dot{2}} = \beta_z \tag{2.25}$$

and from then on it is enough to explicitly treat the undotted case, as the dotted is similar. Next, the matrices $s^\mu{}_\nu$ are arranged in a $2(2k+1) \times 2(2k+1)$ matrix A

$$A = \begin{pmatrix} s^1{}_1 & s^1{}_2 \\ s^2{}_1 & s^2{}_2 \end{pmatrix} \tag{2.26}$$

The drift of the argument is that the A matrices, and the corresponding $2(2l+1) \times 2(2l+1)$ matrices B for dotted indices, will build up the "gamma matrices" of higher spin. Correspondingly, there will be multicomponent spinors ψ_A and ψ_B.

Dirac shows that $A(A+1) = k(k+1)$ from which follows that the eigenvalues of A are k and $-(k+1)$. So although these matrices may be fairly complicated in structure, their diagonal form is simple. Dirac argues (not assuming the α_i to be Hermitian, which they obviously not are from (2.23)) that A can anyway be diagonalized through $A = U^{-1}DU$ with U a nonunitary matrix, and D having the first m diagonal elements as k and the remaining n elements as $-(k+1)$ so that $m+n = 2(2k+1)$. Dirac then makes an ansatz for U and U^{-1} in the form

$$U = (2k+1)^{-\frac{1}{2}} \begin{pmatrix} b_1 & b_2 \\ v_1 & v_2 \end{pmatrix} \qquad U^{-1} = (2k+1)^{-\frac{1}{2}} \begin{pmatrix} a^1 & u^1 \\ a^2 & u^2 \end{pmatrix} \tag{2.27}$$

where the b's are $m\times(2k+1)$, the v's are $n\times(2k+1)$, the a's are $(2k+1)\times m$ and the u's are $(2k+1)\times n$. Then from $UU^{-1} = U^{-1}U = 1$ follows quadratic equations relating the matrices a, b, u and v. Furthermore, the equation $A = U^{-1}DU$ relates these matrices to the s matrices. But the matrices a, b, u and v are not uniquely determined by these equations, although as was elaborated by Fierz, the b matrices are simply related to the v matrices, and the a matrices to the u matrices (see next section). Dirac uses the equations to show that the number of eigenvalues of A are $m = 2k+2$ and $n = 2k$. Then after some further algebra he reduces the equations to the following set (remember $i, j = 1, 2$ and the summation convention is in force)

$$s^{\mu}{}_{\nu} + (k+1)\delta^{\mu}{}_{\nu} = a^{\mu} b_{\nu} \quad \text{and} \quad s^{\mu}{}_{\nu} - k\delta^{\mu}{}_{\nu} = -u^{\mu} v_{\nu} \tag{2.28}$$

$$b_{\mu} a^{\mu} = v_{\mu} u^{\mu} = 2k+1 \quad \text{and} \quad b_{\mu} u^{\mu} = v_{\mu} a^{\mu} \tag{2.29}$$

Dirac does not explicitly write down the following interesting consequences:

$$v_{\mu} s^{\mu}{}_{\nu} = -(k+1) v_{\nu} \quad \text{and} \quad b_{\mu} s^{\mu}{}_{\nu} = k b_{\mu} \tag{2.30}$$

that bring us back to the observation about the eigenvalues of the matrix A. But he does write "[...] we may assume that a^i, b_i, u^i, v_i transform under Lorentz transformations like single-suffix spinors [...]". In any way, equations (2.30) are presumably behind Dirac "assuming" – without any further motivation – the following form for the higher spin wave equations:

$$p^{\dot{\mu}\nu} v_{\nu} \psi_A = m' v^{\dot{\mu}} \psi_B \tag{2.31}$$

$$p_{\dot{\mu}\nu} v^{\dot{\mu}} \psi_B = m'' v_{\nu} \psi_A \tag{2.32}$$

where the matrices v_{μ} and $v^{\dot{\mu}}$ play the role of "gamma" matrices.

The structure of the "spinors" ψ_A and ψ_B are not specified at this stage. However, for the equations to make algebraic sense, the wave vector ψ_A must have $(2k+1)2l$ components, labeled by one undotted index set related to the $2k+1$ columns of v_{μ} and by one dotted set of indices related to the $2l$ rows of $v^{\dot{\mu}}$. Also, the wave vector ψ_B must have $2k(2l+1)$ components, labeled by one dotted index set related to the $2l+1$ columns of $v^{\dot{\mu}}$ and by one undotted set of indices related to the $2k$ rows of v_{μ}. With these choices, left- and right-hand sides of (2.31) and (2.32) can be equated (each side having $4kl$ components).

Clearly, the equations are not unique, as would not be expected from experience with the spin 1/2 equations where there are also choices as to explicit representations of the γ-matrices to use. In the final section of the paper, Dirac argues for "an alternative way of building up the theory", essentially by showing that the effect of the v_{μ} matrices is to convert a spinor with $2k+1$ undotted indices to a spinor with $2k$ undotted indices, and for the $v^{\dot{\mu}}$ matrices to convert a spinor with $2l+1$ dotted indices to a spinor with $2l$ dotted indices (as we indeed saw in the previous paragraph). Based on this, Dirac arrives at equations (2.19) and (2.20) (although still with different masses). We will return to this development when reviewing what M. Fierz did.

2.1.4 M. Fierz

The Fierz paper of 1939 (in German) [47] treats the integer spin fields in tensor language, and the half-integer fields in the Dirac spinor language of [1]. In principle, the spinor language also covers the integer spin cases, and Fierz has a section (3) where the correspondence is outlined.[16] Fierz writes in the "Einleitung" that it is not possible to use tensor language for half-integer spin, and one has to resort to the "recht scwherfälligen Spinorkalkül" of van der Waerden [43]. This, Fierz points out, is however not just a technical problem but has a physical basis in that the half-integer fields obey the exclusion principle and, therefore, the fields are not classically observable.

The stated object of the paper is to quantize the fields and derive expression for the energy-momentum tensors and currents. This is done for integer spin in the first four sections of the paper, the first section discussing the field equations and subsidiary conditions. Although actions are not discussed, it is fair to say that the Fierz paper provides the first substantive treatment of free higher spin fields. For those who wish to go deeper into the very early history of higher spin fields, Fierz has references to the, at that time, contemporary work by Sakata-Yukawa, Jauch, Proca and Kemmer.[17] Let us however move on the second part of the paper (Section 5 and onwards) where the spinor theory of half-integer spin fields is developed. It should also be noted that Fierz is quite critical (not at all unfair) of the Dirac paper, listing several deficiencies, not the least the failure of minimal coupling, an issue that we will return to in the next section.

Fierz starts directly with the spinors (2.17) and (2.18) that we defined above, satisfying the second-order wave equations (2.21) and (2.22). The spinors are demanded to be symmetric in dotted and undotted indices. This can also be expressed as all traces with $\epsilon_{\dot{\alpha}\dot{\beta}}$ and $\epsilon^{\alpha\beta}$ vanish. Fierz then goes on to discuss the A-spinors.

It follows that a spinor $A^{\dot{\beta}_1\dot{\beta}_2\ldots}_{\alpha\gamma_1\gamma_2\ldots}$ has $(2k+1)2l$ independent components. Group theoretically, the A-spinor corresponds to the $D(k, l-\frac{1}{2})$ representation of the Lorentz group (while the B-spinor corresponds to the $D(k-\frac{1}{2}, l)$ representation). To bring down the number of components to the correct number, Fierz imposes the subsidiary, divergence conditions

$$p_{\dot{\beta}_1}{}^{\alpha} A^{\dot{\beta}_1\ldots}_{\alpha\gamma_1\ldots} = 0 \tag{2.33}$$

These are $2k(2l-1)$ in number. The spinor A then describes $2(k+l) = 2s+1$ components as it should. The corresponding argument can be made for the B spinors.

16 Essentially, spinors with an even number of dotted and undotted indices corresponds to symmetric and traceless tensors through translation with the "Pauli" matrices $\sigma_\mu{}^{a\dot{b}}$. See our formulas (1.14) and Section 3.6.4 for more details.

17 We will briefly review some of this work in Section 2.2 below.

Counting field components in Dirac's theory

A totally symmetric object with n indices in two dimensions has $n + 1$ components. The field $A^{\dot\beta_1 \dots \dot\beta_n}_{\alpha\gamma_1\dots\gamma_n}$ therefore has $(n + 1)(n + 2)$ components. The subsidiary condition $p_{\dot\beta_1}{}^{\alpha} A^{\dot\beta_1 \dots \dot\beta_n}_{\alpha\gamma_1\dots\gamma_n} = 0$ removes $n(n + 1)$ components. With $s = n + \frac{1}{2}$, the correct number $2s + 1$ of field components follow. The argument is the same for the B-field. The resulting doubling in field components comes from A and B together describing a particle and an antiparticle.

Now, the second-order field equations (2.21) and (2.22) can be rewritten into the coupled first-order field equations (2.19) and (2.20). Conversely, the first-order equations imply the subsidiary divergence conditions through the index symmetry requirement. However, this is not the way Dirac arrived at the field equations, and Fierz has a long Section (7) where he explicates the connection to the Dirac treatment. Apart from this, Fierz derives the energy-momentum tensor and current vector for half-integer spin fields, as well as performing the quantization.

Fierz elaborates on Dirac

One development of Fierz is to clarify and simplify the structure of the U and U^{-1} matrices (2.27) of Dirac. According to Dirac, the v matrices in (2.27) has dimension $2k(2k + 1)$ and similarly for the a, b and u matrices. Fierz keeps track of this by writing $v(k)$ and likewise for the other matrices. From Dirac's formulae, it is clear that the a and b matrices are not independent of the v and u matrices, and Fierz writes:

> Dabei haben wir das Dirac'sche $b_\nu = u_\nu(k + \frac{1}{2})$ und $a^\nu = v^\nu(k + \frac{1}{2})$ gesetzt; denn diese Matrizen erfüllen genau die gleichen Relationen, die für $u_\nu(k)$, $v^\nu(k)$ gelten, falls man k durch $k + \frac{1}{2}$ ersetzt.

Fierz thus gives U and U^{-1} in the form

$$U = (2k+1)^{-\frac{1}{2}} \begin{pmatrix} u_1(k+\frac{1}{2}) & u_2(k+\frac{1}{2}) \\ v_1(k) & v_2(k) \end{pmatrix} \qquad U^{-1} = (2k+1)^{-\frac{1}{2}} \begin{pmatrix} v^1(k+\frac{1}{2}) & u^1(k) \\ v^2(k+\frac{1}{2}) & u^2(k) \end{pmatrix} \tag{2.34}$$

He then derives, after discussing the matrix dimensions of products of the u and v matrices, the consequences of the requirement $A = U^{-1}DU$ and in particular of the requirements $UU^{-1} = 1$ and $U^{-1}U = 1$. The most interesting of these being the off-diagonal products that follows from $UU^{-1} = 1$.[18] In detail, we get (summation convention in force)[19]

$$u_\mu\left(k + \frac{1}{2}\right)u^\mu(k) = 0 \quad \text{of matrix dimension } (2k+2) \times 2k \tag{2.35}$$

$$v_\mu(k)v^\mu\left(k + \frac{1}{2}\right) = 0 \quad \text{of matrix dimension } 2k \times (2k+2) \tag{2.36}$$

---

18 There is a misprint in the Fierz paper: the reference, on page 24, to formula (6.3) should be to formula (7.3) on the same page.

19 Fierz writes the formula with our k replaced by $k - \frac{1}{2}$.

The interest stems from the fact that, for instance (2.35), implies that

$$u^1\left(k+\frac{1}{2}\right)u^2(k)=u^2\left(k+\frac{1}{2}\right)u^1(k) \tag{2.37}$$

This innocent looking equation tells us that the matrix product $u^\mu(k+\frac{1}{2})u^\nu(k)$ is symmetric in the indices μ,ν. Furthermore, from the matrix dimension of the product, one sees that stringing together several such matrices, in products with the arguments in arithmetic progression with difference $\frac{1}{2}$, one gets rectangular spinor matrices of various dimensions. In particular,

$$P^{\mu_1\mu_2\dots\mu_{2k}}(k)=u^{\mu_1}(k)u^{\mu_2}\left(k-\frac{1}{2}\right)\dots u^{\mu_{2k}}\left(\frac{1}{2}\right) \tag{2.38}$$

is a column matrix with $2k+1$ rows. Likewise

$$R^{\mu_1\mu_2\dots\mu_{2k}}(k)=v^{\mu_1}\left(\frac{1}{2}\right)v^{\mu_2}(1)\dots v^{\mu_{2k}}(k) \tag{2.39}$$

is a row matrix with $2k+1$ columns. That the spinors P and R are symmetric in all indices is guaranteed by the equations (2.35) and (2.36). These spinors can then serve as "wave vectors" for higher spin.

The next step is to observe that a fully symmetric spinor with n undotted indices can be completely characterized by just one number, namely the number s, say, of indices taking the value 1. Call such an object ψ_s with s ranging from 0 to n. Taking $n=2k$ we begin to see that it ought to be possible to set up a $1\leftrightarrow 1$ relation between these objects and spinors $\psi^{\mu_1\dots\mu_{2k}}$. Fierz, based on further "orthonormality" properties of P and R (deriving from the basic $UU^{-1}=U^{-1}U=1$ equations), then writes the identifications

$$P_s^{\mu_1\mu_2\dots\mu_{2k}}(k)\psi_{\mu_1\mu_2\dots\mu_{2k}}=\sqrt{(2k)!}\psi_s \tag{2.40}$$

$$R_{s,\mu_1\mu_2\dots\mu_{2k}}(k)\psi_s=\sqrt{(2k)!}\psi_{\mu_1\mu_2\dots\mu_{2k}} \tag{2.41}$$

Clearly, the corresponding constructions can be carried through for the dotted indices. One then arrives at mixed spinors with both undotted and dotted indices. In this way, the Dirac step from the higher spin wave equations in the form of (2.31) and (2.32) to (2.19) and (2.20) can be taken. The Fierz calculations are also reviewed in Corson, [23], Section 17(ii).

2.1.5 Fierz and Pauli

As we saw in the previous section, Fierz had followed up Dirac's paper on higher spin wave equations with a paper [47] where he second quantized the free massive fields of arbitrary spin – in the manner then well established for low spin fields[20] – and derived expressions for the currents and energy-momentum tensors. This was continued with a paper together with Pauli which can be said to have set the stage for higher spin studies for a very long time. In this paper, the electromagnetic coupling problems,

20 The reference is to P. Jordan and W. Pauli, Z. S. f. Ph. **47** (1928), S. 151.

already mentioned in the Fierz paper, that were to plague the theory into the future, were first analyzed. Dirac wrote in [1] that the higher spin particles could be coupled to electromagnetism through minimal coupling. Apparently, he had not checked the details, otherwise it is unlikely that he would have failed to spot the problems with – in modern parlance – noncommutativity of the covariant derivatives.[21]

The Fierz–Pauli paper starts by considering minimal coupling for the spin 3/2 equations in the two-component spinor formalism of Dirac's paper. They show that, what we now call covariant derivatives, do not commute, and that this leads to inconsistent restrictions on the fields of the theory. An attempt to circumvent the problem by weakening the higher spin equations leads to new problems. They then go on to show that an analogous problem arise for a massive spin 2 field, doing the calculations in tensor notation. The root of the problem is that the field equations and the needed supplementary conditions, become inconsistent when derivatives are replaced by covariant derivatives. New conditions arise, involving the electromagnetic field strength tensor.

The solution to these problems, proposed by Fierz and Pauli, is to derive the field equations and supplementary conditions from an action, and to introduce interactions into the action in a way that ensure that the theory remains consistent in the presence of the interaction.

The main body of the paper treats in detail the Lagrangian field theory of massive and massless fields of spin 3/2 and 2. By introducing suitably auxiliary fields – first in the free theory, then with electromagnetic interaction – all the differential equations of the theory can be derived by varying the action. Interactions are introduced into the Lagrangian so that not only the correct field equations result, but also the right number of subsidiary conditions.

From Fierz and Pauli: the spin $\frac{3}{2}$ electromagnetic inconsistency

We will be using the notation from Section 2.1.3. The force-free Dirac equations for spin 3/2 are then

$$p^{\dot{\alpha}\alpha}A^{\dot{\gamma}}_{\alpha\beta} = -mB^{\dot{\alpha}\dot{\gamma}}_{\beta} \quad \text{and} \quad p_{\alpha\dot{\alpha}}B^{\dot{\alpha}\dot{\beta}}_{\gamma} = -mA^{\dot{\beta}}_{\alpha\gamma} \tag{2.42}$$

Minimal electromagnetic coupling to a vector potential ϕ_μ amounts a replacement $p_\mu \rightarrow p_\mu + e\phi_\mu = \Pi_\mu$. This translates to spinor language as

$$p^{\dot{\alpha}\alpha} \rightarrow \Pi^{\dot{\alpha}\alpha} \quad \text{and} \quad p_{\alpha\dot{\alpha}} \rightarrow \Pi_{\alpha\dot{\alpha}} \tag{2.43}$$

---

21 In the early follow-up literature, the theory is often referred to as the "Fierz–Pauli" theory, perhaps because Dirac did not develop it very far, and the oversight about the coupling problem. The also quite common "Dirac–Fierz–Pauli" (DFP) designation is well motivated by the early history, and will be used occasionally.

The definition of field strengths $f_{\mu\nu} = [\Pi_\mu, \Pi_\nu]$ translates to

$$[\Pi_{\alpha\dot\beta}, \Pi^{\dot\gamma\delta}] = \delta^\delta_\alpha f^{\dot\gamma}_{\dot\beta} + \delta^{\dot\gamma}_{\dot\beta} f^\delta_\alpha \tag{2.44}$$

The number of components of f^δ_α and $f^{\dot\gamma}_{\dot\beta}$ are correct, since they are both traceless in their indices.

The equations (2.42) have two consequences. First, they imply the Klein–Gordon equation for the A and B spinors, respectively. This is seen by applying $p_{\gamma\dot\alpha}$ to the first equation and then using the second. The σ-matrix algebra[22] yields $-p^2 A^{\dot\gamma}_{\gamma\beta} = m^2 A^{\dot\gamma}_{\gamma\beta}$. The corresponding calculation is then done for the second equation. Second, the spinors have zero divergence. This follows from contracting the first equation with $\epsilon_{\dot\alpha\dot\gamma}$ and using index symmetry of the B spinor. The corresponding result holds for the B spinor. For the A spinor, we get

$$p^\alpha_{\dot\gamma} A^{\dot\gamma}_{\alpha\beta} = 0 \tag{2.45}$$

In the presence of a minimally coupled electromagnetic field, the calculation resulting in the Klein–Gordon equation now produces

$$-\Pi^2 A^{\dot\gamma}_{\gamma\beta} + f^\alpha_\gamma A^{\dot\gamma}_{\alpha\beta} = m^2 A^{\dot\gamma}_{\gamma\beta} \tag{2.46}$$

Since the spinor $A^{\dot\gamma}_{\gamma\beta}$ is symmetric in γ and β, we can contract the equation with $\epsilon^{\gamma\beta}$ to get

$$f^{\beta\alpha} A^{\dot\gamma}_{\alpha\beta} = 0 \tag{2.47}$$

This is a new subsidiary condition. It can be viewed as either a restriction on the electromagnetic field, or a restriction on the spinor field, reducing the number of degrees of freedom in the interacting theory as compared to the free theory.

Fierz and Pauli then notes that an analogous problem arises for a spin 2 field, described by a traceless tensor $A_{\mu\nu}$ satisfying the equations

$$\Box A_{\mu\nu} = m^2 A_{\mu\nu} \quad \text{and} \quad \partial^\mu A_{\mu\nu} = 0 \tag{2.48}$$

Now it is incompatibility of the second divergence equation, with the first Klein–Gordon equation, when the derivatives are replaced by the minimally coupled covariant derivatives, that leads to the inconsistency. The spin 3/2 inconsistency can also be seen in this way, so it can be traced back to the appearance of subsidiary conditions on the fields. Such conditions, whether they are implicit consequences of the field equations (as for the Dirac spinor form of the equations), or just supplemented (as in the tensor form of the equations) is the root of the problem, together with the noncommutativity of the covariant derivatives.

In this particular case, electromagnetic coupling, Fierz and Pauli were able cure the problem by deriving the equations of motion and subsidiary conditions from an

---

22 See Section 3.6.4.

action. This involved the introduction of auxiliary fields.[23] In the absence of an electromagnetic field, the auxiliary fields should vanish, and the force-free subsidiary conditions be satisfied. "[...] it is important that a one-to-one correspondence should be possible between the states (eigenfunctions) with the external field and without. This is equivalent to saying that the number of conditions which the field and the auxiliary variables (and their time-derivatives for integral spins) must satisfy *at a definite time* is not diminished by the presence of an external field.". In short, the number of degrees of freedom must remain the same with or without the external field.

The problem that Fierz and Pauli thus identified, and proposed a solution to, was to define the "research program of relativistic wave equations" for a long time. One branch of this program was to construct and investigate first-order Dirac-type wave equations with no subsidiary conditions, thus avoiding the noncommutativity problem. The first such paper actually predated Dirac and Fierz–Pauli by a few years, but its motivation had nothing to do with electromagnetic coupling.

Let us also note that Fierz and Pauli treated the case of massless spin 2 and spin 3/2 fields, for spin 2 noting that the free field equations corresponded to a first approximation of Einstein's equations without matter sources. We end by two quotes. The first one regarding spin 2.

> The gauge transformation [reference to an equation] occurs in gravitational theory as an infinitesimal coordinate transformation. When interactions with matter occur and it is no longer sufficient to restrict oneself to the linear terms the gauge group is altered. This keeps the dimensionality of the possible transformations unchanged; four functions of position always remain arbitrary. It is well known that the existence of an energy-momentum tensor is closely connected with the invariance of gravitational theory under these transformations. Similarly, the gauge invariance of Maxwell's theory is connected with the conservation of charge.

The second quote concerns spin 3/2.

> Whereas the theory for the spin value 2 has an important generalization for force fields, namely the gravitational theory, we here have no connexion with a known theory. To get a generalization of the theory with interactions, one would first of all have to find a physical interpretation of the gauge group, and of the conservation theorem connected with this group.

For spin 3/2, this question has received an answer with the theory of supergravity. For spin higher than 2, the question remains with us to this day.

2.1.6 E. Majorana

Let us go back in time. In 1932, Ettore Majorana published a paper on relativistic particles with arbitrary spin [2]. It may be, and often is, regarded as the very first paper

23 Later it was shown that, despite this, there are other problems with the solutions to the equations.

on “higher spin theory”. However, as discussed in the Introduction to this chapter, we should be careful when we interpret old papers in the light of a present day research program. As such, it is reasonable to date “higher spin theory” in the modern sense back to the Fronsdal work in the 1970s. Before that, it makes more sense to label the research area as “relativistic wave equations”.[24]

Majorana’s work was motivated by trying to avoid the negative energy solutions of the Dirac equation. When the positron was discovered in August 1932, and the C. D. Anderson paper appeared in early 1933, this was not such a pressing need any longer. Dirac had also proposed his hole theory solution for the negative energy states in 1931. The chronology of these developments with regard to the Majorana work is discussed in [49]. Majorana’s paper did not become known at the time, and it did not have any influence on the subsequent development of the subject of relativistic wave equations. The Dirac theory successfully accounted for the fine structure of the hydrogen atom and the magnetic moment of the electron.

Majorana’s paper was reviewed by D. M. Fradkin [50] in 1966, who also discussed other reasons behind the Majorana paper being forgotten. A further review of Majorana’s paper is [49], which also discusses connections with the 1960s work on infinite component wave equations,[25] and [20] which is also a useful history of relativistic wave equations in general. A thorough study of the Majorana representations and their relation to infinite component field theory can be found in [51]. Majorana’s theory was studied and extended by the higher spin community in the 2010s, for this; see [48] and references therein.

As to the actual contents of the Majorana paper, it starts with the Dirac equation, but then proceeds in a different direction. Consider the Dirac equation, written in the form (slightly modernized and with $c = 1$)

$$(E + \alpha \cdot \mathbf{p} - \beta M)\psi = 0 \tag{2.49}$$

with E the energy, $\mathbf{p}$ momentum, M mass, and α^k and β numerical matrices. Majorana did not require the Klein–Gordon equation to hold for the components of the wave function. Indeed, the Dirac–Fierz–Pauli higher spin theories, and many subsequent approaches, can all be seen as based on factorizing the Einstein/Klein–Gordon dispersion relation (2.2) or (2.11) into two coupled first-order wave equations for multicomponent fields, the number of components being related to the spin of the particles. As we have seen, these coupled first-order equations yield extra restrictions on the wavefunctions: the subsidiary conditions. By requiring only the linear form (2.49), these restrictions can be avoided, as pointed out in [50].

24 There is an aura of mystery surrounding Majorana’s life and work. In [48], the authors cite scholarly work indicating that Majorana may have obtained wave equations for single massive spin of the Dirac–Fierz–Pauli-type previous to his 1932 paper.

25 We will discuss such work in our Volume 2.

By not requiring the Klein–Gordon equation, the matrices α^k and β need not satisfy the quadratic equations that allowed Dirac to determine their properties that lead, first to the spin 1/2 equation, and later to the general spin equations. Instead the form of the matrices was determined by only requiring relativistic invariance of the action leading to (2.49) and, this is crucial: that the eigenvalues of β should all be positive. This last requirement leads to unitary representations of the Lorentz group.[26] These are infinite dimensional, and Majorana studied the two simplest representations. Consequently, the wave function ψ has an infinite number of components, and the mass spectrum becomes

$$M_j = \frac{M}{j + 1/2} \quad \text{where } j = j_0, j_0 + 1, \ldots \text{ with } j_0 = 0 \text{ or } 1/2 \tag{2.50}$$

Such a mass spectrum, with mass decreasing as a function of spin, has found no phenomenological application.[27]

2.2 Wave equations of the late 1930s

As the 1930s drew to an end, many authors had written on relativistic wave equations. Apart from the already reviewed works by Majorana, Dirac, Fierz and Pauli, we have papers by L. de Broglie, A. Proca, F. J. Belinfante and N. Kemmer to name a few that we will briefly comment upon. These papers were written toward a backdrop of the theoretical and phenomenological situation at the time: theoretical regarding the problems with quantum electrodynamics which for some authors prompted a search for "new wave equations", phenomenological regarding the problems of understanding and describing the nuclear reactions.[28] In particular, we have the backdrop of the Yukawa meson theory from 1935 for the interaction between protons and neutrons. Yukawa had proposed a wave equation for the nuclear force, and in order to describe the short range of the force, the waves must be governed by a wave equation with a mass term in it. The history of the Yukawa force is very interestingly told by A. Pais in [5] where also Proca, Bhabha and Kemmer played role. Here, we will focus on the wave equations as such, not commenting much more on the phenomenology.[29]

---

26 The detailed reasoning can be found in Majorana's paper and the review papers cited here.

27 The much later studied Regge trajectories concerned increasing linear mass spectra.

28 This must indeed be born in mind. Even though we here review parts of the history from a theoretical-retrospective higher spin point of view, the theoreticians involved in the wave equation research of this era, did their work in order to explain the physics of the laboratories.

29 The reprint volume [52] contains the original papers by Yukawa and other physicists working on the Yukawa model. One paper on wave equations is "On the Wave Equation of Meson" by M. Taketani and S. Sakata, reprinted in pages 84–97. This concerns the so-called DKP-equation to be considered below.

2.2.1 The Proca equation

The wave equation and theory of massive spin 1 particles are due to A. Proca. From today's viewpoint, it may look rather trivial, but that was not so in the mid 1930s. Proca's first papers on the subject are from 1936 [53], contemporary with the Dirac 1936 paper, and predating Fierz 1939 and Fierz–Pauli 1939. It was early times for wave equations and there were a lot of things to do: relativistic invariance, Lagrangian formulation, calculation of the energy-momentum tensor and conserved currents and electromagnetic coupling. So one must not be fooled by the simple appearance of the wave equations themselves.

The *Proca equation* reads

$$\Box\psi_s - \partial_s\partial^r\psi_r = k^2\psi_s \tag{2.51}$$

It can be split up into two first-order equations

$$\partial^r G_{rs} = k^2\psi_s \quad \text{and} \quad G_{rs} = \partial_r\psi_s - \partial_s\psi_r \tag{2.52}$$

Upon contracting the first equation with ∂^s, one gets $\partial^s\psi_s = 0$. Therefore, the field equation (2.51) is equivalent to

$$\Box\psi_s = k^2\psi_s \quad \text{and} \quad \partial^s\psi_s = 0 \tag{2.53}$$

The second equation is needed in order to remove the negative terms in the energy coming from the time component ψ_0 of the field in the first equation.

The similarity of equations (2.51) and (2.52) to electrodynamics, is apparent. With the mass k zero, one gets the electromagnetic wave equations. In that case, however, $\partial^s\psi_s = 0$ is not a consequence of the equations (2.52), and it should not be. It was Pauli in his review article [54] who pointed out that the Proca theory allowed no "gauge transformations of the second kind". Phenomenologically, the Proca equation found application in the Yukawa theory of the nuclear force.

2.2.2 From de Broglie's photon theory to the Bhabha equations

The original thinker Louis de Broglie for a long time entertained a theory for the photon as being a composite system of two spin 1/2 fermions. In case these fermions carried a mass, the so formed photon should also carry a small, indeed very small, mass. The theory had many problems, analyzed at the time by M. H. L. Pryce in [55]. The theory was first formulated in 1934.

One can construe strands of research work on wave equations from de Broglie, via a few other researchers, G. Petiau, J. Géhéniau, R. J. Duffin and N. Kemmer up to the work of H. J. Bhabha in the 1940s. The details of this history is told in [56] and [20]. A crucial ingredient in the work was an algebra of matrices that came to be known as the *DKP-algebra* (after Duffin, Kemmer and Petiau).

L. de Broglie had studied a first-order wave equation of the "Majorana" type, involving four 16×16 matrices that however did not obey the Dirac gamma matrix algebra. The actual algebra for a variant of these matrices was found by G. Petiau.[30] Denoting the matrices with β_μ where μ is a space-time index, the algebra reads

$$\beta_\mu \beta_\nu \beta_\rho + \beta_\rho \beta_\nu \beta_\mu = \beta_\mu \delta_{\nu\rho} + \beta_\rho \delta_{\nu\mu} \tag{2.54}$$

The so formed theory acquired a theoretical life of its own, independent of the de Broglie theory. Phenomenologically, it also became applied to the Yukawa meson theory, as evidenced by papers in the reprint volume [52].

2.2.3 The Kemmer equations

We can start the story with a 1938 Kemmer paper [58], set in the context of the Yukawa theory, that elaborated on the Proca spin 1 theory. Although our focus is not on the phenomenology, it is of some interest to quote, at some length, from the Introduction to the paper, as it gives a flavor of the times, in particular of the thinking of forces as being mediated by particles, an idea that was quite new at the time.

> The description of nuclear interaction in terms of a neutron-proton "exchange force" appears to be well justified and generally accepted. However, there has hitherto been no satisfactory suggestion as to the nature of the field of charged particles, the virtual emission and reabsorption of which, according to Heisenberg's (1932) picture, would give rise to this type of interaction. It may be now considered certain that this field is not identical with the electron-neutrino field of Fermi's theory, the magnitude of nuclear forces being far too large to be compatible with the small empirical value of the constant of β-decay.
>
> As an alternative and simpler description of the nuclear field, Yukawa (1935) put forward the idea that the interaction is transmitted by charged particles obeying Einstein–Bose statistics. He showed that the resulting nuclear potential would be proportional to $r^{-1} exp(-2\pi m_0 cr/h)$, m_0 being the rest mass of the Bose particles. Thus, forces of a correct range would be obtained with a rest mass about 100 times that of the electron.
>
> The apparent discovery of particles ("heavy electrons") with a mass of this order of magnitude in cosmic radiation by Neddermeyer and Anderson (1937) has aroused considerable interest in Yukawa's suggestion, and various aspects of this possibility have been discussed by a number of authors (Yukawa 1935, 1937; Yukawa and Sakata 1937; Oppenheimer and Serber 1937; Stueckelberg 1937; Frohlich and Heitler 1938; Kemmer 1938; Bhabha 1938b).[...]
>
> It is the purpose of this paper to consider this theory from a more general point of view. There are various ways of generalizing Yukawa's treatment, the most important concerning the spin of the new particle. From the mechanism suggested, it is clear that the spin must be taken

30 I have not been able to retrieve this reference. I believe the correct bibliographic data is given in [57].

> to be integral, that is, $2n$ times that of the neutron or proton. Yukawa's equations are one way of describing the case $n = 0$. For this value, there is a second alternative theory, and there are also two independent possibilities for $n = 1$. On the other hand, a consistent theory for higher values of n does not appear possible.

The "heavy electron" discovered in cosmic radiation in 1937, turned out not to be the right meson, one of the staple stories of elementary particle physics. Anyway, let us get started on the wave equations.

After briefly reviewing the quantization of the Proca equations, Kemmer turns to the Dirac spinor equations, which he intends to "abandon" but first use to find the tensor equations that he is actually interested in. These Dirac equations for spin 1 are written as

$$p^{\dot{\alpha}\kappa} A_{\kappa\gamma} = \sqrt{2} m_0 B^{\dot{\alpha}}_{\gamma} \quad \text{and} \quad p_{\dot{\alpha}\kappa} B^{\dot{\alpha}}_{\gamma} = \frac{m_0}{\sqrt{2}} A_{\kappa\gamma} \tag{2.55}$$

Here, $A_{\kappa\gamma}$ is symmetrical in its indices, but Kemmer goes on to study the anti-symmetrical case when $A_{\kappa\gamma} = \epsilon_{\kappa\gamma} A$. Then the equations become

$$p^{\dot{\alpha}}_{\kappa} A = \sqrt{2} m_0 B^{\dot{\alpha}}_{\kappa} \quad \text{and} \quad p^{\kappa}_{\dot{\alpha}} B^{\dot{\alpha}}_{\kappa} = -\sqrt{2} m_0 A \tag{2.56}$$

Kemmer writes that the transition to tensor equations can be done "immediately" but that the result is not unique, in that the spinor equations do not specify how the wave functions transform under "reflexions". Taking this into account, the second pair of equations (2.56) can be written as[31]

$$\frac{\partial \phi}{\partial x_\alpha} = \kappa \chi_\alpha \quad \text{and} \quad \frac{\partial \chi^\alpha}{\partial x_\alpha} = \kappa \phi \tag{2.57}$$

and the first pair (2.55) become

$$\frac{\partial \phi_\beta}{\partial x_\alpha} - \frac{\partial \phi_\alpha}{\partial x_\beta} = \kappa \chi_{\alpha\beta} \quad \text{and} \quad \frac{\partial \chi^{\alpha\beta}}{\partial x_\alpha} = \kappa \phi^\beta \tag{2.58}$$

Alternatively, equations (2.55) can be transcribed as

$$\frac{\partial \phi_{\beta\gamma}}{\partial x_\alpha} - \frac{\partial \phi_{\alpha\gamma}}{\partial x_\beta} + \frac{\partial \phi_{\alpha\beta}}{\partial x_\gamma} = \kappa \chi_{\alpha\beta\gamma} \quad \text{and} \quad \frac{\partial \chi^{\alpha\beta\gamma}}{\partial x_\alpha} = \kappa \phi^{\beta\gamma} \tag{2.59}$$

and the equations (2.57) as

$$\frac{\partial \phi_{\beta\gamma\delta}}{\partial x_\alpha} - \frac{\partial \phi_{\alpha\gamma\delta}}{\partial x_\beta} + \frac{\partial \phi_{\alpha\beta\delta}}{\partial x_\gamma} - \frac{\partial \phi_{\alpha\beta\gamma}}{\partial x_\delta} = \kappa \chi_{\alpha\beta\gamma\delta} \quad \text{and} \quad \frac{\partial \chi^{\alpha\beta\gamma\delta}}{\partial x_\alpha} = \kappa \phi^{\beta\gamma\delta} \tag{2.60}$$

---

31 People did not care that much about index conventions in those early days.

Kemmer then notes that the equations (2.57) are equivalent to the Klein–Gordon equation, while the equations (2.58) are equivalent to the Proca equations. The second set of equations are analogous in all respects, for instance the spin they describe and their quantization, but their properties under reflections differ. In modern parlance, they describe pseudo-scalar and pseudo-vector particles. They were needed for the proper description of the nuclear force, and this constitutes the rest of the paper.

A study in spin one spinors and tensors

?

The translation between the Proca equation and the corresponding Dirac spinor equations ought to be "immediate" but is actually a bit tricky to carry through in detail. Holmes intuition is however the following: the Dirac equations are first-order in derivatives, therefore, it may be suspected that one should compare to the Proca equations written in first-order form as in formulas (2.52).

The next piece of evidence comes from asking what representations of the Lorentz group may correspond to spin 1? Referring back to Dirac's analysis in Section 2.1.3, we may expect that the representations $D(1,0)$, $D(0,1)$ and $D(1/2,1/2)$ are appropriate. These sit between the pairs $(D(1/2,0), D(0,1/2))$ and $(D(1,1/2), D(1/2,1))$ that we know correspond to spin 1/2 and 3/2 respectively. Thus we are looking for symmetric spinors $F_{\alpha\beta}$ and $F_{\dot\alpha\dot\beta}$ corresponding to $D(1,0)$ and $D(0,1)$, respectively. The number of components are $3+3$ suggesting a relation to the anti-symmetric tensor F_{ab}. Deferring details, one may write an abstract correspondence $F_{ab} \leftrightarrow F_{\alpha\dot\alpha\beta\dot\beta}$. Then taking into account that antisymmetry is essentially trivial in two-dimensional spinor space, the correspondence can be refined into $F_{ab} \leftrightarrow F_{\alpha\beta}\epsilon_{\dot\alpha\dot\beta} + F_{\dot\alpha\dot\beta}\epsilon_{\alpha\beta}$ with ϵ the antisymmetric symbol in either undotted or dotted two-dimensional spinor spaces.

Furthermore, the representation $D(1/2,1/2)$ corresponds to the spinor $A_{\alpha}^{\dot\beta}$ which according to standard translation rules represents a space-time vector A_a. Finally, the Dirac equations read

$$p^{\dot\alpha\beta} F_{\beta\gamma} = -mA_{\gamma}^{\dot\alpha} \quad \text{and} \quad p_{\alpha\dot\beta} A_{\gamma}^{\dot\beta} = -mF_{\alpha\gamma} \tag{2.61}$$

Assuming that all spinors are real, we also have the complex conjugates of the equations above

$$p^{\alpha\dot\beta} F_{\dot\beta\dot\gamma} = -mA_{\dot\gamma}^{\alpha} \quad \text{and} \quad p_{\dot\alpha\beta} A_{\dot\gamma}^{\beta} = -mF_{\dot\alpha\dot\gamma} \tag{2.62}$$

For Watson, it is now a matter of algebra to prove that these are the Proca equations in disguise. For the second set of equations, taking "reflexions" into account: treat the spinors as complex.

The story of what happened next is told in [56]. Instead, let us step back and take an anachronistic top-down view of where we are. The kind of first-order wave equations that were studied at the time, and well into the 1950s, could all be written in an abstract Dirac–Majorana way as

$$(B^{\mu}\partial_{\mu} - mA)\psi = 0 \tag{2.63}$$

It is all to easy to think of ψ as a spinor and the B_{μ} and A matrices as some kind of gamma matrices. That is however not a priori necessary, as the Kemmer equations indeed show. Of course, after E. Wigner's work on the representations of the Poincaré group (to be reviewed below), and after the dust had settled, the independent possible choices are indeed delimited by group theory. But that was not clear at the time, and even so, the equations could very well be phenomenologically interesting. With this understanding, let us continue to the next Kemmer paper of 1939, [59]. In this paper, Kemmer starts with the wave equation[32]

$$\partial_\mu \beta_\mu \psi = \kappa \psi \tag{2.64}$$

with the "commutation rules" for the operators β_μ given by the formula (2.54) above.[33] By acting with $\partial_\rho \beta_\rho \beta_\nu$ on the wave equation and using the algebra of the β matrices one gets

$$\partial_\mu \psi = \partial_\nu \beta_\nu \beta_\mu \psi \tag{2.65}$$

Kemmer writes that "[...] the differential relation (2.65) appear as a consequence of the wave equation (2.64) and not as 'initial conditions' to be imposed on the wave function.".[34] This is in contrast to the Dirac formalism where supplementary conditions arise on the wave functions, leading to the Fierz–Pauli problem with minimal electromagnetic coupling. At this stage in reading the paper, this is actually not quite clear, although one may suspect that the equations (2.65) are merely identities, and do not constitute further differential conditions one the wave function components. This, indeed, is the case. But first let us note that the Klein–Gordon equation for ψ can be derived from (2.65) by contracting with ∂_μ and using the wave equation (2.64) twice.

Kemmer has a section on the electromagnetic interaction where he again writes that there can be no inconsistency. He also proves relativistic invariance. Then the algebraic properties of the β_μ matrices are studied.[35] It turns out that there are three inequivalent representations: a ten-dimensional, a five-dimensional and a trivial one-dimensional. There are no more irreducible representations [54]. The explicit form of the matrices are given. As the reader may already have guessed, the ten-dimensional and the five-dimensional representations correspond precisely to the scalar and vector wave equations from the 1938 paper (our equations (2.57) and (2.58)). Equally well, the representations can describe the pseudo-scalar and pseudo-vectors equations. From

32 We use Kemmer's notation.

33 In a footnote, Kemmer acknowledges the work of R. J. Duffin, who in a short "letter to editor" [60], derived the algebra (2.54) by rewriting the Proca spin 1 equations (our equations (2.58)) in terms of a 10-row column matrix ψ containing the vector ϕ_α and tensor $\chi_{\alpha\beta}$ satisfying a Dirac-type wave equation. Duffin also showed that the algebra could be realized in terms of 5×5 matrices, and the spin 0 wave equations (our equations (2.57) were obtained.

34 Our quote is not verbatim.

35 Kemmer acknowledges Pauli's suggestion to perform a detailed study of the matrices, elaborating the analysis of Duffin [60].

the explicit form for the β matrices, it is also seen that the equations (2.65) are indeed differential identities, and does not subject the wave ψ to any further subsidiary constraints.

2.3 E. Wigner and the representations of the Poincaré group

The 1939 paper by Eugene Wigner [4] on the unitary representations of the inhomogeneous Lorentz group is fundamental to modern quantum field theory, and the results of the paper are now deeply worked into the theory. At the time, the paper was also motivated by the problem of constructing a relativistic quantum theory. Wigner writes, in an acknowledgement, that the subject of the paper was suggested to him by Dirac in 1928. A major theme in Dirac's oeuvre was indeed the construction of a relativistic quantum theory, and in this context one could note his 1949 paper [61] on "forms of relativistic dynamics". This paper can be regarded as proposing a research program into finding (all) nonlinear realizations of the Poincaré group. The Dirac paper is mentioned here, because just as the Wigner paper on the linear representations is important for free higher spin theory, so is the Dirac paper for interactions. We will therefore have occasion to return to it in several places, in particular in Chapter 6 on the light-front formulation.

2.3.1 Wigner's 1939 paper

The paper, 56 pages long, can be seen as a foundation for higher spin theory and it makes sense to take the opportunity here to review its "conceptual" contents in some more detail than is usually done.[36] The paper consists of 8 sections.

The first introductory section puts the problem in its quantum mechanical context. Although the contents are by now very familiar, upon reading it, it becomes quite clear that there was nothing at the time of writing, very difficult or strange, with the actual "concept of relativistic quantum mechanics" as far as noninteracting particles went. Technical problems there certainly was and still is, but the basic framework was natural.[37]

36 Reviewing its mathematical details would be too lengthy. Since unitarity of an operator is defined in relation to the inner product of the Hilbert space of states in which it acts, this must be ascertained throughout. This takes up a large part of the Wigner paper. The subject, in the sense of "one-particle states" will be treated in our Section 3.5, but not with the rigor of Wigner.

37 The aura of inconsistency, even mysteriousness, that still surrounds the subject of relativistic quantum mechanics to some extent, probably emanates to a large part from the conceptual and technical problems with the consequences of the theory: negative energy states, anti-particles and infinities, to mention a few. Pictorially speaking, the struggles of the pioneers – up to and including the develop-

The states of the theory, called *wave functions*, form a linear manifold in which a unitary scalar product can be defined. Since the wave function φ, and φ multiplied with a constant, represent the same state, it is possible to normalize the states. Then only the phase of the wave function is arbitrary. The linearity is an expression of the superposition principle. Denoting the scalar product between two normalized states φ and ψ by (ψ, φ), the square of the modulus $|(\psi, \varphi)|^2$ is interpreted as the transition probability between the states. In short, we have a Hilbert space of states.

Relativity enters when the same state φ is described in two different coordinate systems, denoted l and l'. Then φ_l and $\varphi_{l'}$ represent the same state but with different functions. If φ_l is given, all Lorentz transformed[38] states $\varphi_{l'}$ are determined up to a constant factor. Invariance of the transition probability implies $|(\psi_l, \varphi_l)|^2 = |(\psi_{l'}, \varphi_{l'})|^2$. Wigner then argues that the states $\varphi_{l'}$ can be obtained from φ_l by the action of a linear unitary operator through

$$\varphi_{Ll} = D(L)\varphi_l \tag{2.66}$$

where L is the Lorentz transformation that carries the system l into $l' = Ll$. Next, it is shown that the operators $D(L)$ form a representation of the inhomogeneous Lorentz group up to a phase factor ω, that is,

$$D(L_2)D(L_1) = \omega D(L_2L_1) \tag{2.67}$$

The object of the paper is then to determine all such continuous[39] unitary representations. Wigner stresses the generality of the approach, writing: "[...] no assumptions regarding the field nature of the underlying equations are necessary.".

In the second section, after referring to previous work,[40] Wigner states that two representations are physically equivalent if there is a one-to-one correspondence between the states of both representations, which is: (i) invariant under Lorentz transformations and, (ii) such that the transition probabilities between corresponding states are the same. From the second condition follows that there is a unitary operator S connecting states Φ in the two representations[41]

$$\Phi^{(2)} = S\Phi^{(1)} \tag{2.68}$$

ments in the 1940s – with these questions, left behind debris that still may cause confusion for the new-coming student. The tension between first and second quantization can perhaps be felt in the paper, but as Wigner comments elsewhere, the paper is written on the Schrödinger level and it only treats the one-particle theory. The need for second quantization – a many particle theory – or what we now call quantum field theory, can be seen as a consequence of relativistic quantum mechanics. And it was of course in understanding the interaction between electrons and the electromagnetic field that most of the difficulties arose. Standard references treating this intellectual history are [38, 62, 16].

38 Here, Lorentz transformations means inhomogeneous Lorentz transformations.

39 The meaning of this term is defined in the paper.

40 By Majorana 1932, Dirac 1936, Proca 1936 and O. Klein 1936.

41 S could also be antiunitary, and Wigner comments on this case.

Then the first condition means that if $\Phi^{(1)}$ and $\Phi^{(2)}$ correspond to each other in one coordinate system, then the Lorentz transformed states $D^{(1)}(L)\Phi^{(1)}$ and $D^{(2)}(L)\Phi^{(2)}$ correspond each other also. From this, it follows that

$$D^{(2)}(L) = SD^{(1)}(L)S^{-1} \tag{2.69}$$

Therefore, as Wigner writes, the existence of a unitary operator S which transforms $D^{(1)}$ into $D^{(2)}$ is the condition for equivalence of the representations. The rest of the section discusses technical questions. Section 3 is a summary of the contents of the rest of the paper.

The long Section 4 begins with a description of the inhomogeneous Lorentz group. Then follows a study of the properties of the homogeneous group. Its characteristic values and vectors (i. e., eigenvalues and eigenvectors) are determined and it is shown that it has no finite dimensional unitary representations. Next, the decomposition of a homogeneous Lorentz transformation into two rotations and an acceleration in a given direction, is discussed. Finally, it is shown that the group is simple. The long Section 5 is also a technical section on the reduction of the representations from "up to a factor" to "two-valued" representations or "up to a sign". Denoting the translation operators by $T(a)$ and the homogeneous Lorentz transformations by $d(\Lambda)$, the group multiplication laws then read

$$T(a)T(b) = T(a+b) \tag{2.70}$$
$$d(\Lambda)T(a) = T(\Lambda a)d(\Lambda) \tag{2.71}$$
$$d(\Lambda)d(L) = \pm D(\Lambda L) \tag{2.72}$$

These operators act in the Hilbert space of states. After these preliminaries, Section 6 turns to the derivation of the actual representations using the method of the "little group". This is also a technical section, but let us try to capture the ideas.[42]

Since the translation operators $T(a)$ form an Abelian invariant subgroup of the whole inhomogeneous Lorentz group, it is possible to introduce a "coordinate system" in Hilbert space such that the wave functions $\varphi(p,\zeta)$ contain momentum variables p_1, p_2, p_3, p_4 and a discrete variable ζ so that

$$T(a)\varphi(p,\zeta) = e^{ip\cdot a}\varphi(p,\zeta) \tag{2.73}$$

After discussing the unitary scalar product of two wave functions, Wigner defines new operators $P(\Lambda)$ through

$$P(\Lambda)\varphi(p,\zeta) = \varphi(\Lambda^{-1}p,\zeta) \tag{2.74}$$

42 For the reader who wants to localize where we are in relation to S. Weinberg's treatment in [18], this corresponds to Section 2.5 in the reference. See also our Section 3.5.

The properties of these operators are discussed, with the result that one can write $d(\Lambda) = Q(\Lambda)P(\Lambda)$ where $Q(\Lambda)$ is an operator in the space of ζ alone, commuting with all $T(a)$, which can, however, depend on the value of p

$$Q(\Lambda)\varphi(p,\zeta) = \sum_{\eta} Q(p,\Lambda)_{\zeta\eta}\varphi(p,\eta) \tag{2.75}$$

and where $Q(p,\Lambda)_{\zeta\eta}$ are the components of a finite or infinite matrix. Then one gets

$$d(\Lambda)\varphi(p,\zeta) = \sum_{\eta} Q(p,\Lambda)_{\zeta\eta}P(\Lambda)\varphi(p,\eta) = \sum_{\eta} Q(p,\Lambda)_{\zeta\eta}\varphi(\Lambda^{-1}p,\eta) \tag{2.76}$$

Wigner then argues that it is sufficient to consider only representations for which the wave functions vanish except for momenta that can be obtained from any one momenta p_0 (a particular four-momenta) by a homogeneous Lorentz transformation. Such representations can be divided into four classes:

1. $p^2 = P > 0$
2. $p^2 = P = 0;\ p \neq 0$
3. $p = 0$
4. $p^2 = P < 0$

The classes 1 and 2 contain two subclasses each according to whether the time component of the momenta p is positive or negative. The two subclasses of class 1 are denoted by P_+ and P_-, the two subclasses of class 2 by 0_+ and 0_- and the class 3 by 0_0. Class 4 is denoted by $-P$ and has no subclasses.[43]

Wigner then goes on to "[...] give [...] a characterization of the representations with a given P, which is independent of the coordinate system in Hilbert space.". This investigation ends with the result that "[...] when characterizing a representation to the whole inhomogeneous Lorentz group by P and the representation of the little group, it is not necessary to say which p_0 is left invariant by the little group.".

In the last subsection of Section 6, Wigner lists the little groups for the first three classes. In the case 1_+, the little group momentum p_0 is taken as the vector $(0,0,0,1)$. The little group then contains all rotations in the space of the first three coordinates. In the case 0_0, the little group is the whole homogeneous Lorentz group. In both subclasses of the case -1 ($P = -1$), p_0 can be taken as the vector $(1,0,0,0)$ and the little group contains all transformations that leave the form $-x_2^2 - x_3^2 + x_4^2$ invariant. This is the $2+1$ dimensional homogeneous Lorentz group. For the remaining case 0_+, the determination of the little group is somewhat more complicated, Wigner writes. Since this is also the case that is the most interesting from the higher spin perspective, let us follow its determination in some detail.

43 In modern treatments, m^2 is used instead of P.

Wigner's determination of the little group for massless representations

Lorentz transformations[44] can be realized in a two-dimensional complex space by collecting the coordinates into matrices

$$\begin{pmatrix} x_4 + x_3 & x_1 + ix_2 \\ x_1 - ix_2 & x_4 - x_3 \end{pmatrix} \tag{2.77}$$

A Lorentz transformation is represented by a 2×2 complex matrix with unit determinant, and its action on the coordinates is given by

$$\begin{pmatrix} a & b \\ c & d \end{pmatrix} \begin{pmatrix} x_4 + x_3 & x_1 + ix_2 \\ x_1 - ix_2 & x_4 - x_3 \end{pmatrix} \begin{pmatrix} a^* & c^* \\ b^* & d^* \end{pmatrix} = \begin{pmatrix} x_4' + x_3' & x_1' + ix_2' \\ x_1' - ix_2' & x_4' - x_3' \end{pmatrix} \tag{2.78}$$

The little group momentum p_0 is now taken as a vector $(0, 0, 1, 1)$. In two-dimensional notation, this is a matrix with zeros except for the upper left element, which is 2. The conditions for p_0 to be left invariant by a Lorentz transformation is $|a|^2 = 1$ and $c = 0$. Hence, a general element of the little group can be written

$$\begin{pmatrix} e^{-i\beta/2} & (x+iy)e^{i\beta/2} \\ 0 & e^{i\beta/2} \end{pmatrix} = \begin{pmatrix} 1 & (x+iy) \\ 0 & 1 \end{pmatrix} \begin{pmatrix} e^{-i\beta/2} & 0 \\ 0 & e^{i\beta/2} \end{pmatrix} \equiv t(x,y)\delta(\beta) \tag{2.79}$$

which we can interpret as a rotation $\delta(\beta)$ followed by a translation $t(x, y)$ in a two-dimensional space.[45] Here, x, y and β are real and $0 \le \beta < 4\pi$. Wigner then argues that the range of variation for β can be restricted to $0 \le \beta < 2\pi$. The exponents $\pm i\beta/2$ are required by the matrix $\delta(\beta)$ to describe a rotation be an angle β. The group multiplication laws become

$$t(x,y)t(x',y') = t(x+x', y+y') \tag{2.80}$$

$$\delta(\beta)t(x,y) = t(x\cos\beta + y\sin\beta, -x\sin\beta + y\cos\beta)\delta(\beta) \tag{2.81}$$

$$\delta(\beta)\delta(\beta') = \delta(\beta+\beta') \tag{2.82}$$

These equations are analogous to the equations (2.70)–(2.72) and show that the little group is isomorphic to the inhomogeneous rotation group in two dimensions, that is, the two-dimensional Euclidean group.

Section 7 of the paper treats the P (massive) and 0 (massless) classes of representations, the massive case being the well-known representations of the three-dimensional rotation group.

For the massless case, we will follow Wigner's treatment in some detail. Due to the similarity to the inhomogeneous Lorentz group itself, it is a possible to introduce momenta-like variables (ξ, η) and a spin-like variable ν instead of ζ in such a way that

$$t(x,y)\varphi(p_0, \xi, \eta, \nu) = e^{i(x\xi + y\eta)}\varphi(p_0, \xi, \eta, \nu) \tag{2.83}$$

44 See our Section 3.4.3.

45 This will be commented on below in connection with the next Wigner paper to be discussed.

and rotation operators $R(\beta)$, acting only on the ξ, η variables

$$R(\beta)\varphi(p_0, \xi, \eta, \nu) = \varphi(p_0, \xi', \eta', \nu) \tag{2.84}$$

where

$$\xi' = \xi \cos\beta - \eta \sin\beta \quad \text{and} \quad \eta' = \xi \sin\beta + \eta \cos\beta \tag{2.85}$$

Then $\delta(\beta)R(\beta)^{-1} = S(\beta)$ will commute with the $t(x, y)$ operators and will contain the ξ, η variables only as parameters, acting only on the ν variable. This is in analogy with the $Q(p, \Lambda)$ operator of the inhomogeneous Lorentz group itself. The transformation law corresponding to (2.76) is

$$\delta(\beta)\varphi(p_0, \xi, \eta, \nu) = \sum_{\omega} S(\beta)_{\nu\omega}\varphi(p_0, \xi', \eta', \omega) \tag{2.86}$$

Now the little group method can be applied to the little group itself.[46] All vectors (ξ, η) can be obtained from one particular vector (ξ_0, η_0) by a rotation (2.85). Since the metric in the ξ, η space is positive definite, this leads to two disjoint cases

$$\xi^2 + \eta^2 = \Xi = 0 \quad \Rightarrow \quad \xi = \eta = 0 \tag{2.87}$$

$$\xi^2 + \eta^2 = \Xi \neq 0 \tag{2.88}$$

A little group transformation should leave the vector (ξ_0, η_0) invariant. In the first case, any rotation in two dimensions does so, and we are interested in one- or two-valued irreducible representations. Then $S(\beta) = e^{is\beta}$, with s integer or half-integer. These are the standard massless particle representations denoted by O_+.

In the second case, the "little group of the little group" is trivial since the only rotation that leaves a vector in two dimensions invariant, is a rotation with $\beta = 0$. For the little group of the inhomogeneous Lorentz group, written in terms of 2×2 complex matrices, this corresponds to the unit matrix I and $-I$.[47] Of these representations, there is one kind denoted by $O(\Xi)$ which is single valued, and one kind denoted by $O'(\Xi)$ that is double valued. Both are characterized by the positive real number Ξ. In the Introduction to the paper, Wigner writes

> [...] the new representation of the Lorentz group which will be described in Section 7 may interest the physicist also. It describes a particle with continuous spin.

There is nothing specific about wave equations in the 1939 paper. For that, we have to look at the next paper.

---

46 The role played the homogeneous Lorentz transformations leaving p^2 invariant, are now played by the two-dimensional Euclidean rotations, leaving the length of a two-dimensional vector invariant.

47 Due to the map from two-dimensional complex matrices to the Lorentz group being $2 \to 1$. See Section 3.4.3.

2.3.2 Relativistische Wellengleichungen

As we have already seen in Section 2.2, a few authors wrote on relativistic wave equations during the 1930s, and more would do so in the second part of the 1940s as we will see in Section 2.4. These papers were not based on any general analysis of the representations of the Poincaré group, at least as far as can be judged from their explicit content, but rather, the authors wrote down Lorentz invariant wave equations a priori. Exceptions are the Majorana 1932 paper where unitary representations of the Lorentz group are investigated [2] and the 1936 Dirac paper [1] where the finite dimensional nonunitary representations were constructed. This circumstance is however natural enough, since the Wigner paper appeared first in 1939, and perhaps did not become well known until later. It seems that not until the 1960s was the Wigner analysis taken as a baseline for work on wave equations for elementary particles.[48] So it still remained to sort out how relativistic wave equations were related to the representations of the homogeneous and inhomogeneous Lorentz group. Explicit wave equations had not been discussed in the Wigner 1939 paper.

This issue is discussed in the Wigner 1947 paper on "Relativistische Wellengleichungen" [63].[49] In this paper, Wigner discusses two ways of approaching the question of "das relativistische Einkörperproblem in der Quantenmechanik", or relativistically invariant wave equations.

One way is to directly look for relativistic invariant equations by supplementing the continuous configuration space-time variables of the wave function φ with discrete coordinates, such as one does in the Dirac equation with the spin, or employ vectors and tensors as in the Maxwell equations.

Wigner then notes that many workers have followed this road, and gives a list of papers both predating and antedating the 1939 paper.[50] A drawback of this method is that "[...] derselbe physikalische Sachverhalt in verschiedene mathematische Formen gekleidet werden kann.". This nonuniqueness is exemplified by the electromagnetic field that can described by field strengths or by vector potentials.

The other way "invariantentheoretische" – group theoretical – is the one followed in the Wigner 1939 paper. In the 1947 paper, Wigner describes the method as trying to determine a relativistic invariant linear manifold of states. Wigner first outlines the conceptual basis of this approach, roughly as in the first section of the 1939 paper. He notes that the two approaches are, of course, related, but that the "wave equation" approach often is complicated because several irreducible representations are aggre-

48 See Section 2.6.

49 Predating the more often referred Bargmann–Wigner paper on wave equations [64].

50 The authors cited are Majorana, Dirac, Proca, Kemmer, Fierz, Duffin, Belinfante and O. Klein from the 1930s and Bhabha and Harish-Chandra from the 1940s.

gated.[51] On the other hand, the group theoretical way gives no hint how to introduce interactions.[52]

After this, Wigner goes on the present systems of wave equations corresponding to the representations with $m \neq 0$ and $m = 0$. One point must not be overlooked when reading the paper: as quoted above, after equation (2.67), Wigner initially makes no assumption as to the field nature of the "wave functions". In general, any representation can be described by an appropriate list of functions $\varphi_1, \varphi_2, \dots, \varphi_l$, but nothing is yet said about what variables they depend on. However, when going over to discuss wave equations, Wigner writes

> Es laßt sich zeigen, daß die Funktionen $\varphi_1, \varphi_2, \varphi_3, \dots, \varphi_l$ durch Funktionen $\varphi(p_1, p_2, p_3, p_4, \zeta)$ ersetzt werden konnen, so daß die Variablen $p_1, p_2, p_3, p_4, \zeta$ einfachen Index l ersetzen.

The p-variables are interpreted as the momentum of the particle in the standard way.[53] In an irreducible representation, only functions where $p_4^2 - p_1^2 - p_2^2 - p_3^2$ takes a definite value, occur. In the case when this value is positive, there is a representation for ζ taking values $0, 1, 2, 3, \dots$ corresponding to the spin values $0, \frac{1}{2}, 1, \frac{3}{2}, \dots$ (the classes P_+ and P_-). Wigner notes that, since the wave equations in the previous works he referred to, all describe particles with nonzero mass, they all correspond to combinations of representations with various masses and spin of this kind. Wigner then considers wave functions $\psi(x, y, z, t, \sigma_1, \sigma_2, \dots, \sigma_{2s})$ symmetric in $2s$ "Spinkoordinaten" of the Dirac kind, that is, the σ_j are gamma-matrices γ_{jl} satisfying the usual anticommutators and matrices with different index j commute. The equations read[54]

$$\left(\frac{h}{i}\sum_{jl}\gamma_{jl}\frac{\partial}{\partial x_l} - 2smc\right)\psi(x_l, \sigma_j) = 0 \tag{2.89}$$

$$h^2\left(\frac{\partial^2}{\partial x_1^2} + \frac{\partial^2}{\partial x_2^2} + \frac{\partial^2}{\partial x_3^2} + \frac{\partial^2}{\partial x_4^2}\right)\psi = m^2c^2\psi \tag{2.90}$$

For $s = 0$, there is no equation (2.89). For $s = 1/2$, equation (2.90) is a consequence of (2.89), but not so for $s > 1/2$. Then the Klein–Gordon equation must be supplied

---

51 The contrast between the two approaches is sometimes overemphasized in the secondary literature. True, relativistic wave equations constructed ab initio, often turn out to correspond to reducible representations, but on the other hand, there is no way to derive wave equations from the irreducible representations. As we will see, assumptions has to be made as to what spaces to realize the "wave equations" on.

52 Unless, as proposed by Dirac in [61], one looks for nonlinear realizations of the group.

53 The point discussed here may be of some relevance for the higher spin problem. Although in almost all cases one would like to have momentum space or configuration space fields, it may very well be that higher spin interactions must be described in some other kind of space. The representation theory of the Poincaré group leaves that option open.

54 Wigner denotes space-time indices with l here and takes $x_4 = ct$.

separately in order to have an irreducible representation. The discussion then shifts to the massless representations. There is a paragraph in the paper about setting $m = 0$ in the wave equations. We will defer commenting on this issue to the next section in connection with the Bargmann–Wigner paper of 1947.

Much of the paper concerns the so-called *infinite spin* or *continuous spin* representations, which belong to the case $m = 0$, but had not been discussed in any detail before. Wigner notes that for a wave travelling in the z-direction with momenta $(0, 0, p, p)$ there are Lorentz transformations

$$\begin{pmatrix} 1 & 0 & 0 & 0 \\ 0 & 1 & -\gamma & \gamma \\ 0 & \gamma & 1-\frac{1}{2}\gamma^2 & \frac{1}{2}\gamma^2 \\ 0 & \gamma & -\frac{1}{2}\gamma^2 & 1+\frac{1}{2}\gamma^2 \end{pmatrix} \quad \text{and} \quad \begin{pmatrix} 0 & 1 & -\gamma' & \gamma' \\ 0 & 1 & 0 & 0 \\ \gamma' & 0 & 1-\frac{1}{2}\gamma'^2 & \frac{1}{2}\gamma'^2 \\ \gamma' & 0 & -\frac{1}{2}\gamma'^2 & 1+\frac{1}{2}\gamma'^2 \end{pmatrix} \tag{2.91}$$

and their product (the matrices commute)

$$\begin{pmatrix} 1 & 0 & -\gamma' & \gamma' \\ 0 & 1 & -\gamma & \gamma \\ \gamma' & \gamma & 1-\frac{1}{2}\gamma^2-\frac{1}{2}\gamma'^2 & \frac{1}{2}\gamma^2+\frac{1}{2}\gamma'^2 \\ \gamma' & \gamma & -\frac{1}{2}\gamma^2-\frac{1}{2}\gamma'^2 & 1+\frac{1}{2}\gamma^2+\frac{1}{2}\gamma'^2 \end{pmatrix} \tag{2.92}$$

that leave the wave invariant. In this case, the spin coordinate ζ, can be replaced by two continuous variables π and π'. These variables correspond to the variables ξ and η of the 1939 paper. The notation in the Wigner papers is compared in Table 2.1 below. After briefly summarizing the results of the 1939 paper, Wigner turns to the wave equations.

Table 2.1: Notation for the massless little group in Wigner's papers of 1939, 1947 and 1963.

Symbol	Wigner 1939	Wigner 1947	Wigner 1963
Spin coordinates	ξ, η	π, π'	π'', π'
Lorentz parameters	x, y	λ, λ'	β, α
2-dimensional invariant product	$x\xi + y\eta$	$\lambda\pi + \lambda'\pi'$	$\beta\pi'' + \alpha\pi'$
"Casimir" of the little group	$\xi^2 + \eta^2 = \Xi$	$\pi^2 + \pi'^2$	$\pi'^2 + \pi''^2 = \Xi^2$

Rosetta stone for the Wigner 1939, 1947 and 1963 papers

i

The notation regarding the infinite, or continuous spin, representations are different, and a bit confusing, in the three Wigner papers. For the translation part of the little group, the table 2.1 translates between the papers.

The two Lorentz matrices corresponding to little group translations denoted by $T_\xi(\alpha)$ and $T_\eta(\beta)$ in the 1963 paper correspond to the matrices (2.91) of the 1947 paper. The four-dimensional transformation matrix (2.92) corresponds to the two-dimensional complex matrix (see (2.79))

$$\begin{pmatrix} 1 & \gamma + i\gamma' \\ 0 & 1 \end{pmatrix} \tag{2.93}$$

The wave equations are given without much in the way of motivation. Wigner writes

> Es ist unschwer, eine relativistisch invariante Gleichung im Impulsraume aufzuschreiben, deren losungen die obengenannte Mannigfaltigkeit bilden. Die Wellenfunktionen hangen außer von p_1, p_2, p_3, p_4 noch von vier anderen Vektorkomponenten $\xi_1, \xi_2, \xi_3, \xi_4 = c\tau$ ab. Im Falle des ganzzahligen unendlichen Spins haben wir
>
> $$(p_4^2 - p_1^2 - p_2^2 - p_3^2)\varphi = 0 \tag{2.94a}$$
> $$(p_4\xi_4 - p_1\xi_1 - p_2\xi_2 - p_3\xi_3)\varphi = 0 \tag{2.94b}$$
> $$(\xi_1^2 + \xi_2^2 + \xi_3^2 - \xi_4^2 - l^2)\varphi = 0 \tag{2.94c}$$
> $$\left(p_1\frac{\partial}{\partial \xi_1} + p_2\frac{\partial}{\partial \xi_2} + p_3\frac{\partial}{\partial \xi_3} + p_4\frac{\partial}{\partial \xi_4} + ih\Xi\right)\varphi = 0 \tag{2.94d}$$

Wigner, however, discusses how the two variables π and π' subject to a condition on the sum $\pi^2 + \pi'^2$, can be replaced by the four variables $\xi_1, \xi_2, \xi_3, \xi_4$ subject to three conditions on the wave-functions, expressed by the three equations (2.94b)–(2.94d). This way of writing the equations make it easy to read off their relativistic invariance. The equations define four operators Q acting on wave functions Ψ such that

$$Q_1\Psi = 0 \qquad Q_2\Psi = 0 \qquad Q_3\Psi = 0 \qquad Q_4\Psi = 0 \tag{2.95}$$

Wigner then comments, that although the equations may look arbitrary, and one can write down other similar systems of equations, they are all equivalent according to the general representation theory of the 1939 paper, provided that the equations are consistent and do not describe particles with finite spin or imaginary mass.[55] The consistency requirement is expressed as a set of commutation relations between the operators Q that has to be satisfied. The operator Q_1 (i. e., the "Klein–Gordon" operator) commutes with all the rest, for which we have

$$[Q_2, Q_3] = 0 \qquad [Q_2, Q_4] = Q_1 \qquad [Q_3, Q_4] = -2Q_2 \tag{2.96}$$

So far, the equations describe the single valued, integer spin, representations. For the double valued, half integral, representations, the first equation (Klein–Gordon) is replaced by

$$\left(\frac{1}{c}\frac{\partial}{\partial t} - \sum s_k \frac{\partial}{\partial x_k}\right)\Psi = 0 \tag{2.97}$$

where the s_k are "die Paulische Spinmatrizen".

---

55 Wigner states in several places that while the states themselves, the particles, are uniquely determined by the representation theory, the wave equations are not so.

Let us comment here that the "operator algebra" of (2.96) is reminiscent of the kind of the "first class constraint algebra" that later appeared in Dirac's analysis of constrained Hamiltonian systems (see Sections 3.2.4 and 3.3.3).

The rest of the 1947 paper is devoted to questions having to do with the scalar product in the space of wave functions. It ends with a few comments on the physical interpretation of the infinite spin representations.

Second – or third – thoughts on Wigner's "two ways"

Regarding the two ways of approaching the subject of relativistic wave equations that Wigner discussed in the Introduction to the paper: "wave equation first" versus "representation first", it must be pointed out that the situation is not so clear-cut. Three points may clarify the issue. First, Wigner puts the Majorana 1932 and Dirac 1936 papers in the first category. However, although both papers start with wave equations of the Dirac type, Majorana constructs infinite dimensional unitary representations of the Lorentz group, and Dirac constructs finite dimensional nonunitary representations. The categorization is a therefore not unambiguous. Second, one must not see the categorization as a value judgement. Wigner's method leads from irreducible representations to wave equations for particles with a certain mass and definite spin. The other method often lead to equations describing several mass and spin particles. This is not necessarily a drawback of the method, and was not seen as such by the authors, as multiparticle wave equations could be phenomenologically interesting. From the modern higher spin perspective, where we know that we must consider infinite spectra of spin, irreducibility is also not so important. Third, irreducibility of wave equations come with a price: wave equations often must be supplied with subsidiary conditions, or what essentially amounts to sets of equations, as we have seen above, and will see in the sequel. Such subsidiary conditions lead to problems with interactions, even simple interactions such as the electromagnetic, as the Fierz–Pauli paper pointed out. Indeed, Wigner computed the commutators between his field equations to check their internal consistency. Such computations in general lead to the Fierz–Pauli problem, when external fields are coupled.

2.3.3 The Bargmann–Wigner paper of 1948

This is the paper [64] that is most often referred to regarding relativistic wave equations based on the representation theory. The paper summarizes the results of the 1939 paper, and then goes on to a systematic discussion of wave equations for the most interesting representations: the massive with finite spin (now denoted by P_s) and the massless with finite spin (O_s) and infinite – or continuous – spin ($O(\Xi)$). In all cases, there is a detailed discussion about the norm, or invariant scalar product, in the space of wave functions. The form of the spin operators and of the Casimir operators are also treated in all cases. Indeed, the infinitesimal operators of the Poincaré group, that did not play any significant role in the 1939 paper,[56] are now introduced, as well as the Pauli–Lubanski vector. Here, we will be content to focus on the wave equations.

---

56 See Section 2.A in the 1939 paper.

In the case P_s with $s = 0$, the wave equation is $p^k p_k = m^2$ acting on a one-component wave function ψ. In the higher spin cases, with $s = N/2$ and $N = 1, 2, 3, \ldots$, wave functions $\psi(p; \zeta_1, \ldots, \zeta_N)$ are chosen that depend on N four-valued variables $\zeta_1, \ldots, \zeta_N$ just as in the "Wellengleichungen" paper. Again, for each ζ_ν, four-dimensional gamma-matrices γ_ν^k are introduced, commuting for different values of the "spin index" ν. However, in contrast to the "Wellengleichungen" paper, the wave equations are now given as

$$\gamma_\nu^k p_k \psi = m\psi \quad (\nu = 1, 2, \ldots, N) \tag{2.98}$$

From any one of these equations, one gets the Klein–Gordon equation $p^k p_k \psi = m^2 \psi$ by acting with an operator $\gamma_\nu^k p_k$ and using the gamma-matrix algebra and using the wave equation again. These wave equations then correspond to irreducible representations. To investigate that, Bargmann and Wigner defines a little group by choosing a momentum with components $(0, 0, 0, m)$.[57] Then they assume that γ^4 is diagonal with components $(1, 1, -1, -1)$.[58] For each spin coordinate, the wave equation then reduces to

$$\begin{pmatrix} 1,0,0,0 \\ 0,1,0,0 \\ 0,0,-1,0 \\ 0,0,0,-1 \end{pmatrix} \begin{pmatrix} \psi_1 \\ \psi_2 \\ \psi_3 \\ \psi_4 \end{pmatrix} = \begin{pmatrix} 1,0,0,0 \\ 0,1,0,0 \\ 0,0,1,0 \\ 0,0,0,1 \end{pmatrix} \begin{pmatrix} \psi_1 \\ \psi_2 \\ \psi_3 \\ \psi_4 \end{pmatrix} \tag{2.99}$$

Thus only the first two components are nonzero. This reduces the number of components of the wave function from 4^N to 2^N. This number is further reduced to $N + 1 = 2s + 1$ due to the total symmetry in the ζ coordinates as applied to the first two components of each ψ. This shows that the wave equations (2.98) describe a massive particle with spin and correspond to an irreducible representation. These equations are sometimes referred to as the *Bargmann–Wigner equations*. As such, they are equivalent to the Dirac–Fierz–Pauli equations discussed in Section 2.1.3 through the standard transcription between four-component and two-component spinors (see Section 1.4 formula (1.14)). The equivalence to the more common traceless and divergence-free, symmetric tensor formulation or tensor-spinor in the half- integer spin case, will be discussed in Section 2.4.1.

The next item is the class O_s of representations. The wave equations are obtained by setting $m = 0$ in (2.98). It is argued, in the case $m = 0$, that the linear manifold defined by the wave equation can be decomposed into invariant manifolds with definite

57 Corresponding to choosing a frame where the particle is at rest and then counting the number of independent field components.

58 The "Dirac" representation.

values for any one of the operators $\Gamma = i\gamma_\nu^{\,1}\gamma_\nu^{\,2}\gamma_\nu^{\,3}\gamma_\nu^{\,4}$. In particular, the following split is considered:

$$\Gamma\psi = \psi \qquad (\nu = 1, 2, \ldots, N) \tag{2.100}$$

$$\Gamma\psi = -\psi \quad (\nu = 1, 2, \ldots, N) \tag{2.101}$$

"Both manifolds are invariant under proper Lorentz transformations but go over into each other by reflections: they correspond physically to right and left circular polarization.".[59]

In this case, the little group is defined by a momentum of the form $(0, 0, 1, 1)$. The wave equations become, after multiplication with γ_ν^3

$$\gamma_\nu^{\,3}\gamma_\nu^{\,4}\psi = \psi \quad (\nu = 1, 2, \ldots, N) \tag{2.102}$$

The authors then "assume" that the $\gamma_\nu^{\,3}\gamma_\nu^{\,4}$ matrices are diagonal with elements $(1, 1, -1, -1)$. Since the Γ_ν matrices commute with the $\gamma_\nu^{\,3}\gamma_\nu^{\,4}$, but are not identical with them, they may also be assumed to be diagonal with elements $(1, -1, 1, -1)$. Now, in the manifold defined by (2.102) and (2.100) only the first component of each ψ can be nonzero. In the manifold defined by (2.102) and (2.101), only the second component of each ψ can be nonzero.[60] One thus gets two one-dimensional manifolds of states: "For a given momentum, ψ has only two independent components.".

Comparing the Wigner 1947 and Bargmann–Wigner 1948 wave equations

The reader has certainly noticed a curious difference as to how the wave equations are presented in these two papers. Provided that the wave function ψ is the same in the two papers, the wave equation (2.89) of the Wigner paper corresponds to the sum of the N wave equations (2.98) of the Bargmann–Wigner paper. The first wave equation is therefore weaker than the second set of equations. This may also be surmized by the fact that the Klein–Gordon equations cannot be inferred in the first case, but has to be supplied separately. There is, as far as I can tell, only one explicit comment on the contrasting choices of wave equations. In footnote 8 of the Bargmann–Wigner paper, it says that the sum over all ν was postulated in a paper by H. A. Kramers, F. J. Belinfante and J. K. Lubanski (our [65]) and that this form of the equations were used in the Wigner 1947 paper. This type of equation is closely related to Belinfante's "undor" theory [66] where Belinfante studied quantities that transformed as products of Dirac wave-functions under Lorentz transformations (including reflections).

Implicitly, however, there are two comments in the Wigner paper. The first concerns the need to supply the Klein–Gordon equations separately. It is needed in order to avoid complications that are discussed in a paper by O. Klein [67] (as stated by Wigner, but not elaborated), and to make the system irreducible. Then as a second comment, it is mentioned that the equations are just an example,

59 It is what we now call a chirality split.

60 In the chiral representation of the gamma-matrices, $\gamma^3\gamma^0$ is diagonal with elements $(-1, 1, 1, -1)$ and γ^5 is diagonal with elements $(1, 1, -1, -1)$. The corresponding argument goes through.

and that another example is the equations of the Majorana 1932 paper. As we saw in Section 2.1.6, Majorana does neither assume the Klein–Gordon equations, nor the Dirac-type gamma matrices.

Furthermore, regarding going from $m \neq 0$ to $m = 0$, no subtleties connected with this are noted in the Bargmann–Wigner paper, apart from the drop in number of components from $2s + 1$ to 2. No actual limiting process $m \to 0$ is involved. In the "Wellengleichungen" paper, Wigner writes

> Die Sachlage ist die, daß fur $s > 1/2$ die durch [our equations (2.89), (2.90)] beschriebene Mannigfaltigkeit nicht mehr irreduzibel ist, wenn man darin $m = 0$ setzt[removed footnote], sondern in mehrere Mannigfaltigkeiten zerfallt, denen man die Spins $s, s-1, s-2$, zuschrieben kann. Diese Mannigfaltigkeiten enthalten fur gegebene p zwei linear unabhangige Elemente. Nur die Mannigfaltigkeit mit $s = 0$ enthalt nur ein Element, wenn die p gegeben sind.

So in this case there is no drop in the number of components, but rather a redistribution according to $2s + 1 = 2 + 2 + \cdots + 2 + 1$, the sum containing s terms equal to 2.

For the continuous spin representations, the equations for the case $O(\Xi)$ and $O'(\Xi)$ are taken from the Wigner 1947 paper.

2.3.4 The 1963 "review" paper

This paper is quite interesting. It was presented at an IAEA Seminar on Theoretical Physics, in Miramare, Trieste, in 1962 [247]. Wigner writes that A. Salam had asked him to report "[...] on equations for elementary particles which are not believed to exist in nature.". The rationale for this strange request was the interest, at that time, in studies of scattering amplitudes where the momenta of the particles were extended into the complex plane. The focus of the paper is therefore on equations for the continuous spin representation and representations with imaginary mass. Apart from this, the paper spells out explicitly a general method to set up wave equations corresponding to a given representation.

The first section of the paper is introductory and outlines its context. Sections II–V review the unitary representations of the Poincaré group, the little group, infinitesimal and Casimir operators and the case of positive rest mass.

In the case of zero rest mass, the little group again can be considered to consist of rotations in the xy plane as well two sets of commuting operations $T_\xi(\alpha)$ and $T_\eta(\beta)$, which together form a group isomorphic to the two-dimensional Euclidean group. The four-dimensional matrices $T_\xi(\alpha)$ and $T_\eta(\beta)$ correspond to the matrices (2.91) of the 1947 paper (with the translation $\alpha \leftrightarrow y'$ and $\beta \leftrightarrow y$). About the geometrical interpretation of the little group, Wigner writes:

> Clearly, there is no plane in the four-space of momenta in which these transformations could be interpreted directly as displacements and rotations because all transformations considered are homogeneous. The simplest geometrical picture known to me uses two vectors p^ξ and p^η, of length -1 and orthogonal to each other as well as to p^0. These vectors could be unit vectors parallel to the x and y axes. The $T_\xi(\alpha)$ then adds αp^0 to p^ξ, whereas $T_\eta(\beta)$ then adds βp^0 to p^η.

The representation theory of the massless little group is treated at some length in the paper (Section VI.A). In particular, it is again pointed out that the little group method can be applied to the little group itself, so that if π' and π'' are interpreted as "momenta", then one must have $\pi' + \pi''^2 = \Xi^2$.

But let us turn to the question of setting up wave equations (Section VI.B). Since this is a subject that is very seldom treated in the literature, we will take the opportunity to follow Wigner in detail.[61] Wigner first notes that all known zero mass equations permit only solutions belonging to the representation O_s or its complex conjugate: "[...] the negative energy solutions which are then eliminated or reinterpreted in the second quantized form of the theory.". On the other hand, equations for the $O(\Xi)$ and $O'(\Xi)$ cases, were only obtained after the general method for obtaining equations from representations was devised. Only the case $O(\Xi)$ is then treated (the case $O'(\Xi)$ requires spinor wave functions). It is also noted that the term "equation for a representation" is not clearly defined and that several equations may correspond to the same representation. Not even the variables upon which the wave function depend are determined by the representation. The objective of the method is to systematically delimit the choices.

Wigner's method for obtaining wave equations from representations

The wave function, which may have one or more components, satisfying one or more equations, should transform under the Poincare group according to the representation in question. Then the variables upon which the wave function depend, should be of such a nature that they clearly indicate how the wave function transform. This means that the variables must be four vectors, called "ordinary vectors", or differences between four vectors, called "difference vectors". Next, Wigner goes on to determine the number of vectors and difference vectors needed.

The principle is to find precisely the right number of variables so that every Poincaré transformation that changes the representation, also changes the wave function. The intuition behind this can be understood as follows. Suppose we were interested in finding the wave function for the spin 1 massless representation. A scalar field would obviously not do since it is invariant under the spin part of a Lorentz transformation. On the other hand, a tensor field φ_{kl} would be redundant since the symmetric part can be set to zero.

Thus, the variables should be able to completely describe a frame of reference. A frame of reference can be given by an ordinary vector, defining the origin and four difference vectors, defining the coordinate axes. These are too many, but the number can easily be reduced.

Wigner starts with the difference vectors. One of them can naturally be identified with the momentum vector p and given the length m^2. The other three difference vectors are assumed to be mutually orthogonal, orthogonal to the momentum vector and of length 1 or -1, whichever is possible. One of them is therefore completely fixed in terms of the momentum and the other two. Another one of them is fixed up to a single variable. This one variable will also turn out to be unnecessary, although it is not immediately clear, apart form the fact that we want to have quadruplets of vector components. We are

61 The section in the paper is interesting and well worth reading in its entirety.

left with two difference vectors, p and ξ say. The conditions on them give the wave equations

$$(p \cdot p)\psi = m^2\psi \tag{2.103}$$
$$(\xi \cdot \xi)\psi = -\psi \tag{2.104}$$
$$(p \cdot \xi)\psi = 0 \tag{2.105}$$

These equations are so far common to all representations. The wave function ψ depends on the components of the difference vectors p and ξ as well as on the ordinary vector x that allows the wave function to vary under translations. This variation yields one more equation common to all representations. In order to represent the translation part of the Poincaré group according to formula (2.73), the wave functions must satisfy

$$\frac{\partial}{\partial x_k}\psi = -ip^k\psi \tag{2.106}$$

Then the vectors x are "unnecessary variables". If ψ is given as a function of p and ξ for say $x = 0$, equation (2.106) determines ψ for all x according to

$$\psi(x, p, \xi) = e^{-ip\cdot x}\psi(0, p, \xi) \tag{2.107}$$

By integrating over p and x, respectively, this provides us with the familiar Fourier transform pairs of p-space or x-space wave functions. We therefore has the option to work with wave functions that depend on either x, ξ or p, ξ with the transcription $\partial/\partial x = -ip$.

What remains to be done is to find further equations defining the representation in question. Wigner exemplifies for the case P_0, and then goes on to the case $O(\Xi)$. There is first a geometrical interpretation for the constraints (2.104) and (2.105). Then Wigner goes on to calculate an expression for the square $W = -w \cdot w$ of the Pauli–Lubanski vector

$$w^k = \frac{1}{2}\epsilon^{klmn}p_l M_{mn} \tag{2.108}$$

where the angular momentum operators are taken as

$$M_{mn} = i\left(p_m\frac{\partial}{\partial p_n} - p_n\frac{\partial}{\partial p_m} + \xi_m\frac{\partial}{\partial \xi_n} - \xi_n\frac{\partial}{\partial \xi_m}\right) \tag{2.109}$$

The space-time momentum part of M_{mn} does not contribute to w^k (due to the contraction of two momenta into the antisymmetric ϵ tensor). Even so, W is a quite complicated expression, but acting on the wave function and taking the already established wave equations (2.103)–(2.105) into account, one finally arrives at

$$W = -\frac{\partial^2}{\partial \xi^n \partial \xi_l}p^n p_l\psi = \left(i\frac{\partial}{\partial \xi_l}p_l\right)^2\psi \tag{2.110}$$

Since $W\psi = \Xi^2\psi$, the linear space of wave functions can be decomposed into two subspaces with

$$i\frac{\partial}{\partial \xi_l}p_l\psi = \pm\Xi\psi \tag{2.111}$$

Wigner then argues that these two spaces are equivalent, and that one can choose the positive sign. This then is the fourth wave equation for the representation space $O(\Xi)$. We thus arrive at the wave equations (2.94) of the 1947 paper.

There is obviously a lot more that can be said about these representations and the corresponding wave equations, and Wigner does so, but we will defer further discussion to later chapters.

The paper [68], written for a conference on group theoretical concepts in theoretical physics, is a review paper and it does not contain much new on the subject, apart from simplifications and a new formalism. The paper discuss space and time reflection symmetry. As an aside, it can be noted that Wigner did not write more on Poincaré representations. Obviously, he thought that not much more could be said on the topic.

2.4 The 1940s and early 1950s

In very broad terms, there was a hiatus in fundamental research during the Second World War, except military related. After the war, focus continued to be on nuclear physics, but there was also a return to the problems of quantum electrodynamics, which were quite rapidly solved in the well known way. Experimentally, the 1950s was a period of building new powerful particle accelerators. It is in a way paradoxical that the destructive development of bombs and military technology[62] had, as a side effect, a positive effect on fundamental physics research. In 10 years time, after the end of the war, there would be ample reasons to return to higher spin massive particles and fields. But a few new developments occurred just before, during and just after the war.

2.4.1 Rarita and Schwinger

A couple of years after the Fierz and Pauli paper, in 1941, came the short Rarita–Schwinger letter on half-integral spin fields [69]. The paper introduces the now common way of representing higher spin fermions in terms of tensor-spinors, instead of the "complicated" – as the authors write – spinor formalism of Dirac and Fierz–Pauli. The field equations, with the spinor index suppressed, appear as

$$(\gamma^{\tau}\partial_{\tau} + \kappa)\Psi_{\mu_1\ldots\mu_k} = 0 \quad \text{with} \quad \gamma^{\alpha}\Psi_{\alpha\mu_2\ldots\mu_k} = 0 \tag{2.112}$$

As the authors write, the supplementary conditions from integer spin theory

$$\partial^{\alpha}\Psi_{\alpha\mu_2\ldots\mu_k} = 0 \quad \text{and} \quad \Psi^{\alpha}{}_{\alpha\mu_3\ldots\mu_k} = 0 \tag{2.113}$$

follows from the second, *γ-trace* equation of (2.112).

62 An example is Lamb–Retherford of measurement of the Lamb shift using new microwave technology.

Spin 3/2 is then treated as a special case, a Lagrangian is given and it is pointed out that no auxiliary spinors are needed. In the case of zero mass, the theory is gauge invariant. Subsequently, spin 3/2 fields became known as *Rarita–Schwinger fields*.[63] Electromagnetic coupling could be introduced without the Fierz–Pauli inconsistency appearing. However, in 1969, G. Velo and D. Zwanziger [70] showed that a more subtle inconsistency – waves propagating faster than light – arises when charged spin 3/2 fields are coupled to the electromagnetic field.[64]

Jumping ahead a bit, in 1954, a new formulation of the spin 3/2 wave equation was proposed by S. N. Gupta [71]. He combined the two-component spin 3/2 spinors $a^{\dot{\alpha}}_{\beta\nu}$, $b^{\dot{\alpha}\dot{\beta}}_{\nu}$ and the spin 1/2 auxiliary spinors c_α, $d^{\dot{\alpha}}$ of Fierz and Pauli, into a 16- component object ψ. From there on, a Lagrangian theory could be developed yielding a wave equation $\alpha_\mu(\partial\psi/\partial x_\mu) + \kappa\psi = 0$ with the 16×16 matrices α_μ satisfying an algebra $\sum(\alpha_\mu\alpha_\nu - \delta_{\mu\nu})\alpha_\lambda\alpha_\rho = 0$ where $\sum$ denotes a sum over all permutations of the indices μ, ν, λ and ρ. Gupta quantized the theory and introduced electromagnetic interactions in the standard fashion. From a higher spin perspective, this paper underscores the Wigner point: wave equations are not unique. We have seen three, in their appearance at least, different formulations of spin 3/2 theory.[65]

Then in 1955, P. A. Moldauer and K. M. Case published a paper on half-integer spin particles, in particular spin 3/2 and spin 5/2, motivated by the "[...] many new particles whose spins and moments have not yet been measured [...]" [72]. The authors derive the spinor-tensor form of the wave equations (2.112) and (2.113) from the Dirac–Fierz–Pauli equations. They then proceed from there on to study electromagnetic interactions of the particles and compute magnetic moments for spin 3/2 and 5/2.

2.4.2 The representations of the Lorentz group

Toward the end of the 1940s, the problem of the representations of the homogeneous Lorentz group was completely solved by a number of workers, one of them Harish-Chandra, who worked with H. J. Bhabha in India before going to Cambridge to study under Dirac for a PhD.[66] The situation regarding the general representations was very well summed up in the first paragraph of Harish-Chandra's paper [74], that also is his thesis:

> All the finite irreducible representations of the Lorentz group are well known. Every such representation is characterized by an ordered pair of numbers p and q such that $2p$ and $2q$ are integral

63 In the same issue of Physical Review, the Rarita–Schwinger letter is followed by a letter (to which Rarita and Schwinger refer) by S. Kusaka on β-decay with a tentative spin 3/2 neutrino.

64 We will return to this problem in volume 2 of the present work.

65 Gupta does not refer to the Rarita–Schwinger paper.

66 For a biography of Harish-Chandra, see [73].

> and ≥ 0. None of these representations, however, is [sic] unitary. Dirac (1945) has recently drawn attention to the existence of some unitary, though infinite, representations with a view to their possible physical applications. The present paper is concerned with the investigation of the general irreducible representations of the proper Lorentz group.

Then follows a succinct summary of the main, somewhat involved, results of the paper.[67]

> It is found that such a representation can, in general, be characterized by an ordered pair (k, k^*) of complex numbers such that $2(k - k^*)$ is an integer and may therefore be denoted by $\mathcal{D}(\kappa, \kappa^*)$. However, in case both κ and κ^* are of the form $\frac{1}{4}n$ $(n \neq -2)$ there exist two irreducible representations $\mathcal{D}^+(\kappa, \kappa^*)$ and $\mathcal{D}^-(\kappa, \kappa^*)$ corresponding to the pair (κ, κ^*). The finite representations correspond to $\mathcal{D}^-(\kappa, \kappa^*)$ with κ, κ^* both of the form $\frac{1}{2}n$ $(n \neq -1)$, the connexion between (p, q) and (κ, κ^*) being given by $p = |\kappa + \frac{1}{2}| - \frac{1}{2}$, $q = |\kappa^* + \frac{1}{2}| - \frac{1}{2}$. In general $\mathcal{D}(\kappa, \kappa^*)$ is unitary only if either
>
> $$\kappa = -\frac{1}{2} + i\nu + \frac{1}{2}n, \quad \kappa^* = -\frac{1}{2} + i\nu - \frac{1}{2}n \tag{2.114}$$
>
> $$\text{or} \quad \kappa = \kappa^* = -\frac{1}{2} + \nu, \quad |\nu| \leq \frac{1}{2} \tag{2.115}$$
>
> Here, ν is an arbitrary real number. However, in the special case when κ, κ^* are both of the form $\frac{1}{4}n$ $(n \neq -2)$, $\mathcal{D}^-(\kappa, \kappa^*)$ is unitary only if $\frac{1}{2} \geq |\kappa + \frac{1}{2}| = |\kappa^* + \frac{1}{2}|$, while $\mathcal{D}^+(\kappa, \kappa^*)$ is unitary whenever $|\kappa + \frac{1}{2}| = |\kappa^* + \frac{1}{2}|$.

For finite component field theory, it is the finite, nonunitary representations that are interesting. We will return to them in Chapter 3.

Wave functions and/or fields in the 1930s and 1940s

There seems to be a persistent confusion throughout the literature over the concepts of "waves" versus "fields", ranging from a real confusion over concepts to an inattentive use of words (no doubt, sometimes the confusion rests with the reader). Involved in this are the designations "classical" versus "quantum", "nonrelativistic" versus "relativistic", as well as the ideas of first and second quantization. The issues are tied up with the way the theoretical development of quantum mechanics came about, but also with phenomenological questions regarding the nuclear forces and their description. One way of understanding the theoretical problems may be the following.

Although quantum mechanics was developed with clear knowledge of special relativity, what initially worked very well (and still works well) and solved the paradigm crisis from the end of the nineteenth century, was the nonrelativistic theories of Heisenberg and Schrödinger. In particular, the Schrödinger equation, which was a "wave equation". But Schrödinger had already found a relativistic "wave equation", soon found by other authors,[68] and from Dirac's 1936 paper, the search was on

67 In the quote below, I have made typographical changes to notation as well removed a footnote explaining the meaning of "integer". See paragraph 17 of Corson [23], or the original paper by Harish-Chandra [74]. Others workers credited with these results are Bargmann [75] and I. M Gelfand and M. Naimark. For references to work by these authors; see the book [76].

68 for instance, V. A. Fock and T. E. de Donder.

for relativistic wave equations. But the waves of these relativistic wave equations had various related problems, as we have seen, and could not be interpreted as probability waves. The problems were solved by a second round of quantization.

As regards the phenomenological side, in the second half of the 1930s, efficient computational techniques were needed in connection to the Yukawa meson theory of the nuclear force. Ontologically, however, it seems that the meson, thought to mediate the nuclear force, was regarded in two different ways according to whether one was interested in its electromagnetic interaction or nuclear interactions. This is clear from the following quote from [77].

> Up to the present much work on meson theory has been done by considering it as a field theory, and the equation as field equation. This situation has its origin in the fact that the meson was found originally as a field of the heavy particles. However, if we restrict ourselves to the problems of the interaction between the meson and the electromagnetic field, it seems more adequate to treat the meson equation in the form of a wave equation just like the case of other charged particles [...].

From this quote, we can read off a thinking where the meson was considered a "wave" when it interacted with electromagnetism, but as a "field" when it itself mediated the nuclear force. These authors also comment (in a footnote that I have removed from the quote), on a similar discussion in the introduction to the Kemmer 1939 paper (see our Section 2.2.3).

Let us now take up the story where we left it in late 1930s. Much was known about relativistic wave equations, but the subject lacked systematics. A state of the art review [54] regarding lower spin theory, was published by Pauli in 1941.

2.4.3 H. J. Bhabha's general theory of wave equations

In the mid-1940s, much was known about relativistic wave equations, but the subject lacked systematics, and there were still confusing issues having to do with the wave-particle duality, second quantization and unitarity versus nonunitarity of the representations.

As we have had occasion to remark in several places, the origin of the subject of relativistic wave equations within the development of quantum mechanics itself, lead to a confusion about the ontological nature of the wave equations. This confusion seems to have persisted well into the 1960s. From a mathematical point of view, a wave equation is just a partial differential equation, and a wave function is just a function. However, in nonrelativistic quantum mechanics, the wave function, governed by the Schrödinger equation, describes a quantum state and it is interpreted as probability amplitude. It was only natural to take over this ontology to the Dirac equation and Klein–Gordon equation and the further wave equations that were invented and studied. Apart from other problems with this flawed conception – as it turned out to be – the spin 1 wave equation for the electromagnetic field did not fit into this conceptual scheme very well. Since it was a classical field, it could only be "first" quantized, while the "already quantized" relativistic wave equations, must be "second" quantized. The eventual resolution of all this is now well known. Relativistic wave equations do not govern quantum states, that is, quantum wave functions, but quantum field operators. This is the essence of the ontology of quantum field theory. We make this remark here, because, as we will also occasionally remark, it may be helpful to have in the back of one's mind when issues about unitarity or nonunitarity of representation of the Poincaré and Lorentz group come up, as it does regularly. In short: quantum states need to be unitary, operators not.[69]

The Bhabha papers [78–80], as many papers from this time and tradition, more or less explicitly made a direct connection between elementary particles and wave equations. Every elementary particle was thought to have its own governing relativistic wave equation as a free, noninteracting, particle. Investigating the general form of such wave equations therefore was an important research area. The application of various wave equations to calculating properties of elementary particles then became in the 1950s a driving force behind this research, alongside the theoretical interest. When Bhabha wrote, however, no particles with spin exceeding 1 were known.

In the paper from 1945 [79], Bhabha studies linear wave equations of the Dirac-like type

$$(p_k \alpha^k + \chi)\psi = 0 \tag{2.116}$$

where p_k are the usual space-time differential operators multiplied by i and α^k are four $d \times d$ matrices describing the spin properties of the particle. Bhabha works in four dimensions, so k runs from 0 to 3. The goal is to determine the nature of these matrices. He makes no assumption about χ apart from it being a constant.[70] We recognize the equations as having the same form as the Majorana 1932 equation, but since there is no positivity requirement on χ, we will not be lead to infinite dimensional unitary representations. But like Majorana, Bhabha does not require the Klein–Gordon equation to hold for the individual components of ψ; therefore – as he explains clearly – his theory will not be equivalent to the Dirac–Fierz–Pauli theory. For short, there will be no subsidiary conditions.[71] This is one of the main motivations for Bhabha's work: to study linear wave equations without any subsidiary conditions.

69 This is not to say that the Bhabha papers – the subject of the present section – suffer from this conflation of notions more than other then contemporary papers on relativistic wave equations. The Bhabha papers are clearly written, and perhaps therefore the distinctions here discussed come more easily to mind.

70 In principle, χ could be a matrix. However, if it is nonsingular one could multiply the equation by a multiple of χ^{-1}. At this stage, taking χ as number is no strong restriction.

71 A bit more elaborate (according to Bhabha): the Dirac–Fierz–Pauli equations connect two irreducible spinors that can be split by a transformation into two sets, one of which still connects the two irreducible spinors, while the other set of equations only involve one spinor and are subsidiary conditions.

We will now follow Bhabha a bit into his paper so that we can formulate his results more clearly. Denoting Lorentz transformations by matrices $t_k{}^l$, Bhabha starts by stating that form invariance of the wave equation (2.116) requires the matrices α^k to transform according to

$$\alpha^m = t_n{}^m (S\alpha^n S^{-1}) \tag{2.117}$$

for some nonsingular $d \times d$ matrices S to be determined. It may be good to pause and remember the details of how this comes about.[72]

Recapitulating Lorentz invariance of linear wave equations

---

i Consider two systems with coordinates x^μ and x'^μ connected by a Lorentz transformation $x' = \Lambda x$. The wave equations in the two systems are (we are here using our own conventions of Section 3.4)

$$\alpha^\mu \frac{\partial}{\partial x^\mu} \psi(x) + m\psi(x) = 0 \tag{2.118a}$$

$$\alpha^\mu \frac{\partial}{\partial x'^\mu} \psi'(x') + m\psi'(x') = 0 \tag{2.118b}$$

The wave functions (fields really) are assumed to be linearly related through the transformation

$$\psi'(x') = S(\Lambda)\psi(x) \tag{2.119}$$

where, in the general case, $S(\Lambda)$ is a nonsingular $d \times d$ matrix corresponding to d-component fields ψ. Using this transformation and the chain rule, equation (2.118b) can be referred back to the unprimed system of coordinates

$$\alpha^\mu \frac{\partial x^\nu}{\partial x'^\mu} \frac{\partial}{\partial x^\nu} S(\Lambda)\psi(x) + mS(\Lambda)\psi(x) = 0 \tag{2.120}$$

where $\partial x^\nu / \partial x'^\mu = (\Lambda^{-1})^\nu{}_\mu$. Then inserting $S^{-1}S = 1$ in equation (2.118a) and multiplying by S we find that (2.118b) follows from (2.118a) provided that

$$S(\Lambda)\alpha^\mu S^{-1}(\Lambda) = (\Lambda^{-1})^\mu{}_\nu \alpha^\nu \tag{2.121}$$

This is Bhabha's equation (2.117). In this derivation, we have tacitly assumed – as is often done – that the α matrices are the same in the two systems. Though a bit illogical, this assumption can be made, as pointed out in [27] who refer to a proof [82] of the unitary equivalence of all 4×4 gamma matrices.

The matrices S must form a d-dimensional representation of the Lorentz group. The corresponding infinitesimal generators are denoted by I^{mn}. Their algebra is

$$[I^{mn}, I^{rs}] = -g^{mr} I^{ns} + g^{ms} I^{nr} + g^{nr} I^{ms} - g^{ns} I^{mr} \tag{2.122}$$

72 For a textbook reference, see for instance [27], Section 2.2 and [81], Section 2-1-3.

The infinitesimal version of Bhabha's equation (2.117) then becomes

$$[\alpha^m, I^{rs}] = g^{mr}\alpha^s - g^{ms}\alpha^r \tag{2.123}$$

Bhabha then argues that it is consistent to choose matrices α^m so that the following equation holds:

$$[\alpha^m, \alpha^n] = cI^{mn} \tag{2.124}$$

The choice is not a mandatory, but it can be done consistent with equations (2.122) and (2.123). The constant c can take any value, so by redefining the α's it can be set to 1. This changes the parameter χ, but since it is anyway arbitrary at this stage, that is of no consequence, although it clearly related to the rest mass of the particle. Before continuing, let us dig a little deeper into the motivation behind the equation $[\alpha^m, \alpha^n] = I^{mn}$.

Bhabha's principles

In order to understand the historical context of Bhabha's motivation and approach, it is very interesting to read a review type paper [78] of Bhabha that appeared just before [79]. The paper begins with a very crisp summary of the development of atomic, nuclear and then elementary particle physics from the end of the nineteenth century to the 1940s. The well-known low spin wave theories of Klein–Gordon, Dirac, Proca and Kemmer (DKP), that described the known elementary particles, are reviewed. Bhabha then turns to higher spin wave equations and states two general principles that must be common to them all (I quote freely here).

A It can be deduced from the equations that each component of the wave function satisfies the second-order Klein–Gordon equation. This is physically equivalent to the statement that the particle described by the field has in each case only one value of the rest mass (except for sign).

B The particle-field is completely described by an equation of the Dirac form (2.116)[73] without the help of any further subsidiary conditions. The transformation properties of the wave function, and hence the spin of the particle, are determined entirely by the infinitesimal transformations I^{mn} defined by $I^{mn} = [\alpha^m, \alpha^n]$ satisfying (2.123).

The insistence on principle **B** is given in the next paragraph where the DFP theory is reviewed and criticized. First, Bhabha notes that the Dirac theory does not include all equations that are consistent with assumption **A**. There are equations

$$p^{\gamma}_{\dot{\lambda}} A^{\alpha\beta\ldots}_{\dot{\mu}\dot{\nu}\ldots} = \chi B^{\gamma\alpha\beta\ldots}_{\dot{\gamma}\dot{\mu}\dot{\nu}\ldots} \tag{2.125}$$

$$p^{\dot{\gamma}}_{\gamma} B^{\gamma\alpha\beta\ldots}_{\dot{\gamma}\dot{\mu}\dot{\nu}\ldots} = \chi A^{\alpha\beta\ldots}_{\dot{\mu}\dot{\nu}\ldots} \tag{2.126}$$

For low spin values, these equations correspond to the DKP equations. Second, Bhabha takes up the minimal coupling problem of Fierz–Pauli, and third, in connection to this, he points out that not even the free Dirac equations for higher spin can be derived from an action without the introduction of

73 Bhabha notes that it would be allowed to replace the constant mass term χ by a matrix term $\beta\chi$ with β a matrix that commutes with all the I^{rs}.

auxiliary fields. This becomes increasingly cumbersome as the spin increases. These problems then, pave the way for the alternative scheme proposed by Bhabha.[74]

Let us now return to the 1945 paper [79]. There are therefore two alternatives, Bhabha writes: either the equation $[\alpha^m, \alpha^n] = I^{mn}$ holds, or it does not hold. If it does not hold, then the only restriction on the α^m matrices is equation (2.123) governing the correct Lorentz transformation properties of the matrices. In the long technical Section 5 of the paper, Bhabha investigates the general form of the α matrices. It is found that equation (2.124) does not hold in general. In particular, it does not hold for the DFP equations beyond spin 1.

Turning instead to the case where the equation $[\alpha^m, \alpha^n] = I^{mn}$ is assumed to hold, Bhabha shows that the three sets of commutation relations given by (2.122), (2.123) and (2.124) can be viewed as a Lorentz algebra in 5 dimensions. This is achieved by extending the index range to $0, 1, 2, 3, 4$ and putting $I^{m4} = -I^{4m} = \alpha^m$ and defining $g^{44} = -1$ and $g^{m4} = 0$ for $m \neq 4$. With Bhabha's original metric g^{mn} with signature $(+ - --)$, this yields a new metric with signature $(+ - - - -)$ and an SO(1, 4) Lorentz algebra.[75] This means that the known representation theory of the five-dimensional Lorentz group can be used for the α^m matrices.

Such representations can be labelled by the numbers λ_1 and λ_2 with $\lambda_1 \geq \lambda_2 \geq 0$, both integer or half-integer. These representations are closely related to the finite dimensional representations $D(k, l)$ of the four-dimensional Lorentz group, as might be expected. The theory is outlined in Sections 2–4 in Bhabha's paper. Here, we will state the multimass/multispin nature of the equations. A particle of integral spin $\lambda = \lambda_1$ has 2λ possible values for the rest mass, and a particle of half-integral spin has $2\lambda + 1$ possible values for the rest mass according to

$$\text{Integral spin:} \quad \pm\chi, \pm\chi/2, \pm\chi/3, \ldots, \pm\chi/\lambda \tag{2.127}$$

$$\text{Half-integral spin:} \quad \pm 2\chi, \pm 2\chi/3, \pm 2\chi/5, \ldots, \pm\chi/\lambda \tag{2.128}$$

One may notice a close resemblance to the Majorana spectrum of equation (2.50). The difference is only that the Majorana spectrum is infinite, whereas the Bhabha spectrum is finite. The hope, at the time of Bhabha, that such a spectrum might have phenomenological applications, never materialized.

The implication of this result is that the principle **A** can only be satisfied for spin 1/2 and 1. Only in these two cases, can a single mass Klein–Gordon equation be derived form the linear wave equation. Therefore, the two principles **A** and **B** cannot be reconciled for higher spin.

---

74 Approaches similar to Bhabha's was researched by J. K. Lubanski and by B. S. Madhava Rao, referred to by Bhabha.

75 Starting with signature $(-+++)$, one must choose $c = -1$ to get SO(1, 4), otherwise one gets SO(2, 3). Actually, SO(2, 3) also results with the choice $c = -1$ for metric signature $+ - --$.

2.4.4 Directions in mid century: where we have been and where we are heading

Anyone entering the literature on relativistic wave equations will most likely be struck by how extensive it is, a fact remarked on by Bargmann and Wigner in 1947.[76] On second thoughts, this is however not so strange, given that wave equations was the theoretical tool for understanding elementary particles in those years. The reader may benefit from a coarse top-down view before reading on. If we allow ourselves – anachronistically for sure – to refer everything treated here to "higher spin field theory" one could discern one period, from Dirac and Fierz–Pauli in the late 1930s up to Bhabha in the late 1940s, that may be designated the "wave equation period". It roughly split into two branches, one concerned with Lagrangians with auxiliary fields capable to integrate electromagnetic and gravitational interactions, and another one concerned with Dirac-like wave-equations without subsidiary conditions. This second branch ended with the Bhabha papers of the 1940s, when the subject was more or less completely clarified. Then there is a second period from the early 1960s up to the Fronsdal 1978 paper, that could be designated the "Lagrangian period". As such, it can be seen as a continuation of the first branch of the first period.

At the end of the 1970s, the Yang–Mills Standard Model was established as the phenomenological model of elementary particles and forces. One may perhaps say that there was no need for new wave equations any longer, and the ones already researched were anyway sufficient for any conceivable phenomenological need, if such a need would arise. Our particular direction, "higher spin gauge field theory", materialized at this time with the C. Fronsdal and J. Fang papers.

When it comes to the wave equations themselves, one can discern three approaches. First, one that is "postulational" starting from a given wave equation type such as the Bhabha equation (inspired and generalized from Dirac, also studied by Majorana), that is rather easily motivated since it has a quite obvious relativistic invariant form. Second, there is the "Bargmann–Wigner" approach that starts from the representations of the Poincaré group and sets up wave equations as realizations of the abstract representations. Third, there is the "Weinberg" approach of the early 1960s that starts from the physical states of given mass and spin that may appear in scattering experiments and derives the wave equations from there. The Weinberg approach is actually an elaboration of the Bargmann–Wigner approach.

2.4.5 H. Umezawa's general theory of wave equations

Hiroomi Umezawa, in his 1956 textbook on quantum field theory [83], has an extensive discussion of relativistic wave equations, extending over four chapters. The theory

76 Quite a few authors writing at the time and later make similar remarks.

is quite general, and subsumes much of what lay before and came later.[77] The book starts with a historical introductory chapter that is still interesting to read today, as it gives a view of the thinking of those days. Those were days before the explosion of new particle discoveries of the late 1950s and the 1960s. Days when it was not known if the, newly sorted out, renormalized quantum field theory of the electromagnetic interaction, could be extended to the nuclear forces. Second quantized relativistic wave equations, for particles of various mass, spin and other characteristics, were an important theoretical tool for calculations aimed at comparison with experiment.

Let us briefly review the Umezawa text. Umezawa considers[78] wave functions ψ_j with n components and writes a relativistic wave equation as

$$D_{ij}(\partial)\psi_j(x) = 0 \tag{2.129}$$

with the matrix indices $i, j, \ldots$ as yet unspecified. It is not explicitly stated in the book, but the matrix $D_{ij}(\partial)$ may be of finite order in partial derivatives ∂_μ. Requiring the Klein–Gordon equation to hold then prompts the existence of another "derivation operator" (Umezawa's designation) $d(\partial)$ with the property

$$d_{ij}(\partial)D_{jk}(\partial) = (\Box - m^2)\delta_{ik} \tag{2.130}$$

For the $d(\partial)$ operator, Umezawa gives an ansatz of finite order b in ∂_μ:

$$d_{ij}(\partial) = \alpha_{ij} + \alpha_{ij}^{\mu}\partial_\mu + \cdots + \alpha_{ij}^{\mu_1 \ldots \mu_b}\partial_{\mu_1} \ldots \partial_{\mu_b} \tag{2.131}$$

Turning to linear wave equations, the most general form is again given by

$$D(\partial) = \rho^\mu \partial_\mu + m\beta \tag{2.132}$$

with matrices ρ_μ and β. An ansatz of order $b = 1$ for the operator $d(\partial)$ is

$$d(\partial) = \alpha + \alpha^\mu \partial_\mu \tag{2.133}$$

with matrices α and α_μ. One immediate consequence of the Klein–Gordon condition (2.130) is $m\alpha\beta = -m^2 I$, from which we see that β is nonsingular and has an inverse β^{-1}. Multiplying $D(\partial)$ with β^{-1} and defining new matrices $\beta_\mu = \beta^{-1}\rho_\mu$ yields the new wave equation

$$(\beta^\mu \partial_\mu + m)\psi = 0 \tag{2.134}$$

---

77 Umezawa is, in the book, careful with original references, but there are no references cited for the general equations. One may assume that the synthesis given in Chapter II of his book is his own version of theory that at this time was considered "well known". It could also be noted that there is no reference to the Majorana 1932 paper.

78 We depart somewhat from the notation used by Umezawa.

The Klein–Gordon condition now reads

$$(\alpha + \alpha^\mu \partial_\mu)(\beta^\nu \partial_\nu + m) = (\Box - m^2)I \tag{2.135}$$

from which we get the three equations

$$m\alpha = -m^2 I \qquad \alpha\beta_\mu + m\alpha_\mu = 0 \qquad \alpha_\mu\beta_\nu + \alpha_\nu\beta_\mu = 2\eta_{\mu\nu} \tag{2.136}$$

Thus $\alpha = -mI$ and $\alpha_\mu = \beta_\mu$. All in all, we now have

$$D(\partial) = \beta^\nu\partial_\nu + m \quad \text{and} \quad d(\partial) = \beta^\nu\partial_\nu - m \quad \text{with} \quad \beta_\mu\beta_\nu + \beta_\nu\beta_\mu = 2\eta_{\mu\nu} \tag{2.137}$$

The Dirac equation is of course a special case of these equations.

Umezawa then devotes a chapter to the Dirac equation, which he, at the end, reformulates in terms of two-component spinors. In the following Chapter IV, arbitrary spin wave equations are studied within the formalism of Dirac–Fierz–Pauli as well as Rarita–Schwinger. In this chapter, Umezawa discusses massless gauge invariant wave equations in the gauge where the divergence and trace of the gauge fields are set to zero. The system retains an invariance under gauge transformations with parameters subject to the same conditions as the fields. In modern parlance, the transformations are really *re-gauge transformations*, and the gauge is called TT-gauge (see Section 5.1.1).[79]

In Chapter V on wave equations, Umezawa returns to the general case of Chapter II and develops the theory that overlaps with the theory of Bhabha. Umezawa starts from the linear equation (2.134), and using Lorentz invariance, arrives at precisely the equations (2.122) and (2.123) of Bhabha (with β matrices instead of α matrices). He then notes that these equations are not sufficient to completely determine the algebra of the β matrices.[80] They only show that the β matrices may transform in reducible finite dimensional representations of the Lorentz group, these being possible to decompose into irreducible representations of the three-dimensional rotation group. In effect, the particles described by the wave equation are multispin.

Umezawa then considers two cases, Case I where the wave functions are required to satisfy the Klein–Gordon equation, and Case II where a multimass Klein–Gordon equation

$$\prod_{s=1}^{n} (\Box - m_s^2)\psi = 0 \tag{2.138}$$

should be derivable from the linear wave equation $(\beta^\mu\partial_\mu + m)\psi = 0$. He then considers an example of this case which is precisely the Bhabha theory with the β matrices satisfying the equation $[\beta^\mu, \beta^\nu] = S^{\mu\nu}$ with $S^{\mu\nu}$ the infinitesimal Lorentz generators.

---

79 The equations indeed correspond to TT gauge-fixed Fronsdal equations, although the Fronsdal equations were not known at this time.

80 As noted by Bhabha, see our Section 2.4.3.

Umezawa's Case I leads to an inverse wave operator of the form (2.131) with an explicit solution for the $\alpha_{\mu_1 \dots \mu_j}$ matrices in terms of sums of products of matrices β_{μ_i}. These matrices, in their turn, satisfy higher order equations of the form

$$\sum_{\text{perm.}} \beta_{\mu_1} \cdots \beta_{\mu_{b-1}} [\delta_{\mu_b \mu_{b+1}} - \beta_{\mu_b} \beta_{\mu_{b+1}}] = 0 \tag{2.139}$$

where b is the border of derivatives in the inverse wave operator $d(\partial)$.

But enough is enough, and I choose to stop here, leaving the reader to peruse the Umezawa theory at her own leisure. In this connection, a paper by Harish-Chandra [84], to which Umezawa refers, is interesting. Harish-Chandra investigates properties of the β_μ matrices in linear wave equations of the Bhabha type in relation to the question of subsidiary conditions and the demand of a single mass second-order equation. We will briefly return to this question in connection to C. Fronsdal's thesis paper to be discussed below in Section 2.5.1.

2.4.6 Last words on covariant wave equations

During writing the above notes, I have not visited the whole network of research into covariant wave equations, rather I have stopped at some of the major stations. As others have noted, Bhabha's work can be seen as a terminus for the Dirac-type trains. As regards Bhabha's work, a follow-up paper [85] from 1949 can be seen as "rethinking" the theory in quantum field theory terms. There are also a series of papers from the mid-1970s by R. A. Krajcik and M. M. Nieto, investigating the Bhabha theory. The references can be found in the overview paper [56] that also recounts the history of this kind of wave equation. As regards other contributions, there are papers by Harish-Chandra, who with his more mathematical inclination, supplied results and proofs to the general theory. One such paper is [84] to which we have already referred. The reader may note the following stations [67, 86–91], that I did not stop at. I also decided to pass by F. J. Belinfante's "undors" ("unda" = wave) [92, 66, 65], as stopping there might have lured me to take a look at the spin-statistics theorem. Undors figure implicitly in Wigner's 1947 paper (see Section 2.3.2). There are plenty of other stops, directions can be found in [23] and [20].

2.5 The search for Lagrangians for massive fields

In the 1960s, some researchers took up the problem of constructing wave equations and Lagrangians for free higher spin fields and their interactions. The motivation was the higher spin resonances discovered in the new high energy accelerator laboratories. Many new particles had been discovered starting already in the 1950s,[81] but their masses and spins were uncertain at first. It anyway prompted a few authors in the mid-1950s – S. J. Gupta [71], P. A. Moldauer and K. M. Case [72] and C. Fronsdal [93] – to return to the Dirac–Fierz–Pauli theory, in particular to the case of spin 3/2.

A basic problem for the free theory is the mismatch between the number of physical states for massive spin s particles, as stipulated by the irreducible representations of the Poincaré group, and the number of components of covariantly indexed fields that one would like to use to describe them in field theory. Already beyond spin 1, this problem becomes nontrivial. One gets a two-way problem of projecting out physical states from the covariant wave fields, and the inverse problem of building covariant wave fields from one-particle states.

It may be thought that the problem of formulating arbitrary spin wave equations and Lagrangians was already solved by the Dirac–Fierz–Pauli theory of the early 1930s, but as it turned out, much more had to be done and could be done, both theoretically and in relation to applying the theory to high energy particle processes. From the mid-1950s on, the problem became motivated by phenomenology, and not just by curiosity. As we will see, there was a drive from the two-component spinor formalism of Dirac–Fierz–Pauli coupled equations to, a perhaps more easily handled and conventional, tensor and tensor-spinor formalism. No doubt, the severe minimal coupling problems inherent in the DFP theory, with its need for ever more intricate choices of auxiliary fields and Lagrangian terms, as the spin increased, was a major motivation.

Counting massive states

A fully symmetric tensor field $\varphi_{\mu_1 \ldots \mu_s}$ has $\binom{s+3}{3}$ components in four space-time dimensions, whereas a massive spin s particle has $2s + 1$ states. The field satisfies the Klein–Gordon field equation

$$(\Box - m^2)\varphi_{\mu_1 \ldots \mu_s} = 0 \tag{2.140}$$

The number of field components is brought down by requiring the field to be *traceless* and *divergence-free*, that is by imposing the conditions

$$\varphi'_{\mu_3 \ldots \mu_s} = 0 \tag{2.141}$$

$$\partial \cdot \varphi_{\mu_2 \ldots \mu_s} = 0 \tag{2.142}$$

In the case of half-integer spin, with spinor-tensor fields, the equations are replaced by the "Rarita–Schwinger" equations (2.112).

The counting of field components, in the integer spin case, becomes

$$\binom{s+3}{3} - \binom{s+1}{3} - \left(\binom{s+2}{3} - \binom{s}{3}\right) = 2s + 1 \tag{2.143}$$

---

81 The discovery of the neutral pion in 1950 at the Berkeley synchrocyclotron was the first new particle to be discovered by an accelerator laboratory. See [5], Chapter 19(b).

where we have taken into account that the trace of the divergence must not be counted twice. The condition on the divergence is motivated by another consideration: the necessity of positive energy [47]. Consider a plane wave with momentum k_μ satisfying $k^2 = -m^2$. Transform to the rest system where $k_\mu = (m, 0, 0, 0)$. The equation (2.142) then implies $m\varphi_{0\mu_2 \ldots \mu_s=0}$. This means that all field components with at least one time-like index are zero. With only space-like indices, the energy will be positive. Again counting the number of field components (now symmetric traceless tensors in three space dimensions) yields $2s + 1$.

2.5.1 Fronsdal's thesis paper

One paper that can be said to mark the beginnings of the "modern" Lagrangian theory of higher spin, is Fronsdal's thesis paper [93] published in 1958. It takes its motivation partly from the newly discovered higher spin massive resonances in high energy collisions, in particular the hyperons Λ and Σ. At the time, the spin of the particles were still uncertain. Another motivation was a wish to further simplify the formalism of the Fierz–Pauli theory and to develop the structure of arbitrary spin theory. Fronsdal writes that, in spite of simplifications by Rarita–Schwinger, Gupta and Moldauer-Case: "[...] calculations are very lengthy [...]". The aim of the paper is "[...] to present a new simple formulation of the Fierz–Pauli theory, and to analyze in some detail the structure of this theory for arbitrary spin.".

The paper is situated within Wigner's Poincaré invariant theory and works from the tensor-spinor formulation, rather than the Dirac–Fierz–Pauli two-component language. It is noted that the Klein–Gordon equation for each component of the wave function follows from the first invariant of the Poincaré group $p^2 = -m^2$, while the second invariant $S^2 = s(s+1)$ requires that the projections on lower spin values vanish. This yields the conditions of symmetry, vanishing divergence and tracelessness.[82]

Fronsdal therefore introduces projection operators to project out physical components from symmetric tensor fields subject to the subsidiary conditions (2.141) and (2.142), and correspondingly for half-integer spin fields. Denoting the wave equation with $\eta\varphi = 0$ and the subsidiary conditions collectively with $\eta_i\varphi = 0$, Fronsdal requires a projection operator Θ to obey

$$\begin{aligned} \eta_i(\Theta\varphi) &= 0 \\ (\eta_i\varphi = 0) &\to (\varphi = \Theta\varphi) \\ (\eta\varphi = 0) &\to (\eta\Theta\varphi = 0) \end{aligned} \tag{2.144}$$

Such projection operators are constructed but they turn out to be nonlocal in that they involved the inverse of the d'Alembertian operator p^2 (this will be illustrated in the next Section 2.5.2).

82 Compare to the Wigner approach reviewed in Section 2.3.4.

After developing the general theory of the projection operators, Fronsdal continues with a discussion of the Fierz–Pauli problem with electromagnetic interaction of higher spin particles. Writing the wave equations and Lagrangians with projection operators simplifies the deduction of the subsidiary conditions, but creates a new problem with the nonlocal nature of the projectors. The problem becomes one of replacing such nonlocal equations with equations linear in momenta. The spin 3/2 case is then treated in detail. The method amounts to moving the terms involving $1/p^2$ into an auxiliary field that can subsequently be eliminated. For instance, the theory of Moldauer and Case (referenced in the paper) and the Rarita–Schwinger spin 3/2 theories, are obtained as special cases.

After this (in Section 4 of the paper), Fronsdal turns to the general case of arbitrary spin. He starts from a Dirac-like first-order equation

$$(p^\mu \alpha_\mu + im\beta)\varphi = 0 \quad \text{with} \quad \alpha_\mu^\dagger = \alpha_\mu \quad \text{and} \quad \beta^\dagger = \beta \tag{2.145}$$

The wave function transforms according to some reducible representation of the Lorentz group. To get an irreducible representation, it is required that the wave equation is equivalent to the projected equations

$$(1 - \Theta)\varphi = 0 \quad \text{and} \quad (\mathbf{p} + im)\Theta\varphi = 0 \quad \text{where} \quad \mathbf{p} = p^\mu \alpha_\mu \tag{2.146}$$

with the wave equation invariant under reflection (chirality). From these requirements, Fronsdal then derives commutation relations for the matrices α_μ and β. The equations are given in terms of a matrix $\Gamma_\mu = \beta^{-1}\alpha_\mu$. There is a set of equations, but the interesting one can be written as

$$(\Gamma_\mu p^\mu - \mathbf{p})(\Gamma_\nu p^\nu)^{\bar{n}} = 0 \tag{2.147}$$

Here, $\bar{n}$ is the largest number of fields of given spin that occur in the reducible unprojected wave function φ. Similar, higher order, equations had been derived by Harish-Chandra and by Umezawa and Visconti from the requirement of the wave function to satisfy a single mass Klein–Gordon equations (see our Section 2.4.5 above). Fronsdal derives it from the requirement of the solutions of the wave equation to describe a unique spin. The Fronsdal paper ends with three section on applications of the methods.

2.5.2 S-J. Chang and L. P. S. Singh – C. R. Hagen

Then in 1967, S-J. Chang built upon the Fronsdal theory and proceeded to resolve the nonlocalities through the introduction of auxiliary fields [94]. This interplay between nonlocalities and auxiliary fields is a phenomenon that is common in higher spin theory.[83] Chang carried through the program up to and including spin 4. The Chang procedure is to some extent a return to the Fierz–Pauli approach with auxiliary fields, but now in a tensor formulation.

The introduction to the Chang paper is quite interesting, as it clarifies the respective advantages and disadvantages of the, at that time, prevailing approaches to arbitrary spin theory.

> The first approach emphasizes the transformation properties of field variables under the homogeneous Lorentz group. [...] The second approach follows that of Pauli and Fierz and demands that all field equations and subsidiary conditions should be derived from a generalized action principle.

For the first approach, Chang cites work by S. Weinberg [95, 96], D. L. Pursey [97] and W. K. Tung [98, 99]. We now recognize this way of doing quantum field theory as the one advocated and elaborated in great detail in the Weinberg quantum field theory textbook [18]. Parts of this approach will be reviewed in our Sections 3.3 and 3.4. For historical remarks, see Section 2.6.

As to the procedure actually employed by Chang, it is reminiscent of the deliberations of Fronsdal, but the objective is different. Fronsdal's was to show equivalence between different formulations, Chang's is to replace the need for projectors with auxiliary fields. Although the procedure as such seems a bit ad hoc, it does throw some light on the origin of the auxiliary fields.

What Chang did: the spin 2 example

Following Fronsdal, Chang symbolized the conditions $\varphi'_{\mu_3 \dots \mu_s} = 0$ and $\partial \cdot \varphi_{\mu_2 \dots \mu_s} = 0$ by writing $\eta\varphi = 0$. Fronsdal actually included index symmetry in the conditions, but that is not necessary. He then introduces an orthogonal projection operator $\Theta = \Theta^2$ with the properties

$$\begin{aligned} \eta\Theta\varphi &= 0 \\ \eta\varphi = 0 &\Rightarrow \varphi = \Theta\varphi \end{aligned} \tag{2.148}$$

For the almost trivial case of a vector field, Chang quotes the spin 0 and spin 1 projectors as $\Theta_{\mu\nu}(0) = \Box^{-1}\partial_\mu\partial_\nu$ and $\Theta_{\mu\nu}(1) = \eta_{\mu\nu} - \Box^{-1}\partial_\mu\partial_\nu = \Theta_{\mu\nu}$. Indeed, for the spin 1 part $\varphi_\mu(1) = \Theta_{\mu\nu}(1)\varphi^\nu$ of the vector field, we immediately get $\partial \cdot \varphi(1) = 0$. The projector $\Theta(S)$ that projects out a spin S state is then given in terms of $\Theta(1)$.

Chang then works through the spin 2 case. The nonlocal field equation is taken to be

$$m^2\varphi_{\mu\nu} = \Box[\Theta(2)\varphi]_{\mu\nu} \tag{2.149}$$

---

83 It resurfaced again for higher spin gauge fields with the work of D. Francia and S. Sagnotti in the early 2000s. See Section 5.3.2, Chapter 5.

for a symmetric and traceless field $\varphi_{\mu\nu}$. The spin 2 projector is given by

$$\Theta_{\mu\nu,\lambda\sigma} = \frac{1}{2}[\Theta_{\mu\lambda}\Theta_{\nu\sigma} + \Theta_{\mu\sigma}\Theta_{\nu\lambda} - \frac{2}{3}\Theta_{\mu\nu}\Theta_{\lambda\sigma}] \tag{2.150}$$

in terms of the spin 1 projector. It may seem a bit inconsequential to take $\varphi_{\mu\nu}$ as traceless, as the projector is supposed to project out the traceless part. The same can be said about the field equation (2.149). In any way, one finds that $[\Theta(2)\varphi]_{\mu\nu}$ is traceless and divergence-free no matter whether $\varphi_{\mu\nu}$ itself is traceless or not. The field equation (2.149) works out to

$$m^2\varphi_{\mu\nu} = \Box\varphi_{\mu\nu} - \left(\partial_\mu\partial\cdot\varphi_\nu + \partial_\nu\partial\cdot\varphi_\mu\right) - \frac{1}{2}\eta_{\mu\nu}\partial\cdot\partial\cdot\varphi + \left(\partial_\mu\partial_\nu - \frac{1}{4}\eta_{\mu\nu}\Box\right)\Psi \tag{2.151}$$

where Ψ is the nonlocal field $\Psi = \frac{2}{3}\Box^{-1}\partial\cdot\partial\cdot\varphi$. Next, Chang wants to reinterpret Ψ as an auxiliary field, whose field equation should imply $\Psi = 0$ and $\partial\cdot\partial\cdot\varphi = 0$. In order to do that, he contracts the field equation (2.151) with $\partial_\mu\partial_\nu$ and gets

$$\left(\Box + 2m^2\right)\partial\cdot\partial\cdot\varphi = \frac{3}{2}\Box^2\Psi \tag{2.152}$$

Finally, if Ψ is chosen to satisfy the equation

$$\partial\cdot\partial\cdot\varphi = \frac{3}{2}\left(\Box - 2m^2\right)\Psi \tag{2.153}$$

we get $-6m^2\Psi = 0$ and consequently also $\partial\cdot\partial\cdot\varphi = 0$. Chang proceeds with writing an action for the system, and then goes on to some further reworkings of the system.

Chang's paper contains higher spin Lagrange functions for spin ≤ 4 constructed along these lines. One cannot escape the feeling that the procedure is quite cumbersome, and it soon became superseded. Once it is clear that one cannot escape auxiliary fields, they can be introduced much more systematically without prior recourse to the Fronsdal projectors. It should also be said that the Chang paper treats quantization and Greens functions.

The general theory for massive spin s was then constructed by L. P. S. Singh and C. R. Hagen without any recourse to projection operators [100]. Referring to the Chang paper, these authors write that Chang's method "[...] does not yield a closed form for the Lagrangian of a general-spin field.".[84] The main body of the paper is taken up by the construction of the Lagrangians, but the authors also discuss quantization and electromagnetic interactions. They begin by analyzing the field equations.

The spin 1 theory is almost trivial. The wave equation (the Proca equation) reads

$$\partial^\mu(\partial_\mu\phi_\nu - \partial_\nu\phi_\mu) - m^2\phi_\nu = 0 \tag{2.154}$$

---

84 I have not checked this, and seen no other reference to it. It is not clear to me whether the statement means that the Chang method cannot be extended beyond spin 4, or if it means that no general formulas can be written.

Contracting with ∂^ν yields $\partial^\nu \phi_\nu = 0$ and the wave equation reduces to the Klein–Gordon equation. Of course, the mass m must be nonzero. For spin 2, a new phenomenon appears.

A general wave equation for a symmetric traceless field $\phi_{\mu\nu}$ takes the form

$$(\Box - m^2)\phi_{\mu\nu} - a\left(\partial_\mu \partial \cdot \phi_\nu + \partial_\nu \partial \cdot \phi_\mu - \frac{1}{2}\eta_{\mu\nu}\partial \cdot \partial \cdot \phi\right) = 0 \tag{2.155}$$

where a is a constant to be determined. All terms are symmetric and traceless as is appropriate if this is to be an Euler–Lagrange equation. Contracting with ∂^ν yields

$$m^2 \partial \cdot \phi_\mu = \Box(1-a)\partial \cdot \phi_\mu - \frac{1}{2}a\partial_\mu \partial \cdot \partial \cdot \phi \tag{2.156}$$

The best one can do here is to choose $a = 1$, but $\partial \cdot \partial \cdot \phi = 0$ cannot be deduced from (2.154). To impose this condition, Singh and Hagen introduced a scalar auxiliary field $\phi^{(0)}$ satisfying its own wave equation, as well as modifying (2.154). The only way to couple the scalar is through a Lagrangian term $\partial \cdot \phi_\mu \partial^\mu \phi^{(0)}$. The new Euler–Lagrange equation must then take the following form with new parameters b and c to be determined:

$$(\Box - m^2)\phi_{\mu\nu} - \left(\partial_{(\mu}\partial \cdot \phi_{\nu)} - \frac{1}{2}\eta_{\mu\nu}\partial \cdot \partial \cdot \phi\right) - b\left(\partial_\mu \partial_\nu \phi^{(0)} - \frac{1}{4}\eta_{\mu\nu}\phi^{(0)}\right) = 0 \tag{2.157}$$

$$(\Box - cm^2)\phi^{(0)} - \partial \cdot \partial \cdot \phi = 0 \tag{2.158}$$

Then contracting the first equation with $\partial^\mu \partial^\nu$ and using the second equation for substituting $\partial \cdot \partial \cdot \phi$ yields

$$cm^2 \phi^{(0)} = \frac{1}{4}(2 + 3b)\Box\Box\phi^{(0)} - \left(\frac{1}{2}c - 1\right)\phi^{(0)} \tag{2.159}$$

Choosing $b = -2/3$ and $c = 2$ then gives $\phi^{(0)} = 0$. This then, through equation (2.158), implies the desired result $\partial \cdot \partial \cdot \phi = 0$.

Singh and Hagen performed the corresponding analysis for massive spin 3 field $\phi^{(3)}$ with the result that auxiliary fields $\phi^{(1)}$ and $\phi^{(0)}$ were needed in order to ensure the conditions (2.141) and (2.142) as well as the vanishing of the auxiliary fields themselves. A general pattern then emerged. To construct field equations for a massive spin s field $\phi^{(s)}$ that are equivalent to the Klein–Gordon equation with the constraints (2.141) and (2.142), that is, are traceless and divergence-free:

> One must successively obtain $\phi^{(s,\lambda)} = 0$ for $\lambda = s, s-1, s-2, \ldots, 2$, where
>
> $$\phi^{(s,\lambda)}_{\mu_{\lambda+1}\cdots\mu_s} = \partial^{\mu_1}\ldots\partial^{\mu_\lambda}\phi^{(s)}_{\mu_1\cdots\mu_s} \tag{2.160}$$
>
> is a symmetric traceless tensor of rank $s-\lambda$. At each stage [of the derivation] an auxiliary field – a symmetric, traceless tensor of the same rank as $\phi^{(s,\lambda)}$ – is needed. Thus one introduces symmetric, traceless tensors of rank $0, 1, 2, \ldots, s-2$. These will be labeled $\phi^{(0)}, \phi^{(2)}, \ldots, \phi^{(s-2)}$, respectively, and correspond to the representations $D(\frac{1}{2}j, \frac{1}{2}j)$, $j = 0, 1, 2, \ldots, s-2$, of the Lorentz group. Thus the second-order theory requires $(s+1)^2 + \frac{1}{6}s(s-1)(2s-1)$ field components.

Singh and Hagen then writes down the most general second-order quadratic Lagrangian with undetermined coefficients involving these fields, and compute the field equations. The coefficients are solved for in steps so as to yield the constraints (2.160) and set the auxiliary fields to zero. In this way, a Lagrangian is obtained that gives the Klein–Gordon equation for a symmetric traceless spin s field free of divergence. The auxiliary fields themselves vanish – in the free field theory – due to the field equations. The calculations are quite lengthy. The need for the spectrum $\phi^{(0)}, \phi^{(2)}, \ldots, \phi^{(s-2)}$ of auxiliary fields was actually suggested by Fierz and Pauli in [101]. The authors also discuss the need for a first-order formalism "[...] in order that electromagnetic interactions can be introduced in an unambiguous fashion. For this purpose, more fields have to be introduced.".

2.6 Feynman rules for any spin

Let us turn to the approach referred to by Chang. "Feynman rules for any spin" is the title of a 1963 paper by S. Weinberg, the first in a series of three papers. Even though it has had limited influence on the technical development involved in what became *higher spin field theory*, one of its off-shots – and now we are referring to massless fields – the Weinberg "no long-range forces" result [7], did indeed become one of the defining points of the research program. Despite the fact that higher spin gauge fields were ruled out as regards having any long range effects reminiscent of the spin 1 and spin 2 gauge fields, researchers have had to argue that they are interesting to study nevertheless.

Historically, it is interesting to dwell a bit on Weinberg's motivations behind the approach, as well as the basic assumptions behind it. The actual theory itself will be partly reviewed in our Chapter 3 where focus will be moved to massless fields. As already noted, the approach was at the time an alternative to the Lagrangian-field equation-canonical approach.

2.6.1 The Weinberg papers

We quote from the very first lines of Weinberg's paper [95].

> This article will develop the relativistic theory of higher spin, from a point of view midway between that of classical Lagrangian field theories and the more recent S-matrix approach. Our chief aim is to present the explicit Feynman rules for perturbation calculations, in a formalism that varies as little as possible from one spin to another.

Then Weinberg states the assumptions behind the approach.[85]

85 We stay close to Weinberg's own wording, but not exactly so.

The Weinberg assumptions

(1) Perturbation theory. The S-matrix can be calculated from Dyson's formula

$$S = \sum_{n=0}^{\infty} \frac{(-i)^n}{n!} \int_{-\infty}^{\infty} dt_1 \dots t_n T\{H'(t_1) \dots H'(t_n)\} \tag{2.161}$$

where the Hamiltonian H is split into a free-particle part H_0 and an interaction part H' and where the interaction H' is defined in the interaction representation $H'(t) = \exp(iH_0 t)H' \exp(-iH_0 t)$.

(2) Lorentz invariance. The S-matrix is invariant under proper orthochronous Lorentz transformations. Weinberg writes that a "sufficient and probably necessary condition" for the invariance of S is: $H'(t) = \int d^3x \mathcal{H}(\mathbf{x}, t)$ where

– $\mathcal{H}(x)$ is a scalar. To every inhomogeneous Lorentz transformations $x^\mu \to \Lambda^\mu{}_\nu x^\nu + a^\nu$ there corresponds a unitary operator $U[\Lambda, a]$ such that

$$U[\Lambda, a]\mathcal{H}(x)U^{-1}[\Lambda, a] = \mathcal{H}(\Lambda x + a) \tag{2.162}$$

– For $x - y$ space-like: $[\mathcal{H}(x), \mathcal{H}(y)] = 0$.

(3) Particle interpretation. $\mathcal{H}(x)$ is constructed out of creation and annihilation operators for the free particles described by H_0. Weinberg then writes that "the only known way of making sure" that such an $\mathcal{H}(x)$ satisfy the restrictions under item **(2)** is to form it as a function of sets of fields $\psi_n(x)$ which are linear combinations of the creation and annihilation operators with the following properties:

– The fields transforms according to

$$U[\Lambda, a]\psi_n(x)U^{-1}[\Lambda, a] = \sum_m D_{nm}(\Lambda^{-1})\psi_m(\Lambda x + a) \tag{2.163}$$

where D_{nm} is some representation of Λ. This makes it possible to satisfy Lorentz invariance by coupling the fields $\psi_n(x)$ in various invariant combinations.

– For $x - y$ space-like $[\psi_n(x), \psi_m(y)]_\pm = 0$. This guarantees $[\mathcal{H}(x), \mathcal{H}(y)] = 0$.

The reader of the first volume of the Weinberg textbooks on quantum field theory [18] no doubt recognizes these items, and indeed, Chapters 2–5 of that book develops this approach in great detail. The rest of the paper itself, is a detailed implementation of the assumptions, resulting in explicit Feynman rules. As to the choice of interactions, Weinberg writes that the choice of an interaction Hamiltonian is no more difficult than the choice of an interaction Lagrangian in the canonical approach.

Some further aspects of the approach are interesting to note. In the Weinberg paper, there is no explicit relation with covariant field equations, as that is not the aim of the work. Such relations are, however, explored in papers by D. L. Pursey and W-K. Tung (referred to in the Chang paper).

Weinberg instead constructs physical $2j + 1$ fields (and $2(2j + 1)$ component fields for parity conserving interactions) that only obey the Klein–Gordon equation. Nothing else is needed since there are no extra components. Weinberg writes "[...] any field equation except [the Klein–Gordon equation] is nothing but a confession that the field contains superfluous components.". The asymptotic states are "states", or "elementary particles", represented by unitary representations of the inhomogeneous Lorentz group.[86] This means that the states are labelled by – apart from mass and momenta – spin labels of the ordinary three-dimensional rotation group.

In the absence of manifestly covariant fields, Weinberg constructs invariant interactions by coupling fields using angular momentum addition rules. This approach is fully developed in the third paper in the series [102].

In the second paper [96], Weinberg treats the massless case. From a higher spin perspective, this is a very interesting paper and we will review its technical details in several sections in the next chapter. For here, a quote will suffice.

> For massive particles of spin j, we have already seen in [our [95]] that a field $\psi^{(+)}$ can be constructed out of the $2j + 1$ annihilation operators $a(\mathbf{p}, \sigma)$, which will satisfy the transformation requirements [Poincaré transformations], for any representation (A, B) that "contains" j, i. e., such that
>
> $$j = A + B \text{ or } A + B - 1 \text{ or } \cdots \text{ or } |A - B|.$$
>
> [A spin-one field could be a four-vector $(\frac{1}{2}, \frac{1}{2})$, a tensor $(1, 0)$ or $(0, 1)$, etc.] We might expect the same to be true for mass zero, *but this is not the case* [emphasis of the original]. We will prove [...] that a massless particle operator $a(\mathbf{p}, \sigma)$ of helicity λ can only be used to construct fields which transform according to representations (A, B) such that
>
> $$B - A = \lambda.$$
>
> [...] at least until we broaden our notion of what we mean by a Lorentz transformation. It will be seen that the restriction [removed formula number] arises because of the non-semisimple structure of the little group.

The cryptic comment "at least until we broaden our notion of what we mean by a Lorentz transformation." refer to gauge transformations.[87] We will prove the Weinberg restriction $B - A = \lambda$ in Section 3.5.6.

2.6.2 The D. L. Pursey and W-K. Tung papers

There are quite few papers from the mid-1960s treating field theory using techniques similar to the ones employed in the Weinberg papers. The backdrop is the need to do efficient calculations to interpret experimental scattering data in high energy particle

86 In those days, elementary particles were often taken as the Wigner representations of the Poincaré group.

87 In the light-cone gauge, the interplay between Lorentz transformations and gauge transformations is very concrete. See Section 6.1.5.

reactions.[88] Here, we will just briefly mention papers by D. L. Pursey [97] and W-K. Tung [98, 99], from the mid-1960s. Both authors study the relation between canonical versus covariant free particle wave equations. The basic formalism of these papers is similar to Weinberg's in that they start from wave functions corresponding to the Wigner representations of the Poincaré group. Both papers contain quite interesting discussions, in their respective introductions, of the general situation regarding relativistic wave equations and the problems that were at this time well known.

The papers differ (apart from details of notation) in – at least one respect – that is interesting from the higher spin perspective. The Pursey paper aims at giving a method by which all possible manifestly covariant wave equations, together with subsidiary constraints, for particles with given spin and mass can be constructed, at least in principle. The Tung papers, on the other hand, aim at avoiding the need for subsidiary constraints for the covariant equations. In the second paper [99], it is shown that this is only possible for spin less than or equal to one.[89]

2.7 Gupta on gravitation and electromagnetism

Suraj N. Gupta worked on the quantization of gravity in the early 1950s, en route inventing, independently of Bleuler, what became known as the *Gupta–Bleuler indefinite metric quantization* method for gauge fields [103, 104]. In a first paper [103] from 1952, Gupta quantized linearized Einstein gravity, and found that the field excitations – gravitons – had spin 2 with two independent spin states (i. e., helicities) with axis parallel or antiparallel to the motion of the particles.[90]

Then, in 1954, Gupta wrote an article comparing gravitation and electromagnetism [112]. In this paper, Gupta rewrites the Einstein nonlinear field equations

$$R^{\mu\nu} - \frac{1}{2}Rg^{\mu\nu} = -\frac{1}{2}\kappa^2 T^{\mu\nu} \tag{2.164}$$

88 We may have occasion to return to this topic in Volume 2 of the present work in connection with higher spin interactions.

89 The first Tung paper [98], written as a preliminary report, this restriction to low spin is not found. The author, in the second paper, explains this as due to not having considered CPT in detail. I have not checked the details myself.

90 In a follow up article, he studied the full nonlinear theory. As for quantum gravity, it is an enormous subject. Several authors, among them L. Rosenfeld, P. G. Bergmann and Dirac, approached the problem in the early days. This work became subsumed under Bryce DeWitt's comprehensive set of three long articles [105–107]. The book [108] reviews much of the subsequent development of the subject. For a history of quantum gravity, see for instance [109], Appendix B. The proceedings from the two Oxford Symposia on Quantum Gravity in 1974 and 1980 respectively, give an interesting flavor of the subject as it stood by the end of the 1970s and before string theory became a dominant paradigm [110, 111].

where $T^{\mu\nu}$ is the energy-momentum tensor of matter, into the form

$$\eta^{\alpha\beta}\frac{\partial^2 g^{\mu\nu}}{\partial x^\alpha \partial x^\beta} = \kappa^2 \Theta^{\mu\nu} \tag{2.165}$$

$$\frac{\partial g^{\mu\nu}}{\partial x^\nu} = 0 \tag{2.166}$$

with $\Theta^{\mu\nu}$ the Belinfante improved energy-momentum tensor of the combined system of matter and gravity. The second equation is a coordinate condition. As Gupta remarks, this form of the gravitational equations are not manifestly covariant (since neither η nor Θ are general coordinate tensors), still they can be written in this form in any frame of reference. Gupta stresses that in equation (2.165), the left-hand side is linear in $g^{\mu\nu}$, while all the nonlinearities reside in $\Theta^{\mu\nu}$. After these equations, Gupta writes:

> Further, we can regard the flat space as the zeroth-order approximation to the Riemannian space. It can then be shown [Gupta reference to our [103]] that the field quantities, occurring in Einstein's theory, can be expressed as infinite series in the flat space.

This, presumably, refers to linearizing the field equation (2.165), because then Gupta continues:

> Therefore, keeping Einstein's theory mathematically unchanged, we can pass over from the Riemann space to the flat space. After passing over to the flat space, the general covariance of the theory is no longer apparent, but the theory still remains manifestly Lorentz covariant. In this way, Einstein's theory can also be regarded as a theory of gravitation in flat space with a Lagrangian density containing an infinite number of terms.

This passage may very well mark a shift in perspective to regarding gravity not so much as a inherently geometrical theory, but rather as a highly nonlinear field theory in Minkowski space-time, much like the paradigm of particle physics. Gupta had in fact adopted this point of view of linearizing gravity in the 1952 paper on quantization. But now focus shifted from linearizing the nonlinear theory to constructing the nonlinear theory from the linear spin 2 field equations.[91] In the very same year, 1954, the first nonlinear spin 1 theory was constructed by Yang and Mills [114].

This has lead to two distinct approaches to gravity. One approach – *the deformation theoretic* – trying to build up the field theory iteratively, order by order, starting with the free theory and successively adding interactions as well as corrections to the symmetries. The other approach, *gauge theoretic*, was pioneered by Utyiama in a paper [115] where the Yang–Mills procedure was applied to the Lorentz group (see Section 2.9.2).

---

91 Already Fierz and Pauli discuss the massless spin 2 field equations and find that they agree with the linearized Einstein equations in the absence of matter. Furthermore, N. Rosen in a paper from 1940 [113], considers a version of general relativity in flat space, but according to J. Fang and C. Fronsdal in [8]: "[...] the idea was not well received at first.".

The deformation theoretic program – named *the Gupta program* by J. Fang and C. Fronsdal in [8] – then formed the basis for an analogous proposal for a program for higher spin: *the generalized Gupta program*. But let us get back to the Gupta paper.

Gupta compares the gravitational equations with the electromagnetic equations

$$\Box^2 A_\mu = -j_\mu \tag{2.167}$$

$$\partial^\mu A_\mu = 0 \tag{2.168}$$

and notes the similarity in form. Both sets of equations are Lorentz covariant and invariant under gauge transformations that leave the supplementary (divergence equations) conditions invariant. The striking difference is of course that the electromagnetic equations are linear in the fields A^μ (since they are uncharged). With Yang–Mills theory, this difference disappeared.

Next, comes a most interesting part of the paper where Gupta shows that the nonlinearities are a consequence of the fact that the gravitational field carries spin 2. Start with the spin 2 free field equations (Gupta refers to Fierz and Pauli [101])

$$\Box h_{\mu\nu} = 0 \tag{2.169}$$

$$\partial^\mu h_{\mu\nu} = 0 \tag{2.170}$$

In the presence of interactions, the field equation is modified to

$$\Box h_{\mu\nu} = \kappa \Theta_{\mu\nu} \tag{2.171}$$

Taking the divergence, and using (2.170), this equation leads to the condition

$$\partial^\mu \Theta_{\mu\nu} = 0 \tag{2.172}$$

what we now call a *source constraint*. Then Gupta argues that "the only known quantity, which is described by a symmetrical tensor with vanishing divergence, is the total energy momentum tensor of a closed system of fields.".

Now, either $\Theta_{\mu\nu}$ contains contributions from the gravitational field itself, or it does not. The last case is however inconsistent if the closed system of fields provide sources for the gravitational field. Then we must have $\Theta_{\mu\nu} = T_{\mu\nu} + t_{\mu\nu}$, the sum of "matter" energy-momentum and gravitational energy-momentum. Thus in a theory with only gravitational fields, $\Theta_{\mu\nu}$ must be the energy-momentum tensor $t_{\mu\nu}$ of the gravitational field itself, and the pure spin 2 field equation reads

$$\Box h_{\mu\nu} = \kappa t_{\mu\nu} \tag{2.173}$$

Next, one assumes that the field equations can be derived from an action. The source-free field equations (i. e., with $t_{\mu\nu} = 0$) follow from an action quadratic in first-order derivatives of the fields. This action, however, also yields an energy-momentum tensor quadratic in the fields. This tensor must appear in the right-hand side of the field equation. Obtaining it from an action requires adding a term that is cubic in the field and derivatives of the field. This then produces a new contribution to the energy-momentum tensor of cubic order. It adds to the field equation, and obtaining it from the action requires adding still a new term, now quartic in the field and derivatives of the field. Clearly, this procedure iterates indefinitely, and a nonpolynomial theory results.

Gupta argues that this phenomenon is a consequence of the spin 2 of the gravitational particles, basically because energy-momentum sources gravity, and gravity itself carries energy-momentum. The contrast to spin 1 electromagnetism is that the photon carries no charge, so it does not contribute to the electromagnetic current. The iteration then never starts. However, as is well known now, allowing a set of spin 1 fields to carry a non-Abelian charge like SU(2), one does get an iteration that produces Yang–Mills theory. In this case, the iteration stops due to algebraic reasons.

2.8 The generalized Gupta program

It is fairly easy to understand that upon self-coupling a set of fields with themselves, as outlined above, and envisioned by Gupta for spin 2, one will get a potentially infinite iteration yielding a nonpolynomial theory. Carrying it out in practice is an entirely different story. The “Gupta program” for spin 2 – although the designation was not generally adopted and was coined after the program was completed – was undertaken by many authors: R. H. Kraichnan in 1955 [116], W. E. Thirring in 1961 [117], R. P. Feynman in 1962 [118], W. Wyss in 1965 [119], S. Deser in 1970 [120] and D.G Boulware and S. Deser in 1975 [121]. The history of the program is described by Fang and Fronsdal in their paper *Deformations of gauge groups. Gravitation* [8] (see Section 2.8.1).[92] Further historical comments can be found in J. Preskill's and K. S. Thorne's foreword to the *Feynman Lectures on Gravitation* [118].

It may seem that the philosophy behind the program – a deformation theoretic, self-coupling, iterative procedure starting from a free field theory in a flat background – from the outset will run into problems with the interpretation of gravity as a geometric theory. This is something that also irritates in the gauge-theoretic approach, as we will see. However, the authors of the papers mentioned are careful to write that the deformation approach is complementary to the geometric approach, even though it

92 There is a slight anachronism in the history, as Fang and Fronsdal does not refer to the S. Deser 1970 paper [120] where the deformation problem is solved in a first-order formulation (treating $g^{\mu\nu}$ and $\Gamma_{\mu\nu}{}^{\lambda}$ as independent fields), unless the authors consider such an approach as being outside the program. See our Section 2.9.1. Perhaps it should also be pointed out that, most likely, the authors of the papers cited in this “program”, were not thinking of it as a program. Indeed, according to the reference lists in these papers, they were perhaps not aware of all work done previous to their own.

may lead to alternatives to or variations of the Einstein theory.[93] Today, gravity is well understood in this framework. The same cannot be said about the "gauging" approach which suffers from more obstinate issues of principle. For the history of the gauging approach, see Section 2.9 and for the theory itself, see Section 4.6.

The generalized Gupta program for higher spin is still in its infancy. The first major step forward was the F. A. Berends, G. H. J. Burgers and H. van Dam approach to spin 3 [122] and their general analysis of the procedure [123], and the light-cone approach to arbitrary spin self-interactions [124] of I. Bengtsson, L. Brink and myself. Nowadays, the generalized Gupta program goes under the name of the *Fronsdal program*. These are topics to be treated in Volume 2 of the present work.

2.8.1 The Fang and Fronsdal formalization of the Gupta program

A successful completion of the Gupta program resides in utilizing gauge invariance. As Fang and Fronsdal writes, this was pointed out by Wyss in 1965, who showed that under certain assumptions – criticized and improved on by Fang and Fronsdal – the structure of the deformed gauge group coincides with the Lie algebra of vector fields on a differentiable manifold.

In previous attempts at deriving gravity by deforming the free theory, there had always been a model of matter involved. Using the notation $w_{\mu\nu}$ for the free wave equation for the massless spin 2 field, the coupling to matter is through the energy-momentum tensor $t_{\mu\nu}$

$$w_{\mu\nu} = -\kappa t_{\mu\nu} \tag{2.174}$$

The explicit formula for $w_{\mu\nu}$ is divergence-free, that is, $\partial^\mu w_{\mu\nu} = 0$, therefore, the right-hand source term must also be divergence-free (as would indeed be expected for an energy-momentum tensor). This is often referred to as a *source constraint*.

Fang and Fronsdal interpreted Gupta's idea as aiming at classifying all physically acceptable sources $S_{\mu\nu}$ for the wave equation (2.174) (the Fierz–Pauli equation as Fang and Fronsdal writes) as formal power series[94]

$$S_{\mu\nu} = \kappa t_{\mu\nu} + \sum_{n=1}^{\infty} \kappa^n \delta_n t_{\mu\nu} \tag{2.175}$$

where $\delta_n t_{\mu\nu}$ are polynomials in $h_{\mu\nu}$ and in their first and second derivatives, with coefficients constructed from other fields. The first term, $t_{\mu\nu}$, represents matter and does

93 The designation "deformation" is not used in the original papers. It came into use in the higher spin related literature after the Fang and Fronsdal paper. Fang and Fronsdal were familiar with the mathematical theory (from which the notion derives) of deforming groups and algebras.

94 For the right-hand side, we follow the notation of Fang and Fronsdal.

not depend on $h_{\mu\nu}$. Fang and Fronsdal wanted the reconstruction of gravity to be independent of any particular model of matter, so they proposed a "restricted Gupta program". That meant taking $t_{\mu\nu} = 0$ in (2.175). The iteration starts by requiring the field equation to be derivable from an action.

The Gupta and the restricted Gupta program

Denoting the action by the letter f and the free spin 2 action by f_0, Fang and Fronsdal start from

$$f^m + f_0 - \kappa h^{\mu\nu} t_{\mu\nu} + \sum_{n=1}^{\infty} \kappa^n f_n \equiv f \tag{2.176}$$

where $f_k, k = 0, 1, 2, \ldots$ are polynomials in the components of h and their first-order derivatives.[95] The matter energy-momentum $t_{\mu\nu}$ is included, as well as the free matter action f^m. These terms will be dropped in the restricted program. The spin 2 field equations are

$$\frac{\delta f_0}{\delta h^{\mu\nu}} = \kappa t_{\mu\nu} + \sum_{n=1}^{\infty} \kappa^n \delta_n t_{\mu\nu} \quad \text{where } \delta_n t_{\mu\nu} = -\frac{\delta f_n}{\delta h^{\mu\nu}} \text{ by definition} \tag{2.177}$$

The source constraint is still in force, so we get

$$\partial^\mu \left(t_{\mu\nu} + \delta_1 t_{\mu\nu} + \sum_{n=2}^{\infty} \kappa^{n-1} \delta_n t_{\mu\nu} \right) = 0 \tag{2.178}$$

where the first term in the iteration $\delta_1 t_{\mu\nu}$ is separated out by dividing through by κ. Now Fang and Fronsdal argue: the first term is of order κ, since t is divergence free in the limit $t \to 0$ by virtue of the matter field equations. It follows that the second term must also be of order κ, and thus it must have the form

$$\partial^\mu \delta_1 t_{\mu\nu} = -A_{\nu,\alpha\beta} \left(\frac{\delta f_0}{\delta h_{\alpha\beta}} \right) - D_\nu{}^K \left(\frac{\delta f^m}{\delta \phi^K} \right) \tag{2.179}$$

with $A_{\nu,\alpha\beta}$ and $D_\nu{}^K$ first-order differential operators (including nonderivative terms) and ϕ^K denoting the matter fields.

After this, Fang and Fronsdal describe a way to proceed that starts by writing a general ansatz for $\delta_1 t_{\mu\nu}$ and restricting coefficients for the terms so as to make the left-hand side of (2.179) take the form of the right-hand side. Based on this, historical comments are made on previous work by Kraichnan, Thirring, Wyss and Feynman, which were all based on matter models. The suggestion by Wyss, that "[...] the structure of the Lie algebra of infinitesimal gauge transformations may be fixed by the requirement of consistency of the field equations to lowest order in κ." then leads Fang and Fronsdal over to the restricted program.

95 One would perhaps assume that the polynomial order of f_k is $k + 2$, but the authors make no such statement, only writing that f_1 is assumed to have no constant term and no term linear in h. The point is that, in the presence of matter, the terms involve factors of the matter fields (as remarked after the formula (2.175)). For instance, in the work of Wyss, part of f_2 is found to be bilinear in h and bilinear in matter fields.

In the restricted program, the source $S_{\mu\nu}$ and the action f, are to be constructed entirely from $h_{\mu\nu}$ and its first derivatives, and all matter related terms drop out of the equations written above. In particular, this means that consistency of the field equations to order κ reduces to equation (2.179) without the ${D_\nu}^K$ term. Fang and Fronsdal interpret this equation in the sense

$$\left[\frac{\delta f_0}{\delta h_{\alpha\beta}} = 0\right] \Rightarrow \left[\partial^\mu \delta_1 t_{\mu\nu} = 0\right] \tag{2.180}$$

The authors argue for an ansatz for f_1 of homogeneous order 3 in $h_{\mu\nu}$ (reasonable enough) and then discuss equivalence of formal power series in $h_{\mu\nu}$ under nonlinear field redefinitions not involving any derivatives. Then they state a theorem to the effect that with f_0 the free spin 2 action and f_1 a homogeneous polynomial of order 3, then (2.180) is satisfied if and only if f is equivalent to order κ, either to the series defined by expanding Einstein's Lagrangian (with the zero cosmological constant) with $g_{\mu\nu} = \eta_{\mu\nu} + \kappa h_{\mu\nu}$, or to the series with $f_1 = 0$.

The authors describe the proof as a direct computation: from the most general ansatz for f_1, ignoring exact divergences, calculate $\partial^\mu \delta_1 t_{\mu\nu}$. Then eliminate all terms that contain the d'Alembertian $\square$ by the free field equations and require that the resulting expression vanishes identically. This leaves five undetermined coefficients of which four can be adjusted by field redefinitions, turning f_1 into a constant multiple of the corresponding term in Einstein's Lagrangian. Thus, Fang and Fronsdal concludes, the order κ spin 2 energy-momentum tensors and self-interactions previously derived by Gupta, Wyss and Feynman, under more special assumptions, are unique up to field redefinition equivalence.

In order to continue up to the next order in κ, one would have to retain the particular form of the operator $A_{\nu,\alpha\beta}$ and work from (2.179). Fang and Fronsdal, instead, turn to gauge invariance.

The Fang and Fronsdal paper continues with a discussion of the gauge algebra. As the discussion is technically quite involved, we will just state the result here in a simplified way. A generator of a local infinitesimal coordinate transformation can be written as $\xi^\mu \partial_\mu$. Commuting two such generators, acting on a scalar field ϕ for simplicity, one gets[96]

$$[\eta^\nu \partial_\nu, \xi^\mu \partial_\mu]\phi = (\eta^\nu \partial_\nu \xi^\mu - \xi^\nu \partial_\nu \eta^\mu)\partial_\mu \phi = [\eta, \xi]^\mu \partial_\mu \phi \tag{2.181}$$

This defines a Lie algebra, and Fang and Fronsdal prove that this is the Lie algebra resulting from the restricted Gupta program applied to massless a spin 2 field. The result is Einstein's theory of gravity.[97]

2.9 Gauge invariance, interactions and self-interactions

We reviewed the program of deriving Einstein gravity by deforming the Fierz–Pauli, massless free spin 2 theory in some detail above, since it inspired Fang and Fronsdal to formulate the generalized Gupta program of deriving higher spin gauge field inter-

96 See also Section 3.13.1 where the action on general tensor fields is discussed.

97 In the mid-1980s, I. Bengtsson and myself tried to construct higher spin gauge algebras by generalizing (2.181) in an obvious, but naive way, taking for instance gauge generators for spin 3 as $\xi^a_{\mu\nu} T_a \partial^\mu \partial^\nu$ and correspondingly for higher spin [125, 126]. We did not succeed.

actions in an analogous way. As we saw, quite a few authors wrote on this subject in the 1950s and 1960s. There is indeed a very extensive literature devoted to the question of alternative derivations of the theory of gravity, both deformation theoretic and gauge theoretic. Such work has of course been highly interesting as a source of ideas and intuition for work on the higher spin problem, in particular since a "geometric" approach has also proved evasive. We will not go very deeply into this history here, partly because it is not explicitly "higher spin", partly because it belongs more properly to Volume 2, where we will return to it. A few comments will suffice for now.

2.9.1 Stanley Deser's first-order deformation

A paper from 1970 by S. Deser [120] is highly interesting in the context of deformation theoretic approaches to gravity. When expanded in terms of a spin 2 field, gravity becomes a nonpolynomial theory. This fact indicates that an approach to deriving gravity by deforming a free spin 2 theory is difficult to carry through, as the analysis referred to above by Fang and Fronsdal indeed shows. However, it is possible rewrite the Einstein action in a first-order form where the metric and the connection are considered as independent fields. Deser writes the action as

$$I = \int d^4x \mathfrak{g}^{\mu\nu} R_{\mu\nu}(\Gamma) \tag{2.182}$$

where the metric tensor density is $\mathfrak{g}^{\mu\nu} = \sqrt{g} g^{\mu\nu}$ and the Ricci tensor

$$R_{\mu\nu} = \partial_\mu \Gamma_{\nu\alpha}{}^{\alpha} - \partial_\alpha \Gamma_{\mu\nu}{}^{\alpha} + \Gamma_{\mu\beta}{}^{\alpha} \Gamma_{\alpha\nu}{}^{\beta} - \Gamma_{\alpha\beta}{}^{\alpha} \Gamma_{\mu\nu}{}^{\beta} \tag{2.183}$$

is expressed entirely in terms of the connection Γ. Taking $\mathfrak{g}$ and Γ as independent fields, the action is cubic.[98] Varying the action with respect to $\mathfrak{g}^{\mu\nu}$ is trivial, and just gives $R_{\mu\nu} = 0$ with $R_{\mu\nu}$ given by equation (2.183). Varying with respect to $\Gamma_{\mu\nu}{}^{\rho}$ is more complicated, but it eventually leads to the familiar equation

$$\Gamma_{\mu\nu}{}^{\sigma} = \frac{1}{2} g^{\sigma\rho} \left(\frac{\partial g_{\nu\rho}}{\partial x^\mu} + \frac{\partial g_{\mu\rho}}{\partial x^\nu} - \frac{\partial g_{\mu\nu}}{\partial x^\rho} \right) \tag{2.184}$$

The calculations indicated so far are standard, and we review them in Section 4.7.2.

Deser first linearizes the action and field equations by taking $\mathfrak{g}^{\mu\nu} = \eta^{\mu\nu} + h^{\mu\nu}$. The linearized $R_{\mu\nu} = 0$ is then the free spin 2 field equation. He then adds to the right-hand side of this equation the "stress-tensor" of the linear action. This starts an iteration that is expected to continue with an infinite number of terms, as we have seen above.

98 The contractions over indices in the expression for $R_{\mu\nu}$ do not involve the metric. Index contraction is an operation in tensor algebra and have nothing in particular to with any metric on the manifold. Deser writes $R_{\mu\nu}$ so that it is explicitly symmetric.

However, Deser shows that in this particular approach it stops already after the first step, producing the first-order cubic action. Intuitively, the nonpolynomicity of gravity emerges due to the need to invert $g^{\mu\nu} = \eta^{\mu\nu} + h^{\mu\nu}$ where $h^{\mu\nu}$ is taken as the spin 2 field. We will not go any further here with the details of the argument. It is elaborated and discussed in [127].

2.9.2 Utiyama, Sciama and Kibble

Two years after the construction of the SU(2) gauge theory of isotopic spin by C. N. Yang and R. L. Mills in 1954 [114], R. Utiyama [115] proposed a gauge theory of gravity where the Lorentz group played the role of gauge group. According to historical comments in L. O'Raifeartaigh's overview *The Dawning of Gauge Theory* [128], some parts of Utiyama's work was done independently at the same time as the Yang–Mills paper was published. Upon learning about the Yang–Mills paper, he did not at first publish his own work until he realized that it was more general. Utiyama, who thought in terms of a "general gauge theory", was in particular interested in the gravitational example. His paper treated general finite-dimensional (compact or noncompact) Lie groups [115, 128]. His work has however become mostly known for the gauging of the Lorentz group.

As we will see in the technical Section 4.6.1, gauging the Lorentz group naturally leads to the introduction of connections as Lorentz gauge fields, but yields no motivation for the metric or vierbein fields. Consequently, Utiyama's paper was criticized for introducing the vierbeins in an ad hoc manner, and therefore, begging the question of a nongeometrical motivation for gravity.

The problem of gauging the Lorentz group was then considered by D. Sciama [129, 130] from a different viewpoint.[99] Sciama started from the full Einstein general relativity in the vierbein formulation, and then "on that background", so to speak, gauged the Lorentz group. The difference in relation to Utiyama was that Sciama did not aim at deriving general relativity, but rather to clarify the role of "spin" as a kind of "charge" in general relativity. The result was Einstein gravity with torsion coupled to the angular momentum current.

A proper "gauge theory of gravity" was studied by T. W. B. Kibble [131], who undertook the gauging of the in-homogeneous Lorentz group, in that way providing a motivation for the vierbeins. Simply put, the vierbeins are the gauge fields of the local translations. Due to the importance of the problem, a large number of papers have appeared over the years, but I will not try to relate the history of the subject. The problem was actively researched during the late 1970s and the 1980s in connection to supergravity. For the premid-1980s history, see [132–134] and the very useful commented

99 Sciama does not refer to Utiyama.

reprint volume [135] which contains very many references, as well as articles that put the central papers in perspective. In our Section 4.6, we will study Kibble's approach.

2.9.3 Ogievetskij – Polubarinov

In the early 1960s, there are a number of papers by V. I. Ogievetskij and I. V. Polubarinov investigating the interplay between space-time symmetries, internal symmetries (such as isospin) and the spin of the fields in the theory. This was after Yang–Mills but before the establishment of the standard model, and the phenomenology of time had still not received its now well-known systematization. This is reflected in the context of the paper [136], parts of which are relevant for the general problem of finding self-interacting field theories. One theme of the paper is set out in the first, one and a half paragraphs.

> In the strong interactions of elementary particles, quantities such as isotopic spin, strangeness, and number of baryons, are conserved. At first glance, this group of conservation laws and the corresponding invariances are by no means connected with the Minkowski space-time properties. At the same time, conservation of energy momentum and angular momentum is explicitly bound up with the homogeneity and isotropy of space-time.
> In the abstract [of the paper], we have stated that the first group of conservation laws is [sic] intrinsically connected with the space-time property of vector fields having definite spin. This statement is surprising, and we shall attempt to explain it.

The statement is indeed surprising, since we are accustomed to thinking about internal symmetries, global or local, as having nothing to do with space-time symmetries. In fact, the no-go theorems of Coleman-Mandula [137] and O'Raifeartaigh [138] some years later, ruled out the possibility any nontrivial relations between internal and space-time symmetries. So, if the Ogievetskij–Polubarinov paper is not wrong, it is interesting to understand what they are claiming.

The paper concerns field theories for spins 0, 1/2 and 1, as was appropriate for strong interaction physics. To get started, consider field equations for a set of vector fields b^i_μ, indexed by i with possibly different masses m_i (no sum over i)

$$\Box b^i_\mu - \partial_\mu \partial \cdot b^i - m_i^2 b^i_\mu = -j^i_\mu \tag{2.185}$$

Here, j^i_μ are currents constructed out of other fields, as well as of the b^i_μ themselves in the self-interacting case. It is required that it should follow from the field equations that the spin of the each field is 1, that is, superfluous field components should be removed.[100] In the massive case, computing the divergence of the field equation yields

100 The mismatch between the number of components for Lorentz invariant fields and the number of d. o. f. for Poincaré invariant physical states, is briefly discussed in the paper.

$m_i^2 \partial \cdot b^i = \partial \cdot j^i$. Therefore, when the fields b^i_μ are free ($j^i_\mu = 0$), they must have zero divergence. This condition must be maintained in the interacting case in order that the number of field components do not change when the interaction is present (see the Fierz–Pauli analysis reviewed in Section 2.1.5). This, in its turn, leads to current conservation $\partial \cdot j^i = 0$. In the massless case, computing the trace of the field equation, also yields current conservation, but no condition at all on $\partial \cdot b^i$. In conclusion,

$$\partial \cdot b^i = \begin{cases} 0 & \text{if } m^2 \neq 0 \\ \text{arbitrary} & \text{if } m^2 = 0 \end{cases} \tag{2.186}$$

In both cases, the current j^i_μ is conserved, indicating that the theory is invariant under some phase transformation. In both cases, the divergence of the field equation is zero. This leads to the question: "What invariances and what interactions are possible?".

This is the question investigated in the paper. We now cut straight to Section V of the paper that concerns the self-interaction of vector fields, massive or massless. The most general Lagrangian for vector fields b^i_μ with cubic and quartic self-interaction terms, restricted by requiring dimensionless coupling constants, are set up as an ansatz. The authors include parity violating terms, but let us for simplicity neglect these. The ansatz is then[101]

$$\mathcal{L} = -\frac{1}{4} f^i_{\mu\nu} f^i_{\mu\nu} - \frac{1}{2}(m^2)_{ij} b^i_\mu b^j_\mu + \alpha_{ijk} \partial_\nu b^i_\mu b^j_\mu b^k_\nu + \beta_{ijkl} b^i_\mu b^j_\mu b^k_\nu b^l_\nu \tag{2.187}$$

where $f^i_{\mu\nu} = \partial_\mu b^i_\nu - \partial_\nu b^i_\mu$ as usual. Since the mass-matrix $(m^2)_{ij}$ and coupling constants α_{ijk} and β_{ijkl} are completely arbitrary at this stage (except being real and certain symmetry properties of the β's that follow from the definition of the quartic term), these are indeed all possible terms.

Next, the field equations are derived. Then requiring that the divergence of the field equations be zero, and using the field equations again, a sum of different terms arise, equated to zero. Investigating the structure of the terms, it turns out that there is a single term of a particular structure that imposes the condition $\alpha_{jki} = -\alpha_{kji}$. Then the sum of another two terms being zero, impose the condition $\alpha_{ijk} = -\alpha_{ikj}$. Together we find that the α_{ijk} must be totally antisymmetric. Then the rest of terms involving only α_{ijk} sum to zero.

Continuing with the rest of the terms, Ogievetskij and Polubarinov find that

$$8\beta_{ijkl} + \alpha_{mki}\alpha_{mlj} + \alpha_{mkj}\alpha_{mli} = 0 \tag{2.188}$$

$$\alpha_{mij}\alpha_{klm} + \alpha_{mki}\alpha_{jlm} + \alpha_{mjk}\alpha_{ilm} = 0 \tag{2.189}$$

The first of these equations, when inserted back into the Lagrangian $\mathcal{L}$, yields the familiar Yang–Mills quartic interaction term. The second equation is a Jacobi identity.

101 The summation convention is in force for both types of indices.

It thus transpires that what we get is an adjoint representation of a finite dimensional Lie algebra. The Lagrangian $\mathcal{L}$ is a massive Yang–Mills Lagrangian

$$\mathcal{L} = -\frac{1}{4} G^i_{\mu\nu} G^i_{\mu\nu} - \frac{1}{2}(m^2)_{ij} b^i_\mu b^j_\mu \quad \text{where } G^i_{\mu\nu} = \partial_\mu b^i_\nu - \partial_\nu b^i_\mu + \alpha_{ijk} b^j_\mu b^k_\nu \tag{2.190}$$

Further considerations show that all masses within an irreducible representation must be equal. The theory is invariant under global, infinitesimal transformations

$$\delta b^i_\mu = \alpha_{ijk} \omega^j b^k_\mu \tag{2.191}$$

We can now appreciate the claim made at the beginning of the paper. The authors may just as well speak for themselves.

> The form of the vector field interaction obtained by us is in general similar to that obtained by many other authors who treat the vector fields on the basis of the so-called "gauge principle" [references removed]. This similarity is not accidental and is due to the following facts: *From the very beginning they have postulated the symmetry properties* (e. g., the isotopic invariance) and have assumed the vector field masses equal to zero. They require further the "local" symmetry properties, but they are equivalent to the requirement of arbitrariness of $\partial_\mu b_\mu$, and also single out spin 1 [references removed]. *At the same time, we do not suppose any symmetries; we derive these symmetries.*

To conclude, the Ogievetskij–Polubarinov derivation of massive and massless Yang–Mills theory is an early, and successful instance of the "deformation theoretic" approach to interactions. The authors also go on to fix the interactions to spin 1/2 and spin 0 matter in a similar way.

2.10 The 1970s and Fronsdal's theory

Let us take up the story of wave equations where we left it with Singh and Hagen. Up to the mid-1950s, investigations into higher spin field theory had been focused mainly on massive fields. In the 1950s and 1960s, more and more higher spin massive particles were discovered in the accelerator laboratories. Yang–Mills theory was invented in 1954 [114] and investigated thereafter, but the masslessness of the fields initially made them unlikely to mediate short-range forces. Then in about 20 years the mechanism of spontaneous symmetry breaking was developed and asymptotic freedom was discovered. This together with the renormalization proof made Yang–Mills and massless spin 1 fields viable candidates for the fundamental interactions of weak and strong forces.[102]

102 Rather than trying to supply a list of references here, I refer the reader to the second Volume [139] of S. Weinberg's quantum field theory textbook, which contains both the theory and original references.

Partly as a consequence of the invention of supergravity in 1976 by D. Z. Freedman, P. van Nieuwenhuizen and S. Ferrara [140, 141] and by S. Deser and B. Zumino [142], the focus of theoretical studies of higher spin shifted to massless fields. Before this, there were no strong motivations for studying massless gauge fields of spin greater than 2. Utiyama had initiated the gauge theory approach two gravity, continued by Kibble and Sciama, but this did not lead to investigations of massless higher spin until Fronsdal – and collaborators – raised the question. The majority of workers in the supersymmetry and supergravity research programs, were content with, or most interested in, ruling out massless higher spin fields.

2.10.1 Fronsdal's road to the Fronsdal theory

Christian Fronsdal's 1978 paper [3] on massless fields with integer spin is pivotal in the history of higher spin gauge fields. In this paper, several strands of history from the Dirac, Fierz and Pauli papers, the Bargmann and Wigner papers and the later work by Singh and Hagen converge. It is the starting point of almost all modern research into higher spins. Here, the problem of introducing self-interactions is posed as the central problem. The paper is followed by a companion paper on half-integer spin fields, written together with J. Fang [6].[103]

Fronsdal had been publishing on representations of space-time isometry groups since the late 1950s and the long series of papers [143–148] from 1965 to 1978 on elementary particles in curved space, lead to group theoretical insights that have become relevant for the AdS approach to higher spin and to the AdS/CFT correspondence conjectures, although actually not having been involved in the initial stages of those developments.

Fronsdal's classic paper on higher spin fields takes up the theory of massive fields where it had been left by Singh and Hagen a few years earlier. As we have seen, the Singh–Hagen work built on the work of S-J. Chang, which in its turn built on Fronsdal's thesis paper. Fronsdal uses the massive theory as a springboard to the massless theory.[104] This signals the shift of perspective from matter higher spin fields that had indeed been the main focus of the pioneers Dirac, Fierz and Pauli, and had remained so up until the advent of supergravity in the early 1970s. This period is also the fine era of accelerator physics investigating the weak and strong interactions and the higher spin, strongly interacting, mesons and hadrons.

It is clear from the paper, both from the abstract and from the final section on nonlinear theory, that Fronsdal's main interest concerned self-interactions. The call

103 The two papers are consecutive in the Physical Review.

104 The attempts by many authors to construct higher spin theories inspired Fronsdal to "take the massless limit of my thesis", as he wrote to me in answer to questions.

for a such a research program is explicitly stated.[105]

> A generalized Gupta program is proposed, that is, a search for a scheme for generating a theory of interacting, massless particles, consistent to all orders in the coupling constant.

Two motivations for reviving higher spin theory are stated in the Introduction to the paper, the first being the recent discovery of supergravity and the spin 2 barrier that it seemed to impose. The second motivation came from neutrino physics. At this time, the neutrino was thought to be massless and Fronsdal speculated that its properties and the weak interactions could perhaps be understood from some gauge principle, and that higher spin gauge theory might give clues. Nothing of this sort has materialized.[106]

The paper then takes its starting point in the well-known field equations for a massive spin s field, described by symmetric, traceless and divergence-free tensor $\phi_{\mu_1 \ldots \mu_s}$, abbreviated as ϕ^s, each component satisfying the Klein–Gordon equation (see equations (2.140)–(2.142)). As discovered by Fierz and Pauli, attempts to introduce interactions directly in the field equations lead to changes in the number of degrees of freedom of the field. They had therefore suggested deriving the field equations from an action, something which requires the introduction of auxiliary fields "in order to have enough field components to vary.". This was the problem studied in detail by Singh and Hagen in [100]. Fronsdal agrees with these authors that the set of fields $\Phi = \{\phi^s, \phi^{s-2}, \phi^{s-3}, \ldots, \phi^0\}$ is the "simplest viable choice", all of which are supposed to be traceless.

Fronsdal then writes down the most general Lagrangian, second-order in derivatives, with arbitrary coefficients. The requirement is that the Euler–Lagrange equations must yield the correct field equations for ϕ^s and $\phi^k = 0$ for $k \neq s$, that is, all of the auxiliary fields must be set to zero by varying the action. Fronsdal then solves for the coefficients by a particular method, different from the one used by Singh and Hagen, arriving at the same result.[107] We will try to capture the essence of Fronsdal's derivation, without entering into the rather complex details.

Fronsdal's derivation of the massless theory and the emergence of gauge invariance

i

Fronsdal considers the following Lagrangian for a massive free field:

$$\mathcal{L} = \sum_k \left[\frac{1}{2}\alpha_k (\partial\phi^k)\cdot(\partial\phi^k) + \frac{1}{2}\beta_k (\partial\cdot\phi^k)\cdot(\partial\cdot\phi^k) \right.$$

---

105 It is repeated in [8]. See Section 2.8.1.

106 Neutrinos are now known to be massive.

107 Noting a misprint in Singh and Hagen that these authors had informed Fronsdal about.

$$-\gamma_k\phi^{k-2}\cdot(\partial\partial\cdot\phi^k)+\delta_k\phi^{k-1}\cdot(\partial\cdot\phi^k)-\frac{1}{2}\sigma_k m^2\phi^k\phi^k\Big] \tag{2.192}$$

where the sum over k runs over $0,1,2,\dots s-2,s$. All indices are suppressed "contracted in a unique and self-evident manner".[108] The Euler–Lagrange equations, written with $p=-i\partial$ become

$$\begin{aligned}\delta\phi^k\cdot\big[\alpha_k p^2\phi^k+\beta_k p(p\cdot\phi^k)+\gamma_{k+2}pp\cdot\phi^{k+2}+\gamma_k pp\phi^{k-2}\\ +i\delta_{k+1}p\cdot\phi^{k+1}-i\delta_k p\phi^{k-1}-\sigma_k m^2\phi^k\big]=0\end{aligned} \tag{2.193}$$

The aim is to solve for the coefficients. The details will not concern us, except that in the process, the components of the fields ϕ^k are split up according to "spin content" into a sum of objects $\phi^{k,l}$, with particular coefficients, where each $\phi^{k,l}$ contains $2l+1$ components corresponding to representations of rotation group. The count is correct, since the number of components of ϕ^k is $(k+1)^2$ and $\sum_{l=0}^k(2l+1)=2\sum_{l=0}^k l+k+1=(k+1)^2$. Correspondingly, the field equations (2.193) split up into equations for the spin components $\phi^{k,l}$. To repeat, the problem is to solve for the coefficients so that $(p^2-m^2)\phi^{s,s}=0$ and $\phi^{k,l}=0$ for all other components. This is done and the result is in agreement with Hagen and Singh.

Then Fronsdal turns to the case of zero mass. In that case, the coefficients γ_{s-2} vanish, "[...] an unexpected boon." as Fronsdal writes. Furthermore, $\delta_k=0$ in the massless case, and this has the effect that the fields ϕ^s and ϕ^{s-2} become decoupled from the rest $\phi^{s,s-3},\phi^{s,s-4},\dots$, which can be ignored. What remains of the field equations are now $p^2\phi^{s,s}=0$ for the component $\phi^{s,s}$. For the component $\phi^{s,s-1}$, the equation reduces to an identity. For the components $\phi^{s,l}$ and $\phi^{s-2,l}$, a matrix equation remain

$$p^2\begin{pmatrix}\alpha_{s,l} & \gamma_s\\ \gamma_{s,l} & \alpha_{s-2,l}\end{pmatrix}\begin{pmatrix}\phi^{s,l}\\ \phi^{s-2,l}\end{pmatrix} \tag{2.194}$$

for $l\le s-2$. The matrix is singular which implies that the components $\phi^{s-2,l}$ are expressed in terms of the components $\phi^{s,l}$ for $l\le s-2$. This is the source of gauge invariance. Accepting these equations – which we have only stated – we can now perform a d. o. f. count.

First, the fields ϕ^s and ϕ^{s-2} together has $2(s+1)^2$ components, which is the same number as for a single double traceless field ϕ^s. The way one normally reduces this number down to the physical number 2, is to invoke gauge invariance with a traceless gauge parameter ξ^{s-1} which has s^2 components. Subtracting gauge and regauge components, yields the number 2. This can be done here also, after the Lagrangian and field equations are rewritten for the massless case (which Fronsdal does). More true to the present context is however the following simpler argument.

The field $\phi^{s,s}$ has $2s+1$ components. The field $\phi^{s,s-1}$ decouples since its equation is an identity. The field ϕ^{s-2} has $2s-1$ components, but all its spin projections $\phi^{s-2,l}$ with $l\le s-2$ are equal by the matrix equation (2.194) to the lower spin projections $\phi^{s,s-2},\phi^{s,s-3},\dots$ of $\phi^{s,s}$. These are thus all "gauge" and should not be counted as physical components. Therefore, the number of physical components are $2s+1-(2s-1)=2$. Examining this argument in more detail, one will, however, realize that one needs to invoke the "gauge/regauge" procedure here also.

---

108 Readers familiar with modern "condensed notation" for higher spin should be warned here (see Section 5.1). By the expression $\partial\phi^k$, Fronsdal intends $\partial_\mu\phi^k_{\mu_1\dots\mu_k}$ with no symmetrization understood. Fronsdal gives the example $(\partial\phi)\cdot(\partial\phi)=(\partial^\mu\phi^{\nu\dots})(\partial_\mu\phi_{\nu\dots})$. A dot denotes contraction and it should also be noted that in the expression $\partial\partial\cdot\phi^k$ both partial derivatives are contracted (contrary to the convention in modern condensed notation).

The details of the Fronsdal theory will be reviewed in Chapter 5 beginning in Section 5.1. Here, we will just record the Lagrangian as it is finally written in the Fronsdal paper:

$$\begin{aligned} -\mathcal{L} =& \frac{1}{2}(ph)\cdot(ph) - (s/2)(p\cdot h)\cdot(p\cdot h) - (s/2)(s-1)h'\cdot(pp\cdot h) \\ & - (s/4)(s-1)(ph')\cdot(ph') - (s/8)(s-1)(s-2)(p\cdot h')\cdot(p\cdot h') \end{aligned} \tag{2.195}$$

The higher spin field of spin s is denoted by h. A condensed notation is used, that in various versions has become common in much of the later higher spin literature (see Section 5.1). Here, h stands for $h_{\mu_1\dots\mu_s}$ and p for p_μ interpreted as a derivative $p = -i\partial$. A dot "$\cdot$" stands for index contraction and prime "$'$" stands for an index trace. When no dot occurs between p and h, as in the first term in $\mathcal{L}$, then the two p's should be contracted into each so that $\frac{1}{2}(ph)\cdot(ph) = -\frac{1}{2}h\cdot(p\cdot p)h$ upon partial integration.[109]

The action is invariant under gauge transformations

$$h_{\mu_1\dots\mu_s} \to h_{\mu_1\dots\mu_s} + \sum_1 p_{\mu_1}\xi_{\mu_2\dots\mu_s} \tag{2.196}$$

with a traceless gauge parameter $\xi' = 0$, provided that the field is double traceless, that is $h'' = 0$. This somewhat strange condition appeared when Fronsdal combined the two traceless symmetric tensors ϕ^s and ϕ^{s-2} into the symmetric tensor field h.[110] The symbol $\sum_1$ stands for a symmetric sum. This summation notation was later dropped as the condensed notation was further developed.

Within a year, two investigations into higher spin gauge fields were published by T. Curtright [149] and by B. de Wit and D. Z. Freedman [150] that clarified the structure of the new theory.

2.10.2 T. Curtright's derivation of the Fronsdal theory

T. Curtright's paper [149] takes its stated motivation from the newly found arbitrary spin gauge field theories,[111] and the observation by M. Gell-Mann[112] that an irreducible supermultiplet limited to spins less that or equal to 2, "[...] cannot accommodate a local $SU(3)_{color} \times (SO(2) \times U(1))_{electroweak}$ gauge theory with the necessary vector bosons

109 Note that a misprint in the first term of $-\mathcal{L}$, as written in the paper, is corrected here: the prime $'$ should be a dot $\cdot$.

110 As we will see in Chapter 5, the theory can just as well be developed in terms of tensor fields of rank s and $s-2$.

111 It is written in the paper that the higher spin Lagrangians were obtained before learning of the Fronsdal papers.

112 See reference [5] in Curtright [149].

appearing as fundamental fields", thus perhaps pointing toward the need for "structureless" higher spin fields, at least of spin 3 and spin 5/2. This is interesting since it is an example of how supersymmetry, at this time, was injecting new energy into the old higher spin program.

The Fronsdal theory, both for integer and half-integer spin, was rederived by Curtright by an ansatz-verification method. The higher spin gauge transformation law was postulated to be the one found by Fronsdal in equation (2.196), and similarly for the half-integer case.

The explicit calculations are not presented in the paper, but it is said that the most general form for the free Lagrangians were written down, and that invariance up to total derivatives are only possible for traceless gauge parameters and Lagrangians not containing double or higher traces of the fields. What results are precisely the Fang–Fronsdal Lagrangians for integer and half-integer spin.[113]

Having found the Lagrangians for integer and half-integer spin fields, Curtright then goes on to discuss supersymmetric theories with the spectra $(s, s-1/2)$ and $(s, s+1/2)$. Supersymmetry is actually used to show that the massless higher spin theories are free of lower spin modes.

Toward the end of the paper there are remarks on interactions and the known coupling problems. These questions are also further discussed in a review paper by Curtright [151] from 1980 that offers an interesting view of the standing of the subject at that time. For instance, the at the time newly discovered (by E. Cremmer and B. Julia) nonlinear SU(8)-invariance in $N = 8$ supergravity, again raised the hopes to incorporate the standard model group within supergravity.

2.10.3 B. de Wit and D. Z. Freedman

In their paper [150], B. de Wit and D. Z. Freedman clarified the structure of the theory by introducing a hierarchy of *generalized Christoffel symbols*

$$\Gamma^{(m)}_{\rho_1\dots\rho_m;\mu_1\dots\mu_s} \tag{2.197}$$

The μ indices indicate the spin of the fundamental field $\varphi_{\mu_1\dots\mu_s}$, while the ρ indices indicate the number of derivatives ∂_ρ that appear in the definition of the respective symbols. The symbols are symmetric in both index sets separately. For lower spin, the symbols generalize the spin 1 electromagnetic field strength and the spin 2 free theory Christoffel symbol and curvature tensor.

113 In retrospect, it is interesting to note that Curtright comments that the approach "[...] appears not to be very methodological, beginning as it does with a guess, [...]". The guess concerns the free field gauge transformations. This is actually how the Vasiliev theory starts out, as writing down free field gauge transformations, but then not in the "metric-like" formulation, but in a "frame-like" formulation.

The symbols have simple gauge transformation properties deriving from the underlying gauge field transformation. Generalizing the spin 1 and 2 cases, the "highest" symbol $\Gamma^{(s)}$ is gauge invariant. The special case of spin 3 was further investigated by T. Damour and S. Deser in [152], motivated by a possible geometric origin of the higher spin field theories, hinted at by the existence of these generalized Christoffel symbols. This is a possibility noted also by de Wit and Freedman. This theme was later revived, and studied, by a number of authors (see our Section 5.3.2). The last section of the paper discuss interaction problems and possibilities. More technical details of the de Wit–Freedman construction will be reviewed in Section 5.2.

2.10.4 Hypergravities

Following the successful construction of the supergravity theories in 1976,[114] interest in higher spin gauge fields increased, in particular after the appearance of the Fronsdal theory. We have already noted Curtright's study of simple supersymmetric higher spin gauge multiplets and their free field theory. During the period 1979 to 1982, quite a few authors studied coupling of higher spin gauge fields to gravity, and in particular the coupling of spin 5/2 fields. Such tentative interacting theories were called *hypergravities*. Since this is belongs to the theory of higher spin interactions, we will return to its history in the second volume of the present work. Here, we just list a set of references: [153–157]. Hypergravity theories turned out to suffer from consistency problems.

2.11 The mid-1980s: the BRST approach

In the beginning of 1986, quite a few researchers were apparently working on BRST approaches to massless higher spin fields, myself being one of them. The first paper appearing in print [158] was written by S. Ouvry and J. Stern and published in a September issue of Physics Letters B. It was quite a shock to me when I saw their preprint one morning in the physics department at Queen Mary College, just a week or so after I had submitted my own manuscript to the same journal [159]. I cannot refrain from a bit of personal history at this point.

A bit of personal history

I had worked on the problem of finding field equations and actions for higher spin gauge fields using BRST methods, all through the winter and spring. Covariant string field theory were very active subjects at this time and it was natural to look to string theory for inspiration and methods. I went to

114 in [140, 142, 141].

string seminars at the colleges in London, but I was thinking of massless higher spin all of the time. Ingemar Bengtsson and I had worked on and off since our 1983 work with Lars Brink, on extending it to higher order and to covariantize it. From Fang and Fronsdal [8] and from my own work [125], it was clear that a higher spin theory needed to include a spectrum of all spins. The program I set myself after arriving at Queen Mary College in London was to at least unify free field wave equations and Lagrangians. I approached the problem from two sides. First, I experimented with low spin fields. From various preprints [160–162] that I read at the time, I got the idea that I needed to introduce auxiliary fields for divergences and traces of the gauge fields in order to be able to uniformize the field equations. In parallel, I read papers on free string BRST field theory. This together with experimenting with the $\alpha' \to \infty$ limit of the string theory Virasoro algebra – where it was "pictorially" clear that all Regge trajectories would pile up on the spin axis with zero mass for all fields – finally led to the solution. I actually remember seeing it all clear one afternoon walking up Highgate Road to fetch my son Olof. While my wife and I were working, he was looked after by Kerstin, a Swedish lady that my wife had met by chance at the playground. As usual on the way back home, Olof and I stopped at the bridge over the railway tracks in Kentish Town to look at the trains running to and from Kings Cross.[115] The problem was solved and it remained to work out the details.

The reverse engineering of the Fronsdal field equations will be reviewed later in Section 5.3.1. Here, we will start by putting the higher spin problem into the context of the string field theory of the mid-1980s.

2.11.1 String field theory backdrop

Following the so-called "1984 superstring revolution",[116] many authors and groups were busy covariantizing the light-front string field theory (bosonic and fermionic) that had been generalized to superstrings by M. B. Green and J. H. Schwarz also in collaboration with L. Brink. There were two aspects of the problem: first, to clarify the nature of the covariant free field theory, and second, to extend the light-front cubic interaction into covariant cubic interactions. In regard to the interactions, the method applied was invented by W. Siegel in [163]. In short, it consisted of extending the light-front transverse momenta p_i to fully covariant momenta p_μ while compensating with two new ghost momenta. In the fully elaborated method of [164] the light-front dynamical Lorentz generator J_{i-} could be identified with the BRST Q operator. While working well for free strings, and for field theories in general, this method also promised a way to the interactions via the known interactions in the light-cone gauge. The method was largely superseded by E. Witten's covariant open string field theory [165] that worked a priori with a BRST-invariant interaction.

115 Olof apparently picked up two things in London. He became a model railway enthusiast just like me, and he is now running the O/O Brewing craft beer company with his friend Olle.

116 M. B. Green and J. H. Schwarz had shown that a certain chiral anomaly, disastrous for theories with the ambition to unify the fundamental forces – as pointed out by E. Witten – did not occur in superstring theory. This inspired the huge interest in string theory that has lasted up to the present time.

In regard to the free string theory, it was first treated using BRST methods a few years earlier by M. Kato and K. Ogawa [166] and by S. Hwang [167].[117] These authors arrived at the major conclusion that BRST-nilpotency $Q^2 = 0$ for the bosonic string required the critical dimension $D = 26$ and zero intercept $\alpha_0 = 1$, also showing the theory to be ghost-free and unitary. While the Kato–Ogawa paper utilizes a "covariant operator formalism" from a paper by T. Kugo and I. Ojima [168] devoted to Yang–Mills theory, the Hwang paper derives the same result using the somewhat more streamlined Fradkin–Vilkovisky formalism.[118] In this formalism, once one has derived the first class constraints ψ_a of the theory and their first class algebra – for instance using the Dirac procedure – one can write down the BRST operator as

$$Q = \psi_a \eta^a - \frac{1}{2} \mathcal{P}_c U^c{}_{ab} \eta^a \eta^b \tag{2.198}$$

Here, the $U^c{}_{ab}$ are the structure constants of the first class algebra

$$\{\psi_a, \psi_b\}_- = \psi_c U^c{}_{ab} \tag{2.199}$$

and η and $\mathcal{P}$ are ghost coordinates and momenta satisfying the bracket[119]

$$\{\eta^a, \mathcal{P}_b\}_+ = \delta^a_b \tag{2.200}$$

For the bosonic string, the constraints are the Virasoro generators and the algebra the Virasoro algebra. The so found theory is a first quantization of the string considered as a mechanical system. In retrospect, it may seem quite easy to see how this method can be applied to massless higher spin theory, if only the proper first class constraints are known, or if the corresponding mechanical model is known – if such exists. The story was however not so simple. In a series of papers [163, 172, 173], W. Siegel worked on covariantly second quantized string field theory.

String fields

A *string field* Φ is a "function" of the string coordinate $x^\mu(\sigma)$ where σ is runs from end to end of the string. The field is thus written $\Phi(x^\mu(\sigma))$. The string, viewed as a mechanical system, is actually parametrized by two world-sheet coordinates σ and τ playing the roles of two-dimensional space and time, respectively. The free string coordinate x^μ, after certain gauge fixing, obey a two-dimensional

117 Stephen Hwang was then a graduate student with R. Marnelius in Göteborg.

118 The Hwang paper also treats the "sigma model string theory" in detail, discovered by L. Brink, P. Di Vecchia and P. Howe [169] and S. Deser and B. Zumino [170] in 1976. This string model was later studied by A. M. Polyakov [171].

119 We assume that the underlying theory is bosonic so that the ghosts are fermionic. Thus $\{\cdot,\cdot\}_-$ denotes an antisymmetric bracket (Poisson bracket or quantum commutator) and $\{\cdot,\cdot\}_+$ denotes a symmetric bracket.

wave equation which can be solved in the standard mathematical physics way in terms of a Fourier series expansion over oscillating modes. For the string field $\Phi(x^\mu(\sigma))$, this means that it contains a denumerable infinite number of component fields of increasing spin and mass. To see this in a little more detail, consider first-quantizing the string. There is also a string momentum $p^\mu(\sigma)$ that is conjugate to $x^\mu(\sigma)$ and one arrives at the quantization condition

$$[x^\mu(\sigma), p^\nu(\sigma')] = i\eta^{\mu\nu}\delta(\sigma - \sigma') \tag{2.201}$$

There is an indefinite metric problem here that traditionally was solved by going to a light-cone gauge, but in the 1980s started to be treated by the covariant BRST method. The coefficients of the oscillating modes of the string become harmonic oscillator annihilation and creation operators a_n^μ and $a_n^{\mu\dagger}$ satisfying the commutation relations

$$[a_m^\mu, a_n^{\nu\dagger}] = \delta_{mn}\eta^{\mu\nu} \tag{2.202}$$

The string field can now be expanded in the Fock space of these oscillators

$$|\Phi(x)\rangle = \phi(x)|0\rangle + \phi_\mu^n(x)a_n^{\mu\dagger}|0\rangle + +\phi_{\mu\nu}^{mn}(x)a_m^{\mu\dagger}a_n^{\nu\dagger}|0\rangle + \cdots \tag{2.203}$$

with implicit sums over the mode numbers $m, n, \ldots$. The x in the fields is the *zero mode* in the string coordinate Fourier expansion, corresponding to the space-time coordinate.

Siegel realized that the BRST transformations of the first quantized string could be used to set up second quantized gauge field theory of the string. The result was a gauge-fixed covariant string field theory. The action was BRST-invariant but not gauge invariant. The string field employed in the procedure depended on ghost coordinates, apart from $x^\mu(\sigma)$, corresponding to the mechanics BRST ghosts of the string. The field therefore contained enough components to set up a gauge fixed theory, but needed complicated field redefinitions to get a gauge invariant theory.

Then in collaboration with B. Zwiebach, W. Siegel arrived at a gauge invariant formulation of the bosonic string [174]. Quite a few other authors and groups were also working on this problem: D. Friedan [175], T. Banks and M. Peskin [176], T. Banks, M. Peskin, C.R Preitschopf, D. Friedan and E. Martinec [177], A. Neveu and P. C. West [178], A. Neveu, H. Nicolai and P. C. West [179], K. Itoh, T. Kugo, Kunimoto and H. Ooguri [180] and P. Ramond [160]. E. Witten then pointed out, in [165] that the $\langle\Phi|Q|\Phi\rangle$ form of the string action was already gauge invariant without the need to integrate out superfluous auxiliary modes, as had been done in the other approaches. For a review of string field theory, see [181]. Let us briefly outline the method to construct a field theory out of a mechanics theory that was invented by W. Siegel.

From mechanics to field theory

---

i The Siegel algorithm for constructing a classical field theory from an underlying mechanics model is expressed in the paper [182]. The paper is about superstrings. The method is not formulated in general terms, rather Siegel phrases it as a "strategy": classical mechanics → first quantized mechanics BRST

→ field theory BRST → classical field theory. The method is first exemplified for the bosonic strings, then for the superparticle and finally for the superstring. It is clear from these examples how to apply the method in general.

Siegel notes that it is convenient to work in a Hamiltonian, rather than a Lagrangian, formulation of mechanics, as it involves one less step. This is seen already in the most simple example of the point particle, where the procedure is $L = \frac{1}{2}(\dot{x}^2 - m^2) \to H = \frac{1}{2}(p^2 + m^2) \to \mathcal{L} = \frac{1}{2}\phi(\Box^2 - m^2)\phi$. In this example, the algorithm is so well known that it often goes unnoticed. It does however provide the basic intuition behind the general method. What is added in the Siegel method is that the Hamiltonian H corresponds to the reparametrization constraint of the point particle. In the BRST approach, there is then a corresponding ghost coordinate-momentum pair $(\eta, \mathcal{P})$ and the BRST charge become $Q = \eta H$. For a detailed description of the general method, see Section 3.3.3.

One goal that both the Ouvry and Stern paper and my own paper achieved, was to collect all higher spin gauge fields into one object and writing an action that unified all higher spin gauge actions into one single action. In retrospect, that can be done, and has been done quite easily, but somewhat clumsily, by writing a formal string field-like expansion of the type (2.203) over just one oscillator, thus giving a simple spectrum of higher spin fields, and devising an appropriate kinetic operator.

What could have been done ...

?

A list of all integer spin fields can be collected into a string-like field[120]

$$|\Phi\rangle = (\phi + \phi_\mu \alpha^{\mu\dagger} + \phi_{\mu\nu}\alpha^{\mu\dagger}\alpha^{\nu\dagger} + \cdots)|0\rangle \tag{2.204}$$

The list of all Fronsdal actions for fields of spin $s = 0, 1, 2, 3, \ldots$ can then be written

$$S = \langle\Phi|K|\Phi\rangle \tag{2.205}$$

with the kinetic operator K given by

$$K = \frac{1}{2}\left(\Box - \frac{1}{2}\alpha^\dagger \cdot \alpha^\dagger \Box \alpha \cdot \alpha - \alpha^\dagger \cdot \partial\alpha \cdot \partial + \alpha^\dagger \cdot \alpha^\dagger \alpha \cdot \partial\alpha \cdot \partial - \frac{1}{4}\alpha^\dagger \cdot \alpha^\dagger \alpha^\dagger \cdot \partial\alpha \cdot \partial\alpha \cdot \alpha\right) \tag{2.206}$$

This is actually just a term by term transcription of the Fronsdal action (2.195) for arbitrary spin where the action of the oscillators in $\langle\Phi|K|\Phi\rangle$ picks out the individual actions. S is invariant under gauge transformations

$$\delta\Phi\rangle = \alpha^\dagger \cdot \partial|\Xi\rangle \tag{2.207}$$

120 Some modern higher spin theorists refer this kind of expansion over a coordinate z_μ to a paper by V. Bargmann and I. T. Todorov from 1977 [183]. These authors however state that "[this has been] recognized since the early days of representation theory [...]". In the Introduction of the paper, it explicates the mathematics needed to define scalar products such as the one below (2.205). Here, we compute it using first quantized oscillator algebra.

provided that $\alpha \cdot \alpha|\Xi\rangle = 0$. To make the action explicitly real, the fourth term in the operator K should be written $\frac{1}{2}(\alpha^\dagger \cdot \alpha^\dagger \alpha \cdot \partial\alpha \cdot \partial + \alpha \cdot \alpha\alpha^\dagger \cdot \partial\alpha^\dagger \cdot \partial)$. Then the naive Euler–Lagrange equation that follows from the action is gauge-invariant.

That was not how it was done however.[121] The constructions in both papers was based on a more sophisticated "reverse engineering" of the Fronsdal equation; in my own case inspired by level expansions in string theory rather than the Fronsdal theory itself.

2.11.2 The Ouvry and Stern paper

The paper starts with a short introduction putting it in the context of dual resonance models and covariant string models. It then states the problem, independent of this context, of finding actions and gauge transformations for local fields of arbitrarily high spin. The string field would then appear as a particular collection of such fields $A^{\mu_1 \cdots \mu_n}$ represented by a vector $|A(x)\rangle$ in a Fock space spanned by an infinite set of oscillators as described above in formula (2.203). The authors then remark that one may just as well start with massless gauge fields, and obtain the string theory through some kind of Higgs-like effect.[122] The authors note that with an infinite set of oscillators, fields of any spin and permutation symmetry can be accommodated.

The paper first turns to the particular case of just one oscillator. The spectrum then simplifies to one symmetric tensor field at each excitation level. The Fronsdal action[123] and the gauge transformations are written in terms of a Fock space vector $|A\rangle$ of higher spin fields and a vector of supplementary fields $|B\rangle$ whose components play the role of the traces of the components of $|A\rangle$. The Lagrangian they write can be seen as a reversed engineered Fronsdal Lagrangian (2.195)[124]

$$\begin{aligned} L = & - \langle A|\partial^2 + (a^\dagger \cdot \partial)(a \cdot \partial)|A\rangle - \langle B| - \partial^2 + (a \cdot \partial)(a^\dagger \cdot \partial)|B\rangle \\ & + \langle A|(a^\dagger \cdot \partial)^2|B\rangle + \langle B|(a \cdot \partial)^2|A\rangle \end{aligned} \tag{2.208}$$

---

121 I have no recollection of having seen this particular approach published until 1989. Although it works as a formal unification of the Fronsdal actions, there is no deeper rationale for it than doing just that; see, however, Section 2.11.5 below.

122 Such a conjectured connection between higher spin gauge theory and string theory has become a quite standard motivation for working on massless higher spin theory. The remark made by Ouvry and Stern is perhaps the first occurrence in print. It was discussed at length by the authors at the time of writing. At the end of paper, the authors return to the question of mass generation in the context of introducing interactions.

123 The authors here refer to a paper by T. Curtright.

124 There is a misprint in the paper: there is an A in one place where there should be an a. This is corrected as the formula is reproduced here.

It is invariant under the gauge transformations $\delta|A\rangle = a^\dagger \cdot \partial|\Lambda\rangle$ and $\delta|B\rangle = a \cdot \partial|\Lambda\rangle$.

In this formulation of the theory, the gauge parameter is not required to be traceless. The original Fronsdal action can be recovered setting $|B\rangle = -\frac{1}{2}a \cdot a|A\rangle$.[125] The authors want to generalize this action to the case an arbitrary number of oscillators. In order to this, they introduce a pair of Grassmann variables ξ and η with conjugates $\xi^\dagger = \partial/\partial\eta$ and $\eta^\dagger = \partial/\partial\xi$. A scalar product can be defined so that the action (2.208) can be rewritten in terms of a field $|\psi_0\rangle = |A\rangle + \xi\eta|B\rangle$.

In the general case, Grassmann variables ξ_n and η_n are introduced for each oscillator a_n. Then a set of generators are defined and the algebra they satisfy. We collect these in the box below, as well a few more formulas that are used later in the paper.

The Ouvry–Stern generators and algebras

---

i

Associated with the Grassmann variables there are generators

$$T_+ = \sum_n \eta_n \frac{\partial}{\partial\xi_n} \qquad T_- = \sum_n \xi_n \frac{\partial}{\partial\eta_n} \qquad T_3 = \frac{1}{2}\sum_n\left(\eta_n\frac{\partial}{\partial\eta_n} - \xi_n\frac{\partial}{\partial\xi_n}\right) \tag{2.209}$$

satisfying an SU(2) algebra. Then there are generators

$$G_+ = \sum_n (a_n^\dagger \cdot \partial\partial/\partial\xi_n - a_n \cdot \partial\eta_n) \qquad G_- = -\sum_n (a_n^\dagger \cdot \partial\partial/\partial\eta_n + a_n \cdot \partial\xi_n) \tag{2.210}$$

All together, the generators satisfy the algebra

$$\begin{aligned}
&[T_3, G_\pm] = \pm\frac{1}{2}G_\pm \qquad [T_\pm, G_\mp] = G_\pm \\
&[T_\pm, G_\pm] = 0 \\
&\{G_+, G_-\} = -2\partial^2 T_3 \qquad G_\pm^2 = \pm\partial^2 T_\pm
\end{aligned} \tag{2.211}$$

The paper also defines generators

$$l_n = -\sqrt{n}a_n \cdot \partial \qquad l_{-n} = \sqrt{n}a_n^\dagger \cdot \partial \qquad l_0 = -\partial^2 \tag{2.212}$$

satisfying the algebra

$$[l_n, l_m] = f^k{}_{nm} l_k \quad \text{where } f^k{}_{nm} = -n\delta^{k0}\delta_{n+m,0} \tag{2.213}$$

The authors state that the Lagrangian (2.208) can now be generalized to the case of an arbitrary number of oscillators based on the algebra (2.211). The case of a single oscillator is given explicitly

$$L = \langle\psi_0| - \partial^2 + G_+ T_- G_+ |\psi_0\rangle \tag{2.214}$$

125 This should lead to the action (2.205) with kinetic operator (2.206) written above, with a traceless gauge parameter.

invariant under the gauge transformation

$$\delta|\psi_0\rangle = G_+|\Omega_{1/2} \quad \text{where } \Omega_{1/2} = \xi|\Lambda\rangle \tag{2.215}$$

The paper then discusses properties of the theory, among them the “gauge invariance for gauge invariance” that the theory shows. This is taken as a motivation for a BRST-invariant gauge-fixed formulation. Here, the paper follows the Siegel procedure (reviewed above in Section 2.11.1).

Ghost coordinates c_{-n} and c_n and their derivatives, related to the Grassmann variables ξ_n and η_n and associated conjugates, are introduced. These are associated to the generators $I_{\pm n}$ of (2.212). Furthermore, a ghost coordinate θ, and its derivative, is associated to the generator I_0. The BRST generator so constructed, according to the Siegel algorithm, is written

$$Q = -\theta\partial^2 + (\partial/\partial\theta)T_+ + G_+ \tag{2.216}$$

The nilpotency $Q^2 = 0$ follows from the algebra (2.211). A BRST gauge fixed Lagrangian is given by

$$L_{G.F.} = \int d\theta\langle\chi,\theta|[\theta\partial/\partial\theta, Q]|\chi,\theta\rangle \tag{2.217}$$

invariant under BRST transformations $\delta|\chi\rangle = \epsilon Q|\chi\rangle$. The field $|\chi,\theta\rangle = |\psi\rangle + \theta|\phi\rangle$ contains BRST auxiliary fields and Fadeev–Popov ghosts as well as the physical fields.

Toward the end of the paper, the authors turn to the question of interactions, in which context, the fully gauge invariant action

$$L = \int d\theta\langle\chi_0,\theta|Q|\chi_0,\theta\rangle \tag{2.218}$$

is given. In the field $|\chi_0,\theta\rangle = |\psi_0\rangle + \theta|\phi\rangle$, the component $|\psi_0\rangle$ contains the higher spin fields A and B, whereas the component $|\phi\rangle$ contains nonpropagating auxiliary fields. This action is gauge invariant under transformations $\delta|\chi\rangle = Q|\Lambda\rangle$. When the fields $|\phi\rangle$ are substituted via their nondynamical field equations, one recovers the Lagrangian (2.208). The paper ends with some further discussion of interactions, mass generation and connection to string theory.

2.11.3 The Bengtsson paper

My own paper also employs methods from the then active string field theory research, but it is not motivated by string theory. Rather it is situated squarely in the massless higher spin tradition of Fronsdal.[126] Rereading the paper now, it however becomes apparent how much it was written in the context of discovery rather than the con-

126 Reflecting a basic philosophy of mine regarding higher spin.

text of justification. This context of discovery was very much colored by the reading of string field theory papers and the pedestrian experimentation with auxiliary and Stueckelberg fields, just as in many string field theory papers of the time that were expanding the theory level by level. In retrospect, the $\langle\Phi|Q|\Phi\rangle$ action invariant under gauge transformations $\delta|\Phi\rangle = Q|\Xi\rangle$, with Q constructed from the mechanics first class constraints, is shouting for attention.[127]

In my paper, the issue of finding a BRST gauge-fixed action is bypassed, instead the focus is on BRST gauge invariance from the outset. Inspired by string field theory, I introduced bosonic and fermionic oscillators, generators and their algebra.

The Bengtsson generators and algebras

Bosonic and fermionic oscillators obey the commutators and anticommutators

$$[\alpha^\mu_m, \alpha^\nu_n] = m\delta_{m+n,0}\eta^{\mu\nu} \qquad \{\beta_m, \bar\beta_n\} = \delta_{m+n,0} \tag{2.219}$$

From string theory, the zero-mode degenerate vacua $|+\rangle$ and $|-\rangle$, are borrowed, subject to

$$\bar\beta_0|+\rangle = 0 \qquad \beta_0|-\rangle = 0 \qquad \bar\beta_0|-\rangle = |+\rangle \qquad \beta_0|+\rangle = |-\rangle \tag{2.220}$$

Then departing from string theory, new higher spin generators are defined according to

$$K_m = i\alpha^\mu_m\partial_\mu \quad \text{and} \quad K_0 = -\frac{1}{2}\partial^\mu\partial_\mu \tag{2.221}$$

The algebra is

$$[K_m, K_n] = 2m\delta_{m+n,0}K_0 \quad \text{and} \quad [K_0, K_m] = 0 \tag{2.222}$$

Three further, string theory inspired, operators are introduced

$$D\cdot\partial = \sum_{m\neq 0}\beta^\dagger_m K_m \qquad M = -2\sum_{n>0} n\beta^\dagger_n\beta_n \qquad \mathcal{W} = -\frac{1}{2}(1/n)\bar\beta^\dagger_n\bar\beta_n \tag{2.223}$$

After this preliminary work, the paper postulates an action

$$I = -\langle -|\phi^\dagger K_0\phi|+\rangle - \langle -|\phi^\dagger D\cdot\partial\mathcal{W}D\cdot\partial\phi|+\rangle \tag{2.224}$$

The field ϕ is an expansion over the creation modes (negative indices m) of the bosonic α_m and fermionic oscillators β_m and $\bar\beta_m$ (with an equal number of unbarred and barred

---

127 It apparently was not so in the 1984 to early 1985 string field theory research. During 1985, with the research referred to in Section 2.11.1, it gradually became clear that the mechanics BRST charge Q lead to gauge invariant field theory without first having to go through the BRST invariant gauge fixed theory, and then removing the Fadeev–Popov ghosts and auxiliary fields. One of the first clear statements of this is in the Witten October 1985 paper [165].

oscillators) as well as over the zero mode β_0. It is subject to the constraint $M\phi|+\rangle = 0$. It is shown that the action is invariant under the gauge transformations

$$\delta\phi|+\rangle = D \cdot \partial\Omega|+\rangle \tag{2.225}$$

The proof rest on algebraic relations satisfied by the operators defined in (2.223).[128]

The paper then proceeds to working out the details for spin 1, 2 and 3. In the case of spin 2, it is noted that a scalar field C, occurring at the same excitation level as the spin 2 field $h_{\mu\nu}$, can be identified with the trace of $h_{\mu\nu}$ through a constraint $T\phi|+\rangle = 0$ with $T = \frac{1}{2}\alpha_1\alpha_1 + \bar{\beta}_1\beta_1$. This is the repeated at the spin 3 level where a vector field D_μ can be identified with the trace of the spin 3 field $\phi_{\mu\nu\rho}$. The spin 4 level is not done in detail, but it is noted a constraint $T\phi|+\rangle =$ at this level implies double tracelessness of the spin 4 field through the equations $D_{\mu\nu} = 6\phi_{\mu\nu\rho\rho}$ and $D_{\mu\mu} = 0$. A general formula for the T operator is given.

It is noted that the T operator commutes with the BRST operator so that the constraint can be consistently applied to both fields and parameters, reproducing tracelessness for the gauge parameters and double tracelessness for the fields. It is also noted that there is no need to apply these constraints, gauge invariance is assured by the nilpotency of the BRST operator. The reason for this is the presence of extra independent fields of spin $s-2$ for each spin s higher spin field.[129]

The paper ends with a section on the relation to string theory, briefly discussing the limit $\alpha' \to \infty$ (the zero-tension limit).[130] I discussed the limit in the Veneziano amplitude with M. Green, who showed me that it did not make any sense. This was an early indication that any massless higher spin self-interactions, most likely, must be different from string induced interactions.

There was however one way that the zero tension limit could be interesting. I took the open bosonic string theory Virasoro generators and their algebra and performed the limit $\alpha' \to \infty$ (see Section 5.4.4). The result is precisely the generators (2.221) and the algebra (2.222) that I had postulated for massless higher spin. Using the Fradkin–Vilkovisky–Batalin procedure, I could write the BRST operator as

$$Q = -\frac{1}{2}\beta_0\Box + D \cdot \partial + \bar{\beta}_0 M \tag{2.226}$$

128 The structure of the action (2.224) is similar to the Ouvry–Stern action (2.214) and the proof of invariance rests on the similar algebraic relations. The difference is that the action written here is gauge invariant, not just BRST gauge-fixed invariant.

129 In modern higher spin research, such formulations, that do not require trace constraints on fields and parameters, are called “unconstrained”. Such models will be treated in Sections 5.3 and 5.5. Note also the correspondence to the Fronsdal fields ϕ^s and ϕ^{s-2} (see Section 2.10.1) and to the Ouvry–Stern fields A and B above.

130 The Regge trajectory slope is α'.

Nilpotency is easy to check, and the action is

$$I = -\langle\Phi|Q|\Phi\rangle \tag{2.227}$$

with a field expanded as $|\Phi\rangle = \phi|+\rangle + \psi|-\rangle$. The action is gauge invariant under the gauge transformations $\delta|\Phi\rangle = Q|\Lambda\rangle$ by nilpotency only, with no need for any further arguments.

This seemed almost to good to be true, but the fact was that when worked out level by level, the action I and gauge transformations $\delta|\Phi\rangle$ produced the correct formulas for the component fields. The connection between the $\langle\Phi|Q|\Phi\rangle$ action and the action given in formula (2.224), is the constraint $M\phi|+\rangle = 0$, which is actually a partially gauge choice that allows the auxiliary fields in $\psi|-\rangle$ to be integrated out, just as in string field theory [177]. In modern parlance, the $\langle\Phi|Q|\Phi\rangle$ theory yields a "triplet" formulation of higher spin gauge fields, whereas the action (2.224) yields a "doublet" formulation. These distinctions will be clarified in Chapter 5.

There is almost nothing written in the paper about interactions.[131] I do however comment that some "underlying two-dimensional invariance principle" of "similar strength to string theory" might be needed in order to construct interactions.[132]

2.11.4 A few more papers from the 1980s

Some months after the Ouvry–Stern and Bengtsson papers, there appeared a paper by Y. Meurice [184]. It is not widely cited, perhaps because it does not explicitly take its motivation from string theory or higher spin theory. Rather its motivation was to apply the Siegel "mechanics to field theory algorithm" to more systems. Although the method was, by that time, rather well understood, the paper is nevertheless interesting as it provides a clear exposition of the algorithm: starting from point mechanics, adding further coordinate-momentum pairs, defining bilinear constraints, studying their algebra, first quantization, defining the mechanics BRST charge, studying the ghost and vacuum structure, defining fields and finally setting up actions, gauge transformations and field equations. All this is done quite systematically in the paper.

There is however no application of the method to higher spin theory in the paper, even though the generic fields written down include general mixed symmetry tensor fields. The examples given explicitly concern low spin models such as spin 1, spin 2 and rank 2 antisymmetric tensor field. The reason for this is that the author requires one particular constraint to hold in all models. In the particular case of one additional coordinate-momentum pair (y, p_y), it constrains $p_y^2 + y^2$. When this is expressed in

131 I was luckily advised by M. Green to drop much of the text I had written for the preprint about interactions. The referee on the paper then convinced me to drop the rest.

132 This is a conviction that has followed me up to the present.

terms of oscillators, it becomes a number operator constraint $a^\dagger a - n = 0$ restricting to one particular spin, or representation of the Poincaré group. This puts the paper somewhat out of the higher spin theory line of research.

In 1989, there appeared a paper [185] by F. Hussain, G. Thompson and P. D. Jarvis, where the BRST method, as formulated by Ouvry and Stern, was reviewed and applied to massive fields of any spin and symmetry, among a few other examples. In regard to higher spin fields and supersymmetry, there is a follow-up paper to the Ouvry–Stern paper by M. Bellon and S. Ouvry that treats this subject [186].

A comprehensive review of models of point particles of any spin, bosonic as well as fermionic, with and without supersymmetry, can be found in [187].

Apropos early references to a higher spin Higgs effect, there is a paper, preprinted in the spring of 1985, by C. S. Aulakh, I. G. Koh and S. Ouvry, discussing higher spin fields with mixed symmetry [188]. The motivation behind the paper is "[the] recent surge in interest in string theories [that] has refocused attention on the old problem of formulating field theories of particles carrying arbitrary representations of the Lorentz group.". The Singh–Hagen theory for massive higher spin fields is briefly discussed in the Introduction, especially in relation to dimensional reduction from D to $D-1$ dimensions that produces massive field theories from massless. The example of linearized Einstein action is given. This is called a "telescopic Higgs effect". The paper works out the BRS-symmetry in the case of massless fields with Young Tableaux symmetry $(2,1,\ldots,1)_n$. Dimensional reduction and the ensuing massive theory is discussed. Mixed symmetry massless higher spin fields were subsequently studied by J. M. F Labastida.

Mixed symmetry fields

i *Mixed symmetry fields* refer to tensor fields corresponding to Young tableaux with more than one row. They arise as representations of the Poincaré group in dimension higher than $D = 4$. They were studied comprehensively in [189, 190] and reviewed in [191]. They also occur as "connection-type" fields in $D = 4$, in particular in the Vasiliev theory. An early study is by T. Curtright in [192].

2.11.5 The Labastida series of papers

Starting with a paper with T. R. Morris [193], and referring to the Aulakh, Koh and Ouvry paper [188], there is a series of papers by J. M. F. Labastida on mixed symmetry fields of arbitrary spin. The motivation is again the interest in string field theory, but as the authors of [193] argue, "one would prefer a field theoretical description of the massless representations based on the principle of gauge invariance.". At the time, only a few examples of mixed Young tableaux symmetry fields had been investigated. There had been neither any need nor any interest. String field theory changed that. These circumstances are again mentioned as motivations in the second paper by Labastida [194]. Working backwards from the known massive representations from string field theory to massless by some "anticompactification" method is "tedious and inelegant" and "does not teach us anything about the rich physics contained in the description of massless particles.".

The paper [194] is interesting in that it uses a bosonic string-like field $|A(x)\rangle$ with no ghost excitations and the corresponding set of N covariant oscillators a_n^μ in order to find gauge invariant field equations $\mathcal{O}|A(x)\rangle = 0$ for mixed symmetry massless fields. The method can be viewed as an "ansatz-coefficient solving" method although it is not phrased so in the paper. Instead, possible contributions to the operator $\mathcal{O}$, built from derivatives and oscillators are listed, taking into account various natural restrictions on its structure. Then gauge invariance, also formulated in the same language, is imposed to restrict the possible operators occurring in the field equation. What results is a generalization of the Fronsdal equations to mixed symmetry fields. The operator is determined to be

$$\mathcal{O} = \Box + a_m^{\dagger\alpha} a_m^{\dagger\beta} \partial_\alpha \partial_\beta + \frac{1}{2} a_m^{\dagger\alpha} a_n^{\dagger\beta} a_m^{\gamma} a_{n,\gamma} \partial_\alpha \partial_\beta \tag{2.228}$$

The paper ends with a sentence on breaking the symmetry to generate some connection to string theory, and a sentence on interactions. The third paper in the series treats fermionic fields [195].

The fourth paper [196], referring to the Ouvry–Stern and Bengtsson papers treats a BRST formulation along the lines of these two papers. The paper starts out by formally introducing N copies of the string BRST operator; the open string is N=1 and the closed string is N=2. The theory is then restricted to massless fields. It is said to be related to the Ouvry–Stern and Bengtsson formulations, but not identical to, as the field content differs.

The fifth paper [197] is in my opinion the most interesting. It sums up much of the previous work done by the author, and formulates the theory in terms of a string field with only bosonic oscillators (no ghost coordinates) as in the third paper [194]. An action of the form $\langle A(x)|\mathcal{O}|A(x)\rangle$, reproducing the field equation of [194], is constructed. What results is a generalization of Fronsdal's theory to mixed symmetry fields. The author writes in the conclusion that the paper does not prove that the number of physical degrees of freedom is correct in all cases. The problem is the trace constraints that become complicated in the mixed symmetry cases.

Let us end with a computation showing an interesting role played by the double tracelessness constraint in this kind of formulation.

... and was actually done. Labastida's action

In [197], Labastida sets the problem of finding an action for higher spin gauge fields of the form

$$\langle A(x)|(\mathcal{O}+\mathcal{E})|A(x)\rangle \tag{2.229}$$

with a Hermitian operator $(\mathcal{O}+\mathcal{E})^\dagger = \mathcal{O}+\mathcal{E}$. The operator $\mathcal{O}$ is the one given above in formula (2.228). Let us simplify to just one oscillator. Then

$$\mathcal{O} = \Box - \alpha^\dagger \cdot \partial\alpha \cdot \partial + \frac{1}{2}\alpha^\dagger \cdot \partial\alpha \cdot \alpha \tag{2.230}$$

The Fronsdal field equations should be derivable from the action (2.229). As we reviewed in Section 2.10.1, that is not possible without an intermediate step of computing the trace of the Euler–Lagrange equations. Labastida requires equivalence of the field equations $(\mathcal{O}+\mathcal{E})|A(x)\rangle = 0$ and $\mathcal{O}|A(x)\rangle = 0$. The result is that the operator $\mathcal{E}$ is

$$\mathcal{E} = -\frac{1}{4}\alpha^\dagger \cdot \alpha^\dagger \alpha \cdot \alpha\, \mathcal{O} \tag{2.231}$$

It is quite interesting to note that in order to show that the Labastida action (2.229) is the same as the action (2.205) with kinetic operator (2.206), one has to use double tracelessness in the form $\alpha \cdot \alpha\alpha \cdot \alpha|A(x)\rangle = 0$.

From the story told here, it is clear that a small number researchers were interested in massless higher spin field theory in 1985–1989, approaching the subject from different angles and interests. Attempts at interactions were done (see Section 2.12.3). The interest did not last long, and apart from M. Vasiliev's own work on the AdS approach to interacting higher spin gauge fields during the 1990s, the subject of Minkowski higher spin theory was almost dormant for about 10 years.

2.11.6 New BRST papers of the late 1990s

About a decade after the initial construction of the BRST approach to Minkowski higher spin gauge fields, there was a return to the theory by A. Pashnev and M. Tsulaia. Their first two papers on the subject concerned massive higher spin fields, falling on a single Regge trajectory, with or without daughter trajectories, depending on the choice of constraints.

The authors assume a first-class constraint $L_0 = -p^2 - \alpha' a^\dagger \cdot a$ and two pairs of second class constraints $L_1 = p \cdot a$, $L_{-1} = p \cdot a^\dagger$ and $L_2 = \frac{1}{2}a \cdot a$, $L_{-2} = \frac{1}{2}a^\dagger \cdot a^\dagger$. The first pair corresponds to transversality and the second to tracelessness, thus being potentially able to describe massive particles.[133] In the first paper [198], both sets of second

133 Compare to the Ouvry–Stern and Bengtsson constraints in Sections 2.11.2 and 2.11.3.

class constraints are converted to first class by the introduction of new canonical variables.[134] That allows for a BRST formulation of the theory. The authors however deem the result unsatisfactory due to the occurrence of square roots $\sqrt{p^2}$ in the converted constraints. In the second paper [199], the authors employ a modified method to the theory with the constraints L_0, L_1, L_{-1}. By dimensional reduction of a corresponding massless theory in $D+1$ dimension, the massive BRST theory in D dimensions is derived. Since the tracelessness constraints are not imposed, the theory has daughter Regge trajectories.

The constructions become a bit involved, and one may perhaps argue that converting second class constraints to first class, is not so natural. But it is interesting to ponder the comparison to the bosonic string where the spectrum contains an infinite number of Regge trajectories of massive states. However, the string has an underlying infinite-dimensional algebra of first class constraints, namely the Virasoro algebra. This algebra, in its turn, emanates from the two-dimensional reparametrization invariance of the string world sheet. Apparently, the truncation of the string to a single Regge trajectory is not so natural from a mechanical gauge theory perspective since the first class property of the constraints do not survive. On the other hand, the first class property survives the zero-tension limit. Then the truncation to one trajectory can be made.

The third paper [200] concerns the massless theory. The authors treat the second class tracelessness constraints (i. e., the constraints that impose double tracelessness for the gauge fields) by converting them to first class. There is a standard part in the BRST Q operator, imposing the first class constraints $p^2 = 0$, $L_1 = 0$, $L_{-1} = 0$ and the second class constraints $L_2 = 0$, $L_{-2} = 0$ (now treated as first class), and the concomitant structure constant terms. Furthermore, there is an additional term in Q that takes the form of ghost oscillators times square roots of a sum of number operators.

In regard to massive higher spin theories treated using BRST techniques, there is a paper from 2005 that discuss this problem [201]. It also contains further references to the massive problem. The research into BRST constructions of massless and massive higher spin theories, both in Minkowski and AdS space-times of general dimension D has continued into the new millennium. As this falls outside the limits set for this chapter, we will refer the reader to the review [202].

2.12 Positive interaction results of the 1980s and 1990s

During the early 1980s, there appeared the first positive results on self-interactions for massless higher spin fields. We will just mention the papers here, and return to a proper history in Volume 2 of the present work.

134 For references to a general method of converting second class constraints to first class, see the list in the papers [198, 199].

2.12.1 Cubic interaction terms on the light-cone

Lars Brink had worked on light-cone formulations of supersymmetric gauge theories, in particular in connection the proof of finiteness of the $N = 4$ Yang–Mills theory together with O. Lindgren and B. E. W. Nilsson [203, 204], but also on supergravity and superstrings together with M. B. Green and J. H. Schwarz. He introduced the method to Ingemar Bengtsson and me, and suggested that we should do higher spin in that formulation.

Lower spin massless – and massive – field theory can be reformulated "on the light-cone" so to speak. The space-time coordinates x^μ can be recombined into the light-front coordinates

$$\begin{aligned} x^+ &= \frac{1}{\sqrt{2}}(x^0 + x^3) \qquad & x &= \frac{1}{\sqrt{2}}(x^1 + ix^2) \\ x^- &= \frac{1}{\sqrt{2}}(x^0 - x^3) \qquad & \bar{x} &= \frac{1}{\sqrt{2}}(x^1 - ix^2) \end{aligned} \tag{2.232}$$

and similarly for momenta p_μ and other kinds of vectors and tensors. For gauge fields, for instance the spin 1 electromagnetic field A^μ, one can furthermore choose the "light-cone gauge" with $A^+ = 0$. It then turns out that the component A^- can be solved for explicitly as

$$A^- = \frac{1}{\partial^+}(\partial\bar{A} + \bar{\partial}A) \tag{2.233}$$

in terms of the physical transverse components A and $\bar{A}$. The free field equations are simply $\Box A = \Box\bar{A} = 0$. To stay in the light-cone gauge, one must "regauge" the fields. This leads to modifications of the Poincaré transformations (see Section 6.1.4). The modified infinitesimal transformations still satisfy the Poincaré algebra. This analysis is true also for spin 2, and indeed for any spin, integer or half-integer. Regardless of spin, a massless gauge field on the light-cone can always be described by a complex field ϕ and its complex conjugate $\bar{\phi}$, corresponding to the two helicities $\pm\lambda$ in four space-time dimensions.

One may now attempt to construct nonlinear contributions to the free equations of motion and action. In the first instance, this means quadratic contributions to the field equations. Setting up an ansatz for such terms, one then requires the – now nonlinear – Poincaré transformations to close. The result for arbitrary integer spin λ is the following cubic interaction term [124]

$$\int d^4x \sum_{n=0}^{\lambda} (-1)^n \binom{\lambda}{n} (\partial^+)^\lambda \phi \left[\frac{\partial}{\partial^+}\right]^{(\lambda-n)} \bar{\phi} \left[\frac{\partial}{\partial^+}\right]^n \bar{\phi} + \text{c.c.} \tag{2.234}$$

with gauge group structure constants (antisymmetrization) understood for odd spin.

A characteristic feature of the light-front cubic interactions for massless higher helicity fields is the simple binomial expansion form they take. This came out somewhat mysteriously from the computations. The structure became more clear when the interaction terms were reformulated in momentum space in terms of vertex operators a few years later in a paper by I. Bengtsson, N. Linden and myself. In that formulation, the momentum structure for the helicity λ cubic interaction is essentially given by $\mathbb{P}^{\lambda}$ where $\mathbb{P}$ is defined by

$$\mathbb{P} = -\frac{1}{3}\sum_{r=1}^{3}(p^{+}_{r+1} - p^{+}_{r+2})p_r \tag{2.235}$$

where the index r is counted modulo 3.[135]

The light-cone approach to higher spin, in the vertex operator formalism of [205], was taken up my R. R. Metsaev in the early 1990s. The cubic vertices were generalized to arbitrary dimension D by E. S. Fradkin and in R. R. Metsaev in [206] and in a "generating function" formalism in [207]. The first analysis of the quartic level of interaction on the light-front was performed by Metsaev in two papers from 1990 and 1991, [208] and [209], respectively.[136] As this belongs to the theory of interacting higher spin fields, we will defer further discussion of these interesting papers to the second volume of the present work. Suffice it to say that the Metsaev papers on quartic light-front interactions came to the attention of the higher spin community in the mid-2010s in connection to a general resurgence of interest in Minkowski higher spin theory. The situation regarding light-front interactions was largely clarified by D. Ponomarev and E. Skvortsov in [210] and by Ponomarev in [211].

2.12.2 Covariant spin 3 self-interaction

The first positive result in a covariant formulation of massless higher spin gauge theory was the paper [122] by F. A. Berends and G. J. H. Burgers and H. van Dam, published in 1984, concerning cubic self-interactions for spin 3. The authors apply the deformation theoretic approach discussed by Fang and Fronsdal in [8] (see our Section 2.8). As the authors write "[...] it has never been applied successfully to a case, where the theory was not known beforehand.". A gauge invariant cubic interaction term for spin 3 gauge fields was constructed. The interaction is consistent with the corresponding light-front

135 The momentum variable $\mathbb{P}$ was first introduced in light-front superstring theory.

136 It seems that these works went largely unnoticed at the time. Few people was working on higher spin, and for those who did, the Vasiliev approach formed a paradigm. Ingemar Bengtsson sent me a copy of [208], but we did not pay it the attention that it deserved (I had left theoretical physics for other intellectual interests).

result: there are three space-time derivatives in the interaction and the gauge field must be antisymmetrized over an internal index.

In a follow-up paper [123], the same authors investigate general properties of self-interacting higher spin gauge theories in Minkowski space-time. It is shown that a pure spin 3 theory cannot exist (see also [212] and [125]). It is noted that a way out of this negative conclusion may be a theory containing an infinite family of massless higher spin fields. This had been suggested in 1979 by Fronsdal in a conference paper [213].

The spin 3 theory was revisited by X. Bekaert, N. Boulanger and S. Cnokaert in 2006 [214] and by X. Bekaert, N. Boulanger and S. Leclercq in 2010 [215] using modern BRST cohomological methods. As these works are beyond the scope of the present chapter, I will defer discussion to the second volume of the present work; suffice it to say that the nonexistence of a pure spin 3 gauge theory is verified.

2.12.3 Cubic interaction terms in the BRST approach

Soon after the discovery of the BRST approach to free higher spin gauge fields, there appeared a paper by I. G. Koh and S. Ouvry investigating interactions in the BRST formalism [216]. The authors study and construct a cubic string-like vertex of the type constructed by E. Witten.[137] The method is in principle to find a vertex operator $\langle V_{123}|$ coupling three higher spin gauge fields $|\chi_r\rangle$ in a cubic interaction term of the form $\langle V_{123}|\chi_1\rangle|\chi_2\rangle|\chi_3\rangle$. The vertex operator also contributes with nonhomogeneous gauge transformations bilinear in fields and parameters. The free theory contains an infinite number of oscillator modes (as in the underlying Ouvry–Stern model).

Demanding gauge invariance to cubic order, which in this formalism is the same as the nilpotency of the cubic BRST operator, leads to a specific form of the vertex operator. Its general form is the same as the string vertex operator with operators $N_n^{rs}\alpha_n^r \cdot p^s$ bilinear in oscillators and momenta where n summed over an infinite number of oscillator modes, and r, s are numbering the three higher spin fields entering the interaction. There are also the corresponding BRST-ghost operators. The notation N_n^{rs} is borrowed from string theory, but does not denote the same functions. There are no string-like operators of the type $N_{nm}^{rs}\alpha_n^r \cdot \alpha_m^s$ bilinear in oscillator modes.

Regarding the resulting interactions, the authors conclude "Finally, as for the component expansion, our gauge invariant interacting theory is quite different from the string's one. For example, the spin-one local field is shown to couple to other spin fields with higher derivatives but does not have the *AApA* coupling of Yang–Mills theories.". This fact can be understood as an effect of the absence of the operator terms $N_{nm}^{rs}\alpha_n^r \cdot \alpha_m^s$ in the vertex.

137 They also refer to work by A. Neveu and P. C. West, J. L. Gervais, and work by L. Baulieu and S. Ouvry.

General aspects of this type of BRST-invariant interacting theories of higher spin fields were studied in a 1989 paper by L. Cappiello, M. Knecht, S. Ouvry and J. Stern [217] and in a 1991 preprint by F. Fougère, M. Knecht and J. Stern [218].

I worked on a BRST-invariant cubic vertex for higher spin gauge fields during the winter and spring of 1987, inspired by the Witten open string field theory and by papers by D. J. Gross and A. Jevicki [219–221] that expressed the Witten construction more explicitly in terms of vertex operators. The year before I had worked together with Ingemar Bengtsson and Noah Linden on a vertex operator construction of the light-front cubic vertices. This was, in its turn, inspired by Linden's work on light-front string field theory, which apart from vertex operator techniques, also featured the momentum variables $\mathbb{P}$ of formula (2.235) that promised a rationale for the binomial structure of the cubic interactions. However, there was no way we could get even Yang–Mills out of the vertex operators we tried. They were written in terms of string-like bilinears $Y^{rs}\alpha_r\alpha_s$ in light-front oscillators (r, s numbering higher spin fields participating in the vertex) and terms $X^r\alpha_r\mathbb{P}$. The reason was that the nonlinear Poincaré algebra unrelentingly forced $Y^{rs} = \delta^{rs}$. Therefore, no Yang–Mills coupling resulted. It then dawned on us that we could try operators of the form $Y^{rst}\alpha_r\alpha_s\alpha_t\mathbb{P}$. It was clear that this ansatz promised to produce all cubic self-interaction terms. Now the Poincaré algebra did not object, but instead gave precisely the correct structure $Y^{rst} = p_t^+/p_r^+p_s^+$ of p^+ momenta. This resulted in our 1987 paper [205]. This paper also contains a full list of all possible cubic interaction terms among higher spin massless fields.

After this, it was only natural to try the same structure in a covariant BRST-gauge invariant formalism. In that way, it was possible to derive a BRST-invariant cubic vertex that did produce the Yang–Mills interaction as well as higher spin interactions of the correct overall derivative structure [222]. As a by-product, it was verified that even in the BRST formalism, covariant operators of the type $N^{rs}\alpha^r \cdot \alpha^s$ are only possible with $N^{rs} \sim \delta^{rs}$ in massless higher spin theory. What was needed was operators of the form $Y^{rstu}\alpha_r \cdot \alpha_s\alpha_t \cdot p_u$ and corresponding ghost operators.

After this, the project came to a standstill, due to difficulties that I did not have tools at the time to address. We will return to this approach in Volume 2 of the present work.

2.12.4 The Fradkin–Vasiliev and Vasiliev anti-deSitter theory

The Fradkin–Vasiliev approach to higher spin has been the most successful so far. It started out in 1980 with a reformulation of free higher spin field theory by M. Vasiliev in terms of what has become known as the *frame formulation* (see Section 5.7). Then there was a lapse in time in the published record until a "tree" of papers appeared at the end of the 1980s and beginnings of the 1990s. This "tree" of papers is constituted of a sequence of papers treating the free field theory, a sequence of papers treating higher spin algebras, a sequence of papers constructing interactions – in particular cubic interactions, and a sequence of papers introducing the *free differential algebra* and unfolding approach. We will study these sequences of papers in detail in Volume 2 of the present work.[138]

It is very natural to seek guidance from the lower spin gauge field theories when trying to set up an interacting field theory for higher spin gauge fields. For a long time, as we have seen, the focus was on electromagnetic and gravitational interactions of massive (nongauge) higher spin fields. The problems then encountered added to the impression that higher spin field theories are inherently inconsistent. After the shift of focus to massless gauge fields initiated by Fang and Fronsdal, the interaction problem also shifted focus: to self-interactions. Then the guidance from lower spin gauge theory also shifted focus: from minimal coupling to trying to generalize the very "gauge invariance" itself and the "equational form" of the lower spin theories to higher spin.

The free field theory gauge invariance of actions and field equations for higher spin was by now – as the 1970s turned to the 1980s – fairly well understood through [3, 150, 149]. Yang–Mills gauge theory of massless spin 1 was of course well established and gauge theory approaches to gravity as massless spin 2 had been investigated (see Section 2.9). But as pointed out in the review article by Kibble and Stelle [134], treating gravity as a gauge theory of the Poincaré group did not generalize very naturally from the spin 1 gauge theory of a (semisimple) Lie group. There are several ways of understanding this – and we will study the problems in Section 4.6. One source of the difficulties being the fact that the Poincaré group is not semisimple, but rather a semidirect sum of the Lorentz group and the Abelian translation group. It turned out that there was a way out, namely to instead gauge the groups $SO(4,1)$ or $SO(3,2)$, the de Sitter or anti-de Sitter groups, respectively. This was done by MacDowell and Mansouri [223] and Stelle and West [224]. Now the action could be written in terms of the curvatures, generalizing Yang–Mills theory. It was here that E. S. Fradkin and M. A. Vasiliev found a way to promote the theory to a field theory of higher spin [225, 226]. The detailed history of the Vasiliev theory, however, belongs to Volume 2 of the present work. For now, I can only refer the reader to existing reviews of the Vasiliev theory, for instance [227] and [228].

2.13 Chapter 2 epilogue

Toward the end of the twentieth century, the interest in higher spin gauge theory started to grow, but this is where the story stops for now. The limit is set at the millennium and at the very beginning of the interacting field theory. An attempt to tell the story of what has happened during the last 20 years, as well as the story of interactions, will be made in the second volume of the present work.

---

138 A few of the papers are written with collaborators S. E. Konstein and V. E. Lopatin.

To all those authors who cannot find their papers in the list of references, that is most likely because your work has been done after year 2000 and mainly concerned interactions. I started to compile a list of names, but realized that it was doomed to be incomplete, and as the sadness of those not on it would likely be greater than the happiness of those on the list, I gave up. I can only offer my apologies.

Anyway, I hope that these historical notes, if nothing else, can be of some help for the researcher who wants to dig deeper into the literature on higher spin wave equations. It should give a first overview at least.

3 Concepts, mathematical structures and notation

In this chapter, we will recapitulate basic concepts of relativity and quantum mechanics and their fusion into quantum field theory as well as some relevant group theory, algebra and differential geometry. In doing this, we can also establish our notation. However, rather than first developing these subjects in the abstract and then applying them to higher spin theory, we will from the outset be guided higher spin thinking. This saves time and space and makes the enterprise more interesting. So even though the contents are well known, the skipping reader may miss some ideas. On the other hand, the chapter is not a intended to be a substitute for proper study of the topics covered, but may serve as a set of reminders, and perhaps a few alternative perspectives. There is a strong focus on the Poincaré group and its representations as it is central to higher spin theory.

Features of higher spin theory such as the infinite number of fields, arbitrarily high orders of derivatives, raises subtle questions of principle within the theory of fields. For this reason partly, I have also decided to include some discussions that seldom find their way into reviews.

The literature on higher spin theory is naturally written on many different levels of mathematical sophistication depending on the various authors inclinations and interests. That is of course not surprising, but is perhaps not optimal for communication. In the hope of, at least to some extent, alleviate the confusion, I have therefore endeavored to phrase the same mathematical concepts in isomorphic language, so to speak. However, rather than trying to invent some kind of "consistent" notation covering all corners of theoretical physics, I have stayed close to standard notation that the newcomer will find in the major texts off the subject. This will be apparent to the expert reader.[1]

Theoretical physics needs a lot of mathematical concepts and techniques, but seldom the sometimes quite heavy notation needed in mathematics to make the concepts precise. I have therefore chosen to be more "verbal" than "formal" in the definitions, trying to avoid introducing notation that are not subsequently used. Further concepts and mathematical structures, needed for interactions, will be introduced in a sister chapter in Volume 2.

1 As already mentioned in Section 1.4, a "unified consistent" notation throughout would risk making large parts of the subject look baroque.

https://doi.org/10.1515/9783110451771-003

3.1 Lagrangian mechanics

Consider a mechanical system described by the coordinates y^i. The action S is the time integral of the *Lagrangian* L

$$S[y(t)] = \int_{t_1}^{t_2} L dt \tag{3.1}$$

where L is a function of coordinates and their time derivatives $y, \dot{y}^i, \ddot{y}^i, \ldots, y^{i(k)}$ up to some finite order k. The *action* $S[y(t)]$ is a *functional* of the *trajectory* $y(t)$, but as is clear from the formula, not a function of the time t. For most ordinary mechanical systems, only first- and second-order derivatives occur.[2] But in higher spin theory, we will have to consider theories with arbitrarily high orders of derivatives. In this section, we will develop some parts of Lagrangian and Hamiltonian *mechanics*, the generalization to *field theory* will be done in the last section of this chapter.

3.1.1 Locality

The requirement of a finite order of derivatives in the Lagrangian is a kind of *strong locality* assumption. It can be understood in the following way.

Consider a smooth function $y = y(t')$ (see figure 3.1). In order to Taylor expand it around a point t say, we in general need derivatives at that point to all orders. For functions that describe the *trajectory* (or history) of a system, we escape that since the equations of motion give us all the higher order derivatives in terms of the lower ones.[3] In field theory, locality assumptions concern space-time derivatives, not just time derivatives.

In higher spin field theory, this requirement of strong locality must be given up since already at the cubic interaction level there are derivatives of arbitrarily high order, the basic spin s–s–s interaction term having s space-time derivatives. Even the free field theory contain objects – the gauge invariant generalized Christoffel symbols – that are higher order in derivatives. These are however normally not used in the kinetic terms as one can use the two-derivative Fronsdal tensor instead (see Section 5.2).

Therefore, by requiring *locality* in higher spin theory, we do allow arbitrarily high powers of derivatives. This is presumably acceptable, since thinking in terms of Taylor expansions of smooth functions, the whole trajectory can be reconstructed from a denumerable infinite set of data: the values of all the derivatives at a certain point, at

2 Partial integration makes for a certain amount of trading between first and second order in derivatives.

3 Either we have an exact formula for the solution to differentiate, or we can step forward in time, using the equations of motion, from the initial point.

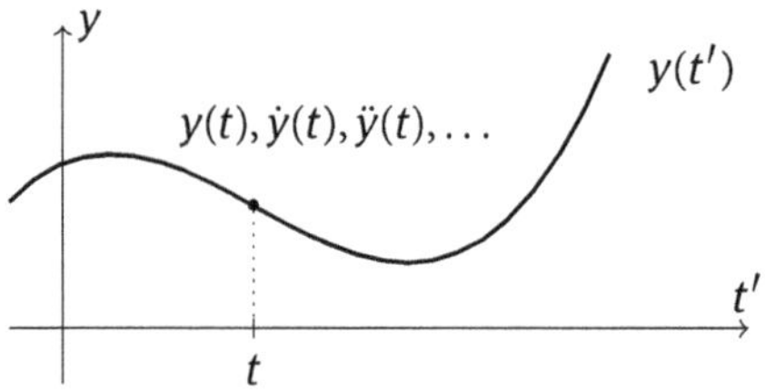

Figure 3.1: Smooth trajectory, Taylor expanded around $t' = t$.

least within certain radii of convergence. There is a mathematical apparatus to handle such situations: infinite jet spaces, which will introduced in Volume 2 where the concept will be needed.

A kind of nonlocality that is much more difficult – perhaps impossible – to accept, is the occurrence of inverse powers of derivatives. Now, an operator $(d/dt)^{-1}$ has no well-defined meaning as it stands, but it can be defined as a integral operator. This makes it nonlocal. Its value cannot be found locally at any point t.[4]

3.1.2 Functional and variational derivatives

The Lagrangian can be viewed either as a function of the coordinates y^i and their time derivatives or as a functional of the trajectory $y(t')$. In the first view, varying the action leads to the Euler–Lagrange equations

$$\begin{aligned}\delta S &= \int \delta L dt = \int dt\Big(\frac{\partial L}{\partial y^i}\delta y^i + \frac{\partial L}{\partial \dot{y}^i}\delta \dot{y}^i + \frac{\partial L}{\partial \ddot{y}^i}\delta \ddot{y}^i + \cdots\Big) \\ &= \int dt\Big(\frac{\partial L}{\partial y^i} - \frac{d}{dt}\frac{\partial L}{\partial \dot{y}^i} + \frac{d^2}{dt^2}\frac{\partial L}{\partial \ddot{y}^i} + \cdots\Big)\delta y^i \end{aligned} \tag{3.2}$$

This calculation is interpreted according to

$$\delta S = \int dt \frac{\delta S}{\delta y^i(t)}\delta y^i(t) = \int dt \frac{\delta L}{\delta y^i}\delta y^i \tag{3.3}$$

so that we can read off

$$\frac{\delta S}{\delta y^i(t)} = \frac{\delta L}{\delta y^i} = \frac{\partial L}{\partial y^i} - \frac{d}{dt}\frac{\partial L}{\partial \dot{y}^i} + \frac{d^2}{dt^2}\frac{\partial L}{\partial \ddot{y}^i} + \cdots \tag{3.4}$$

4 Such operators do occur in the light-front formulation of field theory. There, however, they can be considered as an artefact of the noncovariant formalism. We defer that discussion to Chapter 6.

This equation defines the relation between the *functional derivative* of the integrated Lagrangian (i. e., the action) and the *variational derivative* of the Lagrangian. The difference between variational and functional derivatives should be clear from the following formulas:

$$\frac{\delta y^j}{\delta y^i} = \delta_i^j \qquad \frac{\delta y^j(t)}{\delta y^i(t')} = \delta_i^j \delta(t-t') \tag{3.5}$$

$$\frac{\delta \dot{y}^j}{\delta y^i} = 0 \qquad \frac{\delta \dot{y}^j(t)}{\delta y^i(t')} = \delta_i^j \dot{\delta}(t-t') \tag{3.6}$$

In the second view of the Lagrangian, we therefore have

$$\frac{\delta L(t)}{\delta y^i(t')} = \delta(t-t')\frac{\partial L}{\partial y^i}(t) + \dot{\delta}(t-t')\frac{\partial L}{\partial \dot{y}^i}(t) + \ddot{\delta}(t-t')\frac{\partial L}{\partial \ddot{y}^i}(t) + \cdots \tag{3.7}$$

The two views are consistent as can be seen by computing the functional derivative of the action

$$\begin{aligned} \frac{\delta S}{\delta y^i(t')} &= \int dt \frac{\delta L(t)}{\delta y^i(t')} \\ &= \int dt \left[\delta(t-t')\frac{\partial L}{\partial y^i}(t) + \dot{\delta}(t-t')\frac{\partial L}{\partial \dot{y}^i}(t) + \cdots \right] \\ &= \frac{\partial L}{\partial y^i} - \frac{d}{dt}\frac{\partial L}{\partial \dot{y}^i} + \cdots \end{aligned} \tag{3.8}$$

and we recover the Euler–Lagrange equations. Note that in the just presented formulas, the dots "..." represent a finite number of higher derivatives in standard theories or an infinite series in higher spin-type theories.

3.1.3 General gauge transformations

Gauge transformations depend on arbitrary functions of time in mechanics, and of space-time location in field theory. They can be parametrized by the formula[5]

$$\delta_\xi y^i = \bar{R}^i_{(0)a}\xi^a + \bar{R}^i_{(1)a}\dot{\xi}^a + \bar{R}^i_{(2)a}\ddot{\xi}^a \ldots = R^i_a \xi^a \tag{3.9}$$

where in the last equality, R^i_a acts as derivative operator. The transformations are called *local* when the parameter depends on time, as opposed to *global* when the parameter ξ is constant.

To connect this somewhat abstract formula to something well known, consider the Yang–Mills gauge transformations $\delta A^a_\mu = \partial_\mu \xi^a + g f^{abc} A^b_\mu \xi^c$. The first, inhomogeneous, derivative term $\partial_\mu \xi^a$ corresponds to $\bar{R}^i_{(1)a}\dot{\xi}^a$ in the general formula with

5 We are adopting the notation of [229].

space-time derivatives instead of the time derivative. The second, homogeneous, term $gf^{abc}A^b_\mu\xi^c$ corresponds to abstract term $\bar{R}^i_{(0)a}\xi^a$. Thus with the coefficients $\bar{R}^i_{(k)a}$ taken as appropriate powers (or polynomials) of the coordinates (fields in field theory) the abstract formula (3.9) captures very general types of transformations, local as well as global.

In general discussions, it is customary to use an even more abstract condensed notation, where the time variable (in mechanics) or the space-time variables (in field theory) are subsumed under abstract indices *i* or *a* symbolizing all of the "coordinate" dependence on discrete and continuous variables.[6] Sums over the index is then thought of as including also integrals as appropriate for the context. For the gauge transformations (3.9), the sum over the discrete index *a* is extended to include a time integration. Thus

$$\delta_\xi y^i = R^i_a\xi^a \quad \text{is defined as} \quad \delta_\xi y^i(t) = \int dt' R^i_a(t,t')\xi^a(t') \tag{3.10}$$

To make this reproduce (3.9), we take

$$R^i_a(t,t') = \bar{R}^i_{(0)a}(t)\delta(t-t') + \bar{R}^i_{(1)a}(t)\dot{\delta}(t-t') + \cdots \tag{3.11}$$

where the overdot signifies derivatives with respect to *t*. It has to be kept in mind, when this formalism is employed, that every occurrence of a repeated index may also include integrals as appropriate to context.

3.1.4 Noether identities and gauge algebra

We now compute the variation of the action under a gauge transformation and require it to be zero

$$\delta_\xi S = \frac{\delta S}{\delta y^i}\delta_\xi y^i = \frac{\delta S}{\delta y^i}R^i_a\xi^a = 0 \quad \Rightarrow \quad \frac{\delta S}{\delta y^i}R^i_a = 0 \tag{3.12}$$

since the equality must hold for ξ^a an arbitrary function of time. As we have just discussed, in this expression we have an implicit integration. Using (3.4), we have

$$\begin{aligned}\frac{\delta S}{\delta y^i}R^i_a &= \int \frac{\delta L}{\delta y^i}(t')R^i_a(t,t')dt' \\ &= \frac{\delta L}{\delta y^i}\bar{R}^i_{(0)a} + \frac{d}{dt}\left(\frac{\delta L}{\delta y^i}\bar{R}^i_{(1)a}\right) + \cdots = 0\end{aligned} \tag{3.13}$$

These are the *Noether identities* giving equations connecting various components of the equations of motion. Not all of them can therefore be independent. This is a reflection of the gauge invariance of the system. We will return to these identities, in

6 This notation was introduced by B. DeWitt in [106] in the context of quantum gravity.

Section 3.14 in connection with the deformation theoretic approach to interactions in field theory.

In order to get some grip on the gauge algebra, we compute the commutator of two gauge transformations $\delta_\epsilon y^i = R^i_a \epsilon^a$ and $\delta_\eta y^j = R^j_b \eta^b$, the result of which is

$$[\delta_\epsilon, \delta_\eta]S = \frac{\delta^2 S}{\delta y^i \delta y^j}(R^j_a R^i_b - R^i_a R^j_b)\epsilon^a \eta^b + \frac{\delta S}{\delta y^i}\left(R^j_a \frac{\delta R^i_b}{\delta y^j} - R^j_b \frac{\delta R^i_a}{\delta y^j}\right)\epsilon^a \eta^b \tag{3.14}$$

The first term is zero, and the second we interpret as a new gauge transformation.

To proceed from here, we recognize that one can construct trivial gauge transformation by using the equations of motion. Consider

$$\delta_\eta y^i = \eta^{ij} \frac{\delta S}{\delta y^j} \tag{3.15}$$

with η^{ij} some arbitrary, infinitesimal, antisymmetric function $\eta^{ij} = -\eta^{ji}$ of the y^i. Transformations of this form satisfy the Noether identities by construction, and are present for any action, and should therefore not be regarded as proper gauge transformations. They may, however, appear when commuting proper gauge transformations or when redefining gauge transformations [229]. Such trivial transformations form an ideal of the gauge algebra, and can thus be "factored out" (see Section 3.9).

Furthermore, given gauge transformations $\delta y^i = R^i_a \epsilon^a$ satisfying the Noether identities nontrivially, one can construct new transformations according to

$$\delta_\eta y^i = (R^i_a M^a_b)\eta^b \tag{3.16}$$

When the matrices M^a_b are allowed to depend on the y^i, such transformations are linearly independent of the original transformations $\delta y^i = R^i_a \epsilon^a$. However, they do not lead to independent Noether identities, since

$$\frac{\delta S}{\delta y^i} R^i_a M^a_b = 0 \tag{3.17}$$

are direct consequences of the original Noether identities (3.12).

These circumstances lead to the concept of a *generating set of gauge transformations*, a set of gauge transformations that contain all information about the Noether identities in "minimal way". If such a set of gauge transformations are denoted by $\delta_\epsilon y^i = R^i_a \epsilon^a$, then any other gauge transformation, according to the discussion above, can be written as

$$\delta y^i = m^a R^i_a + M^{ij} \frac{\delta S}{\delta y^j} \tag{3.18}$$

where the coefficients m^a and $M^{ij} = -M^{ji}$ may depend on the y^i. Returning now to the commutator of two gauge transformations in equation (3.14), and taking the transformations to belong the generating set, the commutator must be possible to express as in (3.18), that is, we get

$$R^j_a \frac{\delta R^i_b}{\delta y^j} - R^j_b \frac{\delta R^i_a}{\delta y^j} = C_{ab}{}^c R^i_c + M^{ij}_{ab} \frac{\delta S}{\delta y^j} \tag{3.19}$$

Generating sets are clearly not unique. In case the coefficients M^{ij}_{ab} are nonzero, one speaks of an *open algebra*. On the other hand, in case the coefficients M^{ij}_{ab} are zero, the algebra is called closed. A *closed algebra* is a classical true Lie algebra when the coefficients $C_{ab}{}^c$ are constants.

Higher spin gauge algebras?

Although "higher spin algebras" – taken in a loose sense – are clearly important for the interaction problem and therefore, properly belongs to Volume 2 of the present work, this is a good place for a few general comments.

As pointed out in [229], a generating set of gauge transformations is in general not a basis in the Lie algebra sense. However, on general grounds, the set of all gauge transformations is always a Lie algebra. Gauge theories for which the generating set is a true Lie algebra is therefore quite special in that one may treat the gauge transformations in an abstract way, independent of the field content or the dynamics. As stressed in [229]: "In that case, one can construct the generating set before writing down the action. This is a very lucky instance, however, and many interesting gauge theories are characterized by generating sets that do not form a Lie algebra.".

As we will see in Section 4.2, Yang–Mills theory is the cardinal example of this lucky instance. Indeed, the "gauging paradigm" of taking an abstract global symmetry algebra and making it local in order to perform the kinematical part of the gauging is intuitively based on the successful Yang–Mills example. There is no guarantee that higher spin gauge theory will work out so simple.

3.2 Hamiltonian mechanics

The analysis of constrained dynamical systems is due to Dirac [230–232].[7] We will follow the standard way of describing it for finite dimensional systems, that is mechanics, and then just wave the pen and write that the generalization to infinite dimensional systems, that is field theory, is straightforward, at least in principle. The reader should however be aware of the fact that this straightforwardness is not without its subtleties, complications coming from finite or countable sums being replaced by integrals and the ensuing questions about function spaces to work over. It is at the present not clear if such nontrivialities has any bearing on the higher spin interaction problem. As will be briefly mentioned in Chapter 6 on light-front higher spin theory, there are indeed problems related to the Dirac procedure. For a comprehensive textbook treatment of

7 According to [233], there is a largely unknown precursor work by Léon Rosenfeld.

constrained dynamics; see [229] and for a classic review text, see [234] which also contains some further references to original work on the subject apart from Dirac. A geometrical treatment can be found in [235]. Another classic textbook reference is [236]. For an alternative approach and discussion of specific field theory issues, see [237].

The passage from Lagrangian mechanics to Hamiltonian mechanics is interesting and well motivated in classical generalized mechanics, but it became even more so in view of Dirac's fundamental insight that it provides a natural road for quantization. We have already had several occasions to see the tension between relativity and quantum theory. In this context, it is interesting to quote from Dirac's first paper on the subject.

> With the Lagrangian form the requirements of special relativity can very easily be satisfied, simply by taking the action, that is, the time integral of the Lagrangian, to be Lorentz invariant. There is no such simple way of making the Hamiltonian form relativistic.
> For the purpose of setting up a quantum theory, one must work from the Hamiltonian form. There are well-established rules for passing from Hamilton's dynamics to quantum dynamics, by making the coordinates and momenta into linear operators. [...]
> Thus both forms have their special values at the present time and one must work with both.[8]

Consider an action for a system of N classical degrees of freedom described by generalized coordinates q_n and velocities $\dot{q}_n = dq_n/d\tau$ where τ is some evolution parameter of the system, not necessarily the reference frame time $t = x_0$ in relativistic systems. Write the action as

$$S_L = \int_{\tau_1}^{\tau_2} L(q_n, \dot{q}_n) d\tau \tag{3.20}$$

The Euler–Lagrange equations are (see formula (3.4))

$$\frac{d}{d\tau}\left(\frac{\partial L}{\partial \dot{q}_n}\right) = \frac{\partial L}{\partial q_n} \qquad n = 1, \ldots, N \tag{3.21}$$

If "everything goes as intended", so to speak, the Euler–Lagrange equations yield equations for the accelerations $\ddot{q}_n$. That it may not always turn out "as intended" can be seen by using the chain rule on the left-hand side of (3.21)

$$\ddot{q}_m \frac{\partial^2 L}{\partial \dot{q}_m \partial \dot{q}_n} = \frac{\partial L}{\partial q_n} - \dot{q}_m \frac{\partial^2 L}{\partial q_m \partial \dot{q}_n} \tag{3.22}$$

8 Of course, one may protest by pointing out that one can quantize covariantly using Lagrangian–Feynman path integrals [238] (a method incidentally also going back to Dirac [239]), but that does not make the conceptual tension alluded to here, less interesting.

In order for it to be possible to uniquely express the accelerations in terms of the positions and velocities, the *Hessian matrix*

$$W^{mn} = \frac{\partial^2 L}{\partial \dot{q}_m \partial \dot{q}_n} \tag{3.23}$$

must be invertible. As we will see, in relativistic systems this is in general not the case. In fact, this is the most interesting case, as it leads to the concept of general gauge invariance. In this case, there will be constraints among the coordinates and momenta that require a more elaborate method to pass from Lagrangian mechanics to Hamiltonian mechanics. But let us first do the standard case.

3.2.1 The unconstrained – standard – case

The passage to the Hamiltonian H goes via the definition of the canonical momenta

$$p^n = \frac{\partial L}{\partial \dot{q}_n} \tag{3.24}$$

and the Legendre transformation[9]

$$H = p^n \dot{q}_n - L \tag{3.25}$$

The rationale for introducing H is that it only depends on the velocities $\dot{q}$ through the canonical momenta $p(q, \dot{q})$, and therefore can be expressed as function of positions and momenta only. This can be seen from computing of the variation δH of the Hamiltonian with independent variations δq_n and $\delta \dot{q}_n$ of positions and velocities

$$\delta H = \dot{q}_n \delta p^n + \delta \dot{q}_n p^n - \frac{\partial L}{\partial \dot{q}_n} \delta \dot{q}_n - \frac{\partial L}{\partial q_n} \delta q_n = \dot{q}_n \delta p^n - \frac{\partial L}{\partial q_n} \delta q_n \tag{3.26}$$

Here, δp^n is not an independent variation, but follows from equations (3.24). In fact,

$$\delta p^n = W^{nm} \delta \dot{q}_m + \frac{\partial p^n}{\partial q_m} \delta q_m \tag{3.27}$$

Thus, a variation in the $\dot{q}_n$ that preserves the definition of the momenta while keeping $\delta q_n = 0$ and $\delta p^n = 0$, leaves H invariant. Hence, H can be expressed as a function of the q_n and p^n only. These variables, collectively called *canonical variables*, together form the *phase space* of the system. It is a $2N$-dimensional space.

9 The summation convention is used.

The Euler–Lagrange equations of the motion can now be replaced by the *Hamilton equations* of the motion

$$\dot{p}^n = -\frac{\partial H}{\partial q_n} \quad \text{and} \quad \dot{q}_n = \frac{\partial H}{\partial p^n} \tag{3.28}$$

At this stage one can introduce *Poisson brackets*. Let f and g be functions of the phase space variables q_n and p^n. Then their Poisson bracket $\{f, g\}$ is defined through

$$\{f, g\} = \frac{\partial f}{\partial q_n}\frac{\partial g}{\partial p^n} - \frac{\partial f}{\partial p^n}\frac{\partial g}{\partial q_n} \tag{3.29}$$

In terms of Possion brackets, the Hamilton equations become

$$\dot{p}^n = \{p^n, H\} \quad \text{and} \quad \dot{q}_n = \{q_n, H\} \tag{3.30}$$

The time evolution of the system can, in principle, be computed from these equations.[10] In order to do that, one needs *initial conditions* in the form of values for all the canonical variables at a certain initial time. Pictorially, the initial state is a point in phase space that subsequently moves around as the clock ticks away.

Properties of the Poisson bracket

i

The Poisson bracket is by definition antisymmetric in the sense that $\{f, g\} = -\{g, f\}$. It is linear in either of its members: $\{f + h, g\} = \{f, g\} + \{h, g\}$ and correspondingly for the second member. More interestingly, we have the Leibniz-type rule

$$\{fh, g\} = f\{h, g\} + \{f, g\}h \tag{3.31}$$

and the Jacobi identity

$$\{f, \{g, h\}\} + \{g, \{h, f\}\} + \{h, \{f, g\}\} = 0 \tag{3.32}$$

These properties translate to quantum commutators upon quantization according to the rule (1.1).

3.2.2 The constrained case

When the Hessian matrix W^{mn} in (3.23) is singular, it is not possible to invert the defining equations (3.24) for the momenta, and express all the velocities in terms of the coordinates and momenta. Nevertheless, the Legendre transformation anyhow makes it possible to express the Hamiltonian as a function of positions and a momenta, but

10 Or at least be numerically simulated.

not in a unique way. The argument at the beginning of Section 3.2.1 is still valid even though the Hessian matrix in formula (3.27) is not of maximal rank. There will now be constraints among the momenta and positions. Suppose there are M such constraints $\phi_m(p,q) = 0$ corresponding to the matrix W_{mn} being of maximal rank $N - M$.

A quick way[11] to develop the theory in this case is to state that the Hamiltonian H in equation (3.25) is not unique, rather one can replace it with an "effective" Hamiltonian $\tilde{H}$ defined by

$$\tilde{H} = H + u^m \phi_m(p,q) \approx H \tag{3.33}$$

where the $\approx$ equality sign was introduced by Dirac [230] to signify *weakly equal*. Indeed, the constraints should be written

$$\phi_m(p,q) \approx 0 \quad m = 1, \ldots M \tag{3.34}$$

In [231], Dirac introduces the weak equality to remind us that equalities such as $\phi_m(p,q) = 0$ should not be used before working out Poisson brackets. The reason being that the Poisson brackets of (3.29) presuppose that the q_n and p^n can be considered as independent variables, which is not the case here. In short, the Poisson brackets and the constraints are incompatible. The coefficients u^m are indeterminates with no definite dependence on positions and momenta.

Continuing, with this preliminary understanding, one can compute the equations of motion

$$\dot{q}_n = \{q_n, \tilde{H}\} \approx \frac{\partial H}{\partial p^n} + u^m \frac{\partial \phi_m}{\partial p^n} \tag{3.35}$$

$$\dot{p}^n = \{p^n, \tilde{H}\} \approx -\frac{\partial H}{\partial q_n} - u^m \frac{\partial \phi_m}{\partial q_n} \tag{3.36}$$

where the constraints have indeed been used after the computation of the Poisson brackets. Likewise, the time evolution of any variable f is computed according to[12]

$$\dot{f} = \{f, \tilde{H}\} \approx \{f, H\} + u^m \{f, \phi_m\} \tag{3.37}$$

The constraints so far considered are called *primary constraints* since they follow from the form of the Lagrangian and the equations of motion have not been used. The presence of the constraints in the Hamiltonian equations of motion means that the time evolution contains arbitrary functions of time. This is a manifestation of the original impossibility to solve for all the accelerations in equation (3.22) when the Hessian matrix is singular. However, the story does not end here, because the analysis of this arbitrariness is not yet complete.

---

11 As it is done, for instance, in [234]; this is actually the common method in the secondary and tertiary literature.

12 Note that when performing these computations, one is relying on the abstract properties of the Poisson bracket collected in the box above, in conjunction with the weak equality. For instance, $\{f, u^m \phi_m\} = u^m \{f, \phi_m\} + \{f, u^m\} \phi_m \approx u^m \{f, \phi_m\}$.

Understanding the multipliers u^m

The terms $\{f, u^m\}\phi_m$, that one would perhaps expect to occur in the computations (3.35) and (3.36) are weakly zero. That is just as well, because factors such as $\{f, u^m\}$ cannot be computed since the u^m are not well-defined functions of q_n and p^n, rather they can be interpreted as new canonical variables [230]. A more elaborate derivation of the time development formulas is to return to the variation of H in the computation (3.26) and write it as

$$\delta H = \frac{\partial H}{\partial q_n}\delta q_n + \frac{\partial H}{\partial p^n}\delta p^n = \dot{q}_n \delta p^n - \frac{\partial L}{\partial q_n}\delta q_n \tag{3.38}$$

from which follows

$$\left(\frac{\partial H}{\partial q_n} + \frac{\partial L}{\partial q_n}\right)\delta q_n + \left(\frac{\partial H}{\partial p^n} - \dot{q}_n\right)\delta p^n = 0 \tag{3.39}$$

In the unconstrained case, where q_n and p^n can be varied freely, both coefficients of δq_n and δp^n must be zero. This returns us to the Hamilton equations (3.28). In the constrained case, where we can think of the constraints $\phi(p, q)_m$ as defining a surface in phase space, the "tangential variations" cannot be made independently. It may then be shown that any solution of an equation of the type

$$\lambda^n \delta q_n + \mu_n \delta p^n = 0 \tag{3.40}$$

is of the form

$$\lambda^n = u^m \frac{\partial \phi_m}{\partial q_n} \quad \text{and} \quad \mu_n = u^m \frac{\partial \phi_m}{\partial p^n} \tag{3.41}$$

Combining this with (3.39) again yields the equations (3.35) and (3.36). For details of this argument, see [229]. For Dirac's own argument, see [230].

The Hamiltonian equations of motion (3.35)–(3.36) and the constraints (3.34) can be derived from the action

$$S_H = \int_{\tau_1}^{\tau_2} (\dot{q}_n p^n - H - u^m \phi_m) d\tau \tag{3.42}$$

by independent varying q_n, p^n and u^m. The variables u^m may therefore be interpreted as Lagrange multipliers enforcing the primary constraints. As such, they are indeed indeterminate at this stage of the discussion.

The next step is to require that the τ derivatives of the primary constraints are zero, that is, the constraints should be maintained in time. This gives us the equations

$$\dot{\phi}_n \approx \{\phi_n, H\} + u^m \{\phi_n, \phi_m\} \approx 0 \tag{3.43}$$

These equations may reduce to equations among the q_n and p^n independent of the u^m, or may impose restrictions on the u^m. In the first case, we get new constraints, called *secondary constraints*. They are so-called because they follow from the equations of motion. This procedure is then repeated, computing the time derivatives of the secondary constraints and requiring them to be zero, which again may result in new constraints or conditions on the u^m.[13] The process is repeated until no more secondary constraints result or no more conditions on the u^m occur. At this stage, the set of M initial primary constraints may be enlarged by an additional number K of secondary constraints. Since there is from now on no particular reason to discriminate between primary and secondary constraints, they will all denoted by $\phi_j(q,p)$ where j runs from 1 to $M + K = J$.

Finally, assuming that all constraints have been found, the next step is to investigate if there are any restrictions on the multipliers u^m. Letting m run over the primary constraints and j index any constraint, the restrictions are

$$\{\phi_j, H\} + u^m\{\phi_j, \phi_m\} \approx 0 \tag{3.44}$$

We have here a system of J linear equations in the M undetermined variables u^m with coefficients that are definite functions of the q_n and p^n. These equations must have solutions[14] that we may write as a sum of a particular solution U^m to the inhomogeneous equation, and a general solution V^m to the corresponding homogeneous equation

$$V^m\{\phi_j, \phi_m\} \approx 0 \tag{3.45}$$

Now denote by V_a^m, $a = 1,\ldots,A$ the linearly independent solutions to the system (3.45). Then the general solution the system (3.44) is

$$u^m = U^m + v^a V_a^m \tag{3.46}$$

with undetermined arbitrary coefficients v^a. This means that we have split of the multipliers u^m into a fixed part U^m and an arbitrary part. The split is however not unique in that the particular solutions, as always, are just any particular solutions.

At this stage, it is customary to define what, somewhat unimaginatively, is called the total Hamiltonian. The idea is however to take advantage of the split (3.46) to separate the piece of the time evolution that is completely arbitrary. Returning to the effective Hamiltonian of formula(3.33), we can now write

$$\tilde{H} = H + u^m\phi_m = H + U_m\phi_m + v^a V_a^m \phi_m \tag{3.47}$$

13 Note that in doing this, one should still use the “effective” Hamiltonian in formula (3.33), that is, one does not add the secondary constraints to H. Dirac comments on this in [231]. Dirac writes that, from the Hamiltonian point of view, the essential difference between primary and secondary constraints, is that the primary constraints occur in the equation of motion, while the secondary do not.

14 Unless the original Lagrangian in the action (3.20) is inconsistent for some reason.

This may prompt the following two definitions:

$$H' = H + U^m \phi_m \qquad \text{(The “fixed part” of the Hamiltonian)} \tag{3.48}$$

$$\phi_a = V_a^m \phi_m \qquad \text{(A linear combination of the primary constraints)} \tag{3.49}$$

The *total Hamiltonian* H_T is the defined according to

$$H_T = H' + v^a \phi_a \tag{3.50}$$

thus clearly separating out the part of the time evolution that is arbitrary.

3.2.3 Systematics of the constraints

It remains to understand the structure of the constraints in some more detail. Up to now, we have the split into primary constraints and secondary constraints. The primary constraints follow from the Lagrangian and the definition of the momenta (3.24). The secondary constraints follow from the consistency of the Hamiltonian equations of motion. Then we have the linear combination of primary constraints $\phi_a = V_a^m \phi_m$ coming from the consistency conditions. We will now define the concepts of first class and second class constraints. We will then have three kinds of constraints that we have to understand how they relate to each other.

A function $F(q, p)$ is said to be *first class* if its Poisson bracket with every constraint ϕ_j is weakly zero, that is if

$$\{F, \phi_j\} \approx 0 \quad \text{for } j = 1, \ldots, J \tag{3.51}$$

If the function is not first class, then it is said to be *second class*. That means that it has a nonzero Poisson bracket with at least one of the constraints. This definition allows for a division of all the constraints into two disjoint sets: *first class constraints*, all of which have weakly vanishing Poisson brackets among themselves, and the *second class constraints* where each constraint has a nonvanishing bracket with at least one other constraint in the set.

Focusing on first-class functions, a number of consequences of the definition can be derived. If two functions F and G are first class, then their Poisson bracket $\{F, G\}$ is also first class. This follows from the Jacobi identity (3.32) for the Poisson bracket. This means that the set of first-class functions are closed under the Poisson bracket operation. In particular, this is true for the first-class constraints themselves. Furthermore, it follows that the Hamiltonian H' is first class as well as all the ϕ_a. Thus referring back to (3.49), we learn that the ϕ_a constitute a complete set of first- class primary constraints.

Now, one must exercise a bit of thinking. The set of first-class constraints may derive both from primary constraints and secondary constraints. That means, among other things, that their number is somewhere between 0 and J. The fact that they form a closed set under the Poisson bracket hints at an algebraic structure. To continue, we choose to denote the first-class constraints by γ_r with $r = 1, \ldots, R$, and the second-class constraints by χ_f with $f = 1, \ldots, F$. Clearly, $R + F = J$.

We may tentatively write

$$\{\gamma_r, \gamma_s\} \approx \sum_t c_{rst}\gamma_t \tag{3.52}$$

where the range of the sum is yet unspecified. This is suggestive of a Lie algebra structure (see Section 3.11). This equation can be motivated from the fact that since the Poisson bracket of any two first-class constraints is weakly zero, then it must be a linear combination of first-class constraints. There is no risk that second-class constraints appear on the right-hand side because that would violate the Jacobi identity.

3.2.4 First-class constraints and gauge transformations

Let us take stock of where we are. We have the constraints, sorted into the kinds of Section 3.2.3. We have the first-class total Hamiltonian H_T of formula (3.50). It generates the time evolution of the system, that is, of any dynamical variable F, according to $\dot{F} = \{F, H_T\}$. Due to the presence of the term $v^a\phi_a$ in H_T, the time evolution contains arbitrary functions, parametrized by the coefficients v_a. This arbitrariness we would like to interpret as gauge transformations. But how does that come about?

A brief reminder of canonical transformations

---

i Remember that (not explicitly time dependent) canonical transformations in Hamiltonian dynamics are transformations of the phase space variables

$$Q_m = Q_m(q_n, p_n) \quad \text{and} \quad P_m = P_m(q_n, p_n) \tag{3.53}$$

such that in terms of the new variables, the equations of motion keep their form

$$\dot{Q}_m = \{Q_m, K\} \quad \text{and} \quad \dot{P}_m = \{P_m, K\} \tag{3.54}$$

in terms of a new Hamiltonian $K(Q, P)$ which is the transformation of $H(q, p)$. For infinitesimal canonical transformations, close to the identity transformation, one can show that they can be represented as

$$\delta q_n = \epsilon\{q_n, g\} \quad \text{and} \quad \delta p^n = \epsilon\{p^n, g\} \tag{3.55}$$

where the *generator* $g = g(q, p)$ is some function of the phase space variables, and ϵ an infinitesimal parameter. For other dynamical variables, the transformation reads

$$\delta F = \epsilon\{F, g\} \tag{3.56}$$

For the full story, consult the Chapters 5 and 6 in the book [236].

A fundamental notion of classical mechanics is that if we know the values of all the canonical variables at a certain time τ (the state of the system at that time) then the equations of motion shall determine the state completely at a later time $\tau + \delta\tau$. Therefore, by decree, ambiguities in the state at the later time should be physically irrelevant. Consider then the time evolution generated by the total Hamiltonian in (3.50)

$$\delta F = \delta\tau\{F, H'\} + \delta\tau v^a\{F, \phi_a\} = \delta\tau\{F, H'\} + \epsilon^a\{F, \phi_a\} \tag{3.57}$$

where in the last equality we have absorbed the time increment $\delta\tau$ into the arbitrary parameters v^a to get the arbitrary infinitesimal increments ϵ^a. Thus, on top of the well-defined time evolution given by $\delta\tau\{F, H'\}$, we have arbitrary transformations given by $\epsilon^a\{F, \phi_a\}$. These are called *gauge transformations*.[15]

Now, it would be very nice if the complete set of first class primary constraints ϕ_a occurring in the first-class total Hamiltonian H_T closed on itself, as the full set of first-class constraints must do in (3.52). However, there is nothing in the general theory that guarantees that, and in concrete systems, secondary first-class constraints do occur as the result of Poisson brackets between first-class primary constraints. This raises the question of whether secondary first-class constraints should also be considered as gauge generators or not. Dirac "conjectured" that they should.[16]

> I think it may be that all the first class secondary constraints should be included among the transformations which do not change the physical state, but I have not been able to prove it.

There are counterexamples to the Dirac conjecture. Quite a few simple examples are given in [229], but they appear rather contrived. Instead, as the authors write, there are good reasons for including the secondary first-class constraints among the gauge generators:

- The distinction between first class and second class is natural from the Hamiltonian point of view, while the Lagrangian distinction between primary and secondary is not so.
- The Poisson bracket of two first-class constraints is again a combination of first-class constraints, leading an algebra of gauge transformations.
- The conjecture is true for the major gauge theories of physics.
- The conjecture can be proved under certain regularity conditions.

One may now go on to define an *extended Hamiltonian* that includes all the first- class constraints

$$H_E = H' + v^r \gamma_r \tag{3.58}$$

15 As far as I understand, the physical irrelevance of gauge transformations, cannot be proved. It is rather a phenomenological observation elevated to a principle.

16 See page 23 of [232].

where the index r runs over all the first-class constraints, and the coefficients v^a are supplemented with extra arbitrary variables corresponding to the first-class secondary constraints.

So far, we have tacitly assumed that the set of constraints is irreducible, that is, that they are all independent. For the reducible case, we refer the reader to [229] that treats this case in detail.

3.2.5 Second-class constraints, Dirac brackets and gauge fixing

Second-class constraints, if they are present, can be treated by the introduction of the Dirac bracket. This is a bracket that replaces the Poisson bracket in such a way that the second-class constraints can be set strongly equal to zero either before or after the evaluating a Dirac bracket. The definition of the Dirac bracket hinges on the fact that the matrix of all Poisson brackets

$$C_{\alpha\beta} = \{\chi_\alpha, \chi_\beta\} \tag{3.59}$$

between second-class constraints is invertible. The matrix is antisymmetric, and this incidentally imply that the number of second-class constraints must be even, since the determinant of an antisymmetric matrix is zero if the dimension is odd.

The *Dirac bracket* between two phase space functions F and G is then defined in terms of the Poisson bracket and the inverse of C

$$\{F, G\}_D = \{F, \chi_\alpha\} C^{-1}_{\alpha\beta} \{\chi_\beta, G\} \tag{3.60}$$

The Dirac bracket has all the desirable properties, such as antisymmetry, linearity, the Leibniz rule and the Jacobi identity, that one could wish for.

Since the second-class constraints become identities, there are simple cases where they can be used to completely eliminate some phase space variables from the system. On the other hand, there are examples where the equations cannot be explicitly solved, and the Dirac brackets are quite awkward to work with. We refer the reader to the references cited above for more details. We will employ the Dirac bracket in light-front field theory in Section 6.2.1.

In this context, we may also briefly mention gauge-fixing. One may view the first class constraints, and the concomitant arbitrariness, as a nuisance (although we will not take that point of view). One way of getting rid of this ambiguity is to introduce a number of gauge conditions, equal to the number of first-class constraints. These *gauge conditions* – in practice functions of the phase space variables set weakly to zero – should be such that the full set of first-class constraints and gauge conditions together become second class. Then the theory of Dirac brackets can be employed.

Let us end by noting a very useful heuristic for counting physical degrees of freedom in a theory with constraints. Every second-class constraint removes one phase space degree of freedom, regardless of whether it can be done explicitly or not. Every first-class constraint removes two phase space degrees of freedom. This follows if one imagines supplying a gauge condition for every first-class constraint, thus rendering the whole set second class.

$$\#\text{physical d. o. f.} = \#\text{phase sp. d. o. f.} - \#\text{2nd cl. constr.'s} - 2(\#\text{1st cl. constr.'s}) \quad (3.61)$$

This removal of twice the number of first-class constraints will come back when gauge-fixing gauge field theories, as the heuristic: "gauge-fixing and regauging" (see Section 5.1.1).

One of the most conspicuous examples

A most conspicuous example in field theory occurs already in electrodynamics. The Lagrangian density is proportional to $F_{\mu\nu}F^{\mu\nu}$, with $F_{\mu\nu}$ antisymmetric in its components. Therefore, the A^0 component, the Coulomb field, has no time derivative in L. Thus, without performing any computation, it is immediately clear that the corresponding field momentum Π_0 must be zero. This is a primary constraint. The momenta conjugate to $\mathbf{A}$ are the electric fields $\mathbf{E} = \partial_t \mathbf{A} + \nabla A^0$. The Dirac analysis then turns up the Colulomb law as a secondary constraint $\nabla \cdot \mathbf{E} \approx 0$. Together with $\Pi_0 \approx 0$, it is first class.

3.3 Quantum mechanics and quantum field theory

The topics of this section are book size subjects in their own right. We will confine ourselves to drawing a few baselines, useful for the application to higher spin field theory.

It may be a little confusing, when encountering it for the first time, to accept that the nonrelativistic *Schrödinger equation* in its abstract form

$$i\hbar \frac{\partial}{\partial t}\Psi = H\Psi \quad (3.62)$$

still holds in quantum field theory. The explanation is, perhaps not simple, but certainly illuminating.[17] The observable *time* is normally not represented by an Hermitian operator. In quantum mechanics, time t is a mere parameter labeling the states. What we can do to pave the way for relativity is to treat also the position vector $\mathbf{x}$ as a parameter labeling states and consider quantum fields $\varphi(\mathbf{x}, t)$. Alternatively, the coordinate time t could be considered as an observable corresponding to an operator T. Then the proper time τ can be used as evolution parameter in the abstract Schrödinger equation (3.62) instead of t. This is possible since relativity allows for a reparametrization invariance under transformations $\tau \to \tau'$.

17 For a textbook explanation, see for instance [240].

Quantum mechanics do treat space and time in an unsymmetrical way. This apparent conflict with relativity was a driving force in the theoretical evolution of quantum theory, and the history of the subject actually throws light on the question. Very briefly, de Broglie's intuitive particle wave-mechanics of 1923 was based on special relativistic reasoning (see Section 2.1). However, de Broglie's theory was still within the "old" quantum mechanics. The next step, Heisenberg's matrix mechanics was nonrelativistic, as was Schrödinger's wave mechanics, after he had rejected the relativistic equation. Then came Dirac's insight, that unified the Heisenberg and Schrödinger approaches within the transformation theory. Dirac realized that quantum mechanics could be seen as the equations of classical Hamiltonian mechanics being reinterpreted as quantum equations. The point is that the equations of Hamiltonian mechanics – whether it is nonrelativistic or relativistic – are always linear in time derivatives. This "explains" the general applicability of the Schrödinger equation even for relativistic quantum theories. Or rather, nonrelativistic quantum mechanics becomes the special case it really is. We have seen another aspect of this in the historical chapter. The relativistic wave equations are not equations governing states, they are equations governing quantum operators.

Perhaps significantly, the major theory that does not sit comfortably in this framework is quantum general relativity. One of the major difficulties with quantum gravity is "the problem of time". In quantizing general relativity along conventional lines (canonical quantization), the problem of time does not yield to the above mentioned method of replacing time by another evolution parameter and then relying on reparametrization invariance. Time is too deeply embedded in the kinematics and dynamics of general relativity. For a discussion of these matters, see [108].

Even without invoking classical or quantum mechanics, nonrelativistic or relativistic physics, it is in modern physics thinking clear that in order to perform experiments and do theoretical calculations, we need to set up a grid of spatial coordinates and clocks. We need to put our laboratory, experimental or theoretical, in a coordinate system. That can be done classically, or quantum mechanically, or nonrelativistically, or relativistically, or in any combinations thereof – exactly or approximately. What the relativity theories say (among other things) is that the equations that we either employ or discover, do not depend in any essential way on any particular choice of coordinates. As the theory is developed, gauge symmetries and gauge independence, has to be worked into the picture of relativistic physics.

3.3.1 Baseline quantum mechanics

Two basic ingredients in any quantum theory are the *operators* and the *states*, neither of which are themselves generally accessible to direct observation. Together they constitute what we mean by a *quantum system*. Closer to measurable quantities are the *matrix elements* of operators evaluated between pairs of states. Another, almost defining, property of quantum systems is the linear *superposition* of states into new states. The mathematical structure that has turned out to encode these features of quantum systems in general in a successful way, are the Hilbert spaces (see Section 3.7.6).

Hilbert space in few lines

i

A Hilbert space is a complex linear vector space (superposition possible) with a metric (distance measure between vectors) that derives from an inner product (matrix elements). Linear operators can be applied to the vectors and matrix elements can be computed using the inner product. There is a notion of continuity in the sense of nearness of vectors as measured by the metric.

Relying on the Hilbert space concept, *states of a quantum system* are represented by equivalence classes of vectors Ψ, called *rays*. The inner product between two rays Ψ and Φ is denoted by $\langle\Phi,\Psi\rangle$ and it evaluates to a complex number. The rays are normalized in the sense that $\langle\Psi,\Psi\rangle = 1$ and Ψ and Ψ' belong to the same ray if and only if $\Psi' = c\Psi$ with c a complex number with $|c| = 1$. A ray, that is, a state, can be represented by any of its vectors Ψ belonging to the ray.

The following properties for the inner product between states, are fundamental

$$\langle\Phi,\Psi\rangle^* = \langle\Psi,\Phi\rangle \tag{3.63}$$

$$\langle\Psi,\Psi\rangle \geq 0 \quad \text{with} \quad \langle\Psi,\Psi\rangle = 0 \Leftrightarrow \Psi = 0 \tag{3.64}$$

$$\langle\Phi, c_1\Psi_1 + c_2\Psi_2\rangle = c_1\langle\Phi,\Psi_1\rangle + c_2\langle\Phi,\Psi_2\rangle \tag{3.65}$$

$$\langle c_1\Phi_1 + c_2\Phi_2, \Psi\rangle = c_1^*\langle\Phi_1,\Psi\rangle + c_2^*\langle\Phi_2,\Psi\rangle \tag{3.66}$$

where the two last equations express how the norm behaves under *superposition* of states. States are superposed by summing them with complex coefficients. This is the vector space property of the Hilbert space of states.

Operators A of a quantum system are linear mappings $\Psi \rightarrow A\Psi$ of the Hilbert space into itself. *Matrix elements* of an operator A are given by inner products $\langle\Phi, A\Psi\rangle$ between states Φ and Ψ. The *adjoint* $A^\dagger$ of an operator A is defined by

$$\langle\Phi, A^\dagger\Psi\rangle \stackrel{\text{def}}{=} \langle A\Phi,\Psi\rangle = \langle\Psi, A\Phi\rangle^* \tag{3.67}$$

The first equality is the actual definition and the second equality follows from (3.63). Next, *self-adjoint* or *Hermitian operators* A are such that $A^\dagger = A$.

Observables for a quantum system are represented by Hermitian operators. The motivation is the following. Perform the computation

$$\langle\Phi, A\Psi\rangle^* = \langle A\Psi,\Phi\rangle = \langle\Psi, A^\dagger\Phi\rangle = \langle\Psi, A\Phi\rangle \tag{3.68}$$

where we have used the norm property, adjoint definition and Hermiticity in that order. It is clear that nothing can be said about the reality properties for matrix elements between two different states Ψ and Φ. Such matrix elements are called *transition elements*. However, if the two states are equal, we get

$$\langle \Phi, A\Phi \rangle^* = \langle \Phi, A\Phi \rangle \tag{3.69}$$

This means that Hermitian operators have real diagonal matrix elements. Furthermore, considering the eigenvalue equation $A\Psi = a\Psi$ for a Hermitian operator A, a theorem of linear algebra tells us that the eigenvalues a are all real and that the eigenvectors Ψ are orthogonal. *Orthogonality* of two states is defined as their inner product being zero. Since measurements of any kind are always real numbers (in practice rational numbers), it makes sense to represent observable quantities with Hermitian operators.

Probability interpretation of quantum mechanics

Since states can be in superposition with other states, a quantum system can, in a specific sense, be in several states simultaneously. Consider therefore a system that is in a state represented by a vector Ψ. An experiment is done to measure if it is any one of a set of mutually orthogonal states $\{\Psi_i\}_i$ corresponding to some observable. The *probability* of finding the system in the particular state Ψ_k is $|\langle \Psi, \Psi_k \rangle|^2$. Probabilities sum to 1.

Symmetries of a quantum system are represented by *unitary* operators, with one exception, to be mentioned shortly. The motivation is the following. In order for a linear transformation with an operator U to be a symmetry, "something" must be invariant. This something are the probabilities $|\langle \Phi, \Psi \rangle|^2$ computed between states of the quantum system. Consider a transformation effected by an operator U so that $\Phi' = U\Phi$ and $\Psi' = U\Psi$. Invariance of the probabilities then means demanding

$$|\langle \Phi', \Psi' \rangle|^2 = |\langle \Phi, \Psi \rangle|^2 \tag{3.70}$$

This demand can be satisfied by linear unitary operators

$$\langle U\Phi, U\Psi \rangle = \langle \Phi, \Psi \rangle \tag{3.71}$$

Referring back to the definition (3.67) of an operator, we have for the adjoint of U the equation $\langle \Phi, U^\dagger \Psi \rangle = \langle U\Phi, \Psi \rangle$. The requirement (3.71) then implies for a unitary operator

$$U^\dagger = U^{-1} \tag{3.72}$$

The invariance condition (3.70) can also be satisfied by antilinear and antiunitary operators (for details, see [18]). Time reversal symmetries are represented by such operators.

3.3.2 Simple phase space quantization

Quantization is the process of passing from a classical description of a system to a quantum description. It is not always explicitly recognized, but for a classical system, the states of the system and the physical variables are one and the same. More precisely, if the states are described in phase space by points (q_n, p^n), then all physical variables of the system may be computed as functions of the phase space variables. In the quantum description, there is bifurcation of concepts: we have states and operators. It is clearly seen in the original Schrödinger formulation with wave functions as states, and operators acting on states. It is formalized in the Dirac formulation of quantum mechanics, as outlined in Section 3.3.1 above.

Quantization itself may be a quite complicated procedure, and there is an extensive theory – and philosophy – on the subject. Here, we just repeat the basic scheme already briefly stated in Section 1.4. If a classical mechanical theory is given in the Hamiltonian formulation, then the transition from classical Poisson brackets $\{\cdot,\cdot\}$ to quantum commutators $[\cdot,\cdot]$ is done through the convention:

$$\text{If classically: } \{A, B\} = C, \quad \text{then quantum mechanically: } [\hat{A}, \hat{B}] = i\hbar\hat{C} \tag{3.73}$$

where the Poisson bracket is defined in formula (3.29). In terms of the phase space variables q_n and p^n obeying $\{q_n, p^m\} = \delta_n^m$, we have

$$[\hat{q}_n, \hat{p}^m] = i\hbar\delta_n^m \tag{3.74}$$

In wave mechanics à la Schrödinger, the operators are realized as

$$\hat{p}^n = -i\hbar\frac{\partial}{\partial q_n} \quad \text{and} \quad \hat{q}_n = q_n \tag{3.75}$$

The time evolution of a classical dynamical variable F turns into the time evolution of the corresponding quantum operator F according to

$$\dot{F} = \{F, H\} \quad \rightarrow \quad i\hbar\frac{dF}{d\tau} = [F, H] \tag{3.76}$$

This corresponds to the *Heisenberg picture* where the time evolution of the quantum system is carried by the operators, and the states are constant in time. The Schrödinger equation, given above in formula (3.62), corresponds to the *Schrödinger picture* where the time evolution is carried by the states, and the operators are constant in time.

3.3.3 The Siegel mechanics to field theory algorithm

We will very briefly review a simple instance of a method – based on BRST-symmetry – of passing from a mechanical model to a corresponding field theory. It was invented by W. Siegel, and clarified by E. Witten, in connection with work on string field theory (see Section 2.11.1). The basic intuition behind the method may be argued to go back to L. de Broglie and E. Schrödinger. It is also implicit in Wigner's work on wave equations discussed above (see in particular Section 2.3.2). The method is further developed in [241, 242] and reviewed in [187].

Indeed, referring back to the discussion at the beginning of Section 2.1, a classical free relativistic massless particle "moves" subject to the constraint $p^2 = 0$. Upon first quantization, we get the Klein–Gordon wave equation $\Box\Psi(x) = 0$. This we may regard as a classical field equation. If the particle sports some internal structure such as spin and the corresponding extra variables, then there may be further constraints. If these are of first class, it is possible to construct a field theory via the BRST quantization method, so that the first class constraints generate gauge transformations not just in the mechanics theory, but also in the field theory.

The BRST method itself arose in the covariant quantization and renormalization of the Yang–Mills gauge theory. It was discovered by Becchi–Rouet–Stora and Tyutin independently (hence the name of the method). It was found that the covariantly gauge-fixed – and thus not gauge invariant – action, together with the Faddeev–Popov ghost term, still retained a symmetry closely related to the underlying gauge invariance of the theory. A description of the method can be found in the Weinberg textbook [139]. The method was later generalized to general gauge theories. We will return to this in more detail in our Volume 2 in connection with interacting higher spin theory. Here, we will just outline enough for the BRST approach to be applied to free higher spin gauge field theory in Section 5.4.

The Siegel algorithm starts from a mechanical gauge theory, constructs its BRST operator, and from there writes down a gauge invariant field theory. Let us assume that the mechanical theory only involves bosonic degrees of freedom.[18] The Hamiltonian theory developed above in Section 3.2 may serve as a foundation. We will conform to general notation that has become quite common in the literature.

The Poisson brackets of the mechanical theory will be denoted by $[\cdot,\cdot]$. This allows for a certain flexibility in that one may think of the brackets as quantum commutators. This is actually quite common, and is done in the Siegel algorithm. Once one has derived the first class constraints γ_a of the theory and their first-class algebra – for instance using the Dirac procedure – one can write down the *classical BRST generator* as

$$Q = \gamma_a \eta^a - \frac{1}{2}\mathcal{P}_c U^c{}_{ab}\eta^a\eta^b \tag{3.77}$$

Here, the $U^c{}_{ab}$ are the structure constants of the first class algebra

$$[\gamma_a, \gamma_b] = \psi_c U^c{}_{ab} \tag{3.78}$$

18 Fermionic degrees of freedom can be treated at the price of a more elaborate formalism. It adds no further conceptual depth beyond the existence of the fermionic variables themselves.

and η and $\mathcal{P}$ are ghost coordinates and momenta satisfying the brackets[19]

$$\{\eta^a, \eta^b\} = 0 \qquad \{\mathcal{P}_a, \mathcal{P}_b\} = 0 \qquad \{\eta^a, \mathcal{P}_b\} = \delta^a_b \tag{3.79}$$

This bracket is symmetric, as is appropriate for Grassmann variables. A general *Grassmann variable* θ is one that satisfies $\theta^2 = 0$.[20] The first two brackets in (3.79) show that the ghost phase space variables are Grassmann.

The BRST generator Q is *nilpotent* under these brackets, that is,

$$Q^2 = \frac{1}{2}\{Q, Q\} = 0 \tag{3.80}$$

To show this, one has to use the first class constraint algebra, as well as the Jacobi identities for the structure constants of the algebra. Thus Q records all information about the first class structure of the mechanical theory. To make this statement a little bit more exact, we may also provide information about the ghost extended phase space of the theory $(q_n, p^n, \eta^a, \mathcal{P}_b)$. Let us now focus on the intuition behind the Siegel algorithm.

Intuitive approach to the Siegel algorithm

We have a classical BRST generator Q that the records the structure of a mechanical gauge theory. Upon quantization, the phase space variables become operators that act in a Hilbert space of wave functions. The BRST generator Q becomes a *BRST operator* Q (for which we use the same notation). The wave functions Ψ may be expected to be functions $\Psi(q_n, \eta^a)$. Since the ghost variables are Grassmann, we may expand Ψ in a short polynomial over the ghost variables. Among the coefficient wave functions so obtained, we may suspect that some are "physical" and some are "auxiliary".

Then, among the quantized first class constraints, there may be one constraint that is recognizable as part of a kinetic operator, such as the d'Alembertian $\square$, and some constraints that may be recognizable as generating gauge transformations of the wave functions. Taking advantage of the nilpotency of the BRST operator, one may then try to make sense of the following equations:

$$\text{Action} \quad S = \int \langle \Psi^\dagger, Q\Psi \rangle \quad \Rightarrow \quad \text{Wave equations} \quad W = Q\Psi = 0 \tag{3.81}$$

$$\delta S = 0 \quad \text{and} \quad \delta W = 0 \quad \text{under gauge transformations} \quad \delta\Psi = Q\Xi \tag{3.82}$$

In order to make such a scheme consistent, certain requirements must be met. As a matter of principle, it must be possible to construct an inner product in the Hilbert space of wave functions. This involves not just the original phase space variables (q_n, p^n) but also the ghost extension $(\eta^a, \mathcal{P}_b)$. This inner product must be such that

19 We assume that the underlying theory is bosonic so that the ghosts are fermionic.

20 This is no more strange than $i^2 = -1$ (or just as strange, one might perhaps venture).

the BRST operator is nilpotent not just algebraically (as the classical BRST operator is) but also as acting in the Hilbert space. This may involve issues about normal ordering. Furthermore, properties under Hermitian conjugation must be defined so that Q is self-adjoint as a Hilbert space operator. The corresponding reality of the classical BRST generator (which is simple to arrange if not already checked) is not sufficient in general. All this must be arranged so that the action S is real. All occurrences of ghosts, must be "integrated out" so to speak. This is in practice kept track of by defining *ghost numbers* distinguishing different types of ghost coordinates and momenta from the physical coordinates and momenta.

All this can be done in great generality (see [241]), but a general formalism tends to be somewhat opaque, and it may obscure the underlying idea, which is quite simple. However, the general formalism provides limits to the applicability of the method, for instance on the important question of existence of a Lagrangian for the field theory. Field equations are often more easy to come by than Lagrangians, as they do not require the existence of an inner product on the extended phase space. This is a nontrivial problem in the Vasiliev higher spin theory.

The algorithm is often not too difficult to carry through in concrete cases where the structure of the constraints may serve as a guide toward a consistent implementation. We will see an example in Section 5.4 where the algorithm is used to derive the Fronsdal equations from a simple mechanical model in the way of the original references [158, 159].

3.4 Elements of special relativity

Special relativity, in theoretical physics, and vector spaces, in mathematics, are the archetypes out of which general relativity and manifold theory grew, respectively. Since higher spin theory relies heavily upon abstractions and generalizations of these basic concepts – perhaps eventually going beyond them – we will start with special relativity here and vector spaces in Section 3.7.5.

There are three classes of constant curvature space-times: de Sitter space-time (dS) with positive constant scalar curvature, anti-de Sitter (AdS) space-time with negative curvature and Minkowski space-time (Mi) which falls in between with zero curvature. Considered as vacuum solutions to Einstein's equations of general relativity, these space-times correspond to positive, negative and zero cosmological constant, respectively. The isometry groups, that is, the groups of coordinate transformations that leave the metric invariant[21] are $SO(3,2)$, $SO(4,1)$ and the Poincaré (inhomogeneous Lorentz) group $ISO(3,1)$, respectively. The space-times AdS and dS will be treated in Volume 2. Here, we will focus on Minkowski space-time.

21 Really, *form invariant*; the metric is the same function of the new coordinates as of the initial coordinates. See, for instance, [243], Chapter 13.

A *Poincaré transformation* is a change of coordinates from one system x^μ to another x'^μ given by the formula

$$x'^\mu = \Lambda^\mu{}_\nu x^\nu + a^\mu \tag{3.83}$$

that leaves the *proper time* interval $d\tau$, defined through

$$d\tau^2 = -\eta_{\mu\nu} dx^\mu dx^\nu = dt^2 - d\mathbf{x}^2 \tag{3.84}$$

invariant. The proper time is expressed in terms of the *coordinate differentials* dx^μ, which according to (3.83) transform as

$$dx'^\mu = \Lambda^\mu{}_\nu dx^\nu \tag{3.85}$$

The condition $d\tau'^2 = d\tau^2$ allows a^μ to be arbitrary translations while the Lorentz transformation matrices must satisfy

$$\eta_{\mu\nu}\Lambda^\mu{}_\rho \Lambda^\nu{}_\sigma = \eta_{\rho\sigma} \tag{3.86}$$

From this equation follows that $|\det \Lambda^\mu{}_\nu| = \pm 1$. The Lorentz transformations are therefore invertible.

One can, for practical purposes such as solving field equations in special situations, consider curvilinear coordinates in Minkowski space-time. The preferred systems where the metric takes the form $\mathrm{diag}(-1, 1, 1, 1)$ are called *inertial* coordinates.

We want to represent Poincaré transformations on other, more abstract spaces, than space-time itself, such as, for instance states of physical systems, vector and tensor fields. For that purpose, it is convenient to introduce abstract transformation operators $T_{\Lambda,a}$ implementing the transformations on some, as yet unspecified set, but most often on a vector space. Then on space-time itself, we have

$$x^\mu \mapsto T_{\Lambda,a}(x^\mu) = \Lambda^\mu{}_\nu x^\nu + a^\mu \tag{3.87}$$

By performing two consecutive Poincaré transformations with parameters Λ_1, a_1 and Λ_2, a_2

$$\begin{aligned} x''^\mu &= \Lambda^\mu_{2\ \rho} x'^\rho + a_2^\mu = \Lambda^\mu_{2\ \rho}(\Lambda^\rho_{1\ \nu} x^\nu + a_1^\rho) + a_2^\mu \\ &= (\Lambda^\mu_{2\ \rho}\Lambda^\rho_{1\ \nu}) x^\nu + (\Lambda^\mu_{2\ \rho} a_1^\rho + a_2^\mu) \end{aligned} \tag{3.88}$$

we can read of the abstract composition rule for the Poincaré group

$$T_{\Lambda_2,a_2} T_{\Lambda_1,a_1} = T_{\Lambda_2\Lambda_1,\Lambda_2 a_1 + a_2} \tag{3.89}$$

This formula is an expression of the fact that the Poincaré algebra is a semidirect sum of the Lorentz algebra $\mathfrak{so}(3, 1)$ and the Abelian algebra of translations. The Poincaré group is not semisimple. Any representation of the Poincaré group must conform to this equation (see Section 3.5).

Active versus passive transformations

A space-time transformation $x \rightarrow x'$ and its effects on functions $f(x)$ can be viewed in two ways: either passively or actively. *Passively*, a transformation $x \rightarrow x'$ is viewed as a coordinate change. The actual point p labeled by coordinates x, is after the transformation labeled by new coordinates x'. *Actively*, a transformation $x \rightarrow x'$ is viewed as actually moving the point p, labeled by coordinates x, to a new location, labeled by new coordinates x', in the same coordinate system. The two views are complementary, and convenient depending on context.

3.4.1 Vectors and tensors and Lorentz transformations

Focusing on the Lorentz part of the Poincaré transformations offers a good opportunity to introduce concepts and notation that will be used throughout, while at the same time gaining helpful intuition. Remember that a *contravariant vector* V^ν is any object that transforms – under Lorentz transformations – in the same way as the coordinate differentials dx^μ do, that is,

$$V^\mu \mapsto V'^\mu = \Lambda^\mu{}_\nu V^\nu \tag{3.90}$$

On the other hand, a *covariant vector* V_μ transforms as

$$V_\mu \mapsto V'_\mu = \Lambda_\mu{}^\nu V_\nu \tag{3.91}$$

where $\Lambda_\mu{}^\nu = \eta_{\mu\alpha}\eta^{\nu\beta}\Lambda^\alpha{}_\beta$. The Lorentz matrices in these transformation formulas are each others inverses, as can be directly calculated, or understood, from the fact that $V_\mu V^\mu$ must transform as a scalar. One can write

$$(\Lambda^{-1})^\mu{}_\nu = \Lambda_\nu{}^\mu \tag{3.92}$$

for the inverse.

The distinction between contravariant and covariant vectors becomes really effective first in nonflat space-times (and spaces), but even in Minkowski space-time – its constant metric notwithstanding – the coordinate differentials are natural to treat as contravariant. Furthermore, the partial derivatives or gradients are natural to write as covariant vectors. By differentiating the inverse of the Lorentz transformation of the coordinates $x^\mu = \Lambda_\mu{}^\nu x'_\nu$ we get, using the chain rule,

$$\frac{\partial}{\partial x'^\mu} = \frac{\partial x^\nu}{\partial x'^\mu}\frac{\partial}{\partial x^\nu} \quad \Rightarrow \quad \frac{\partial}{\partial x'^\mu} = \Lambda_\mu{}^\nu \frac{\partial}{\partial x^\nu}$$

This can be interpreted as a transformation rule

$$\partial_\mu \mapsto \partial'_\mu = \Lambda_\mu{}^\nu \partial_\nu \tag{3.93}$$

Tensors are objects with several covariant and contravariant indices. The transformation rule generalizes from formulas (3.90) and (3.91). For instance,

$$T^{\rho\sigma}{}_{\mu\nu} \mapsto T'^{\rho\sigma}{}_{\mu\nu} = \Lambda^{\rho}{}_{\alpha}\Lambda^{\sigma}{}_{\beta}\Lambda_{\mu}{}^{\gamma}\Lambda_{\nu}{}^{\delta}T^{\alpha\beta}{}_{\gamma\delta} \tag{3.94}$$

The general rule is that "all tensor indices transform": contravariant according to (3.90) and covariant according to (3.91).

Tensors with p covariant and q contravariant indices, sometimes abbreviated to T_p^q and designated (p,q)-tensors, form a vector space by themselves: linear combinations of (p,q)-tensors are (p,q)-tensors. Tensors can furthermore be multiplied and the result is a new tensor. In more detail, the product of a (p,q)-tensor and an (m,n)-tensor is a $(p+m,q+n)$-tensor. How does one know that the product is a tensor? Well, the rule is: an object is a tensor if it transforms as a tensor. That a product of any two tensors transform as a tensor therefore follows from the transformation rule applied the factors. The idea is captured by the heuristic formula

$$T'^{q}_{p}T'^{n}_{m} = (\Lambda^{-1})^{p}(\Lambda)^{q}T^{q}_{p}(\Lambda^{-1})^{m}(\Lambda)^{n}T^{n}_{m} = (\Lambda^{-1})^{p+m}(\Lambda)^{q+n}T^{q}_{p}T^{n}_{m}$$

The possibility to multiply tensors in well-defined ways, offers the prospect of promoting the set of all vector spaces of tensors to tensor algebras. That leads to very interesting mathematics that is useful in the theory of higher spin fields. We will explore this mathematics in Sections 3.7.8–3.7.11.

3.4.2 The Poincaré algebra

The Poincaré group is a Lie symmetry group, although not of one of the particularly nice types: simple or semisimple. Properties of Lie groups can to a large extent be analyzed by studying the group elements near the identity, and such an analysis leads to the Lie algebra of the group. We will glimpse the theory in Section 3.11.

For the particular theory of the Poincaré group, it suffices to note that the translation part a^{μ} of a transformation is clearly continuous and we can choose to consider an infinitesimal transformation with parameter ϵ^{μ}. Some more thought makes it clear that also the Lorentz transformation matrices $\Lambda_{\mu}{}^{\nu}$ form a continuous group.[22] The corresponding infinitesimal parameter will be denoted by $\lambda_{\mu\nu}$. It is antisymmetric in its indices and thus parametrizes six independent transformations as is appropriate in four space-time dimensions.

There is a standard way of deriving the Poincaré Lie algebra that is often employed in particle physics and quantum field theory.[23] In very brief outline, the procedure is

22 A good down to earth argument can be found in [244], Chapter 1, Sections 1–3.

23 See [18], Chapter 2 for a detailed derivation, or [245], Chapter 10.

the following. Write the transformation operators $T_{\Lambda,a}$ as unitary operators $U(\Lambda, a)$ – thought of as acting on quantum states (see Section 3.3) – obeying the composition rule (3.89). An *infinitesimal Poincaré transformation* is

$$U(1 + \lambda, \epsilon) = 1 + \frac{i}{2}\lambda_{\mu\nu}J^{\mu\nu} - i\epsilon_\mu P^\mu \tag{3.95}$$

in terms of the generators of Lorentz transformations $J^{\mu\nu}$ and translations P^μ. Then one considers the product $U(\Lambda, a)U(\Lambda', a')U^{-1}(\Lambda, a)$ for infinitesimal $U(\Lambda', a')$. This leads to the transformation rules for the generators themselves

$$U(\Lambda, a)J^{\mu\nu}U^{-1}(\Lambda, a) = \Lambda_\rho{}^\mu\Lambda_\sigma{}^\nu(J^{\rho\sigma} - a^\rho P^\sigma + a^\sigma P^\rho) \tag{3.96}$$

$$U(\Lambda, a)P^\mu U^{-1}(\Lambda, a) = \Lambda_\rho{}^\mu P^\rho \tag{3.97}$$

Thus, under Lorentz transformations ($\epsilon^\rho = 0$), $J^{\rho\sigma}$ is a tensor and P^ρ a vector. Under pure translations ($\lambda^{\rho\sigma} = 0$), P^ρ is invariant, but $J^{\rho\sigma}$ transforms as an angular momentum is expected to do under a change of origin.

Next, letting $U(\Lambda, a)$ itself become infinitesimal, one derives the commutators of the Poincaré generators

$$[J_{\mu\nu}, J_{\rho\sigma}] = i(\eta_{\mu\rho}J_{\nu\sigma} - \eta_{\nu\rho}J_{\mu\sigma} + \eta_{\sigma\mu}J_{\rho\nu} - \eta_{\sigma\nu}J_{\rho\mu}) \tag{3.98}$$

$$[J_{\mu\nu}, P_\rho] = i(\eta_{\mu\rho}P_\nu - \eta_{\nu\rho}P_\mu) \tag{3.99}$$

$$[P_\mu, P_\nu] = 0 \tag{3.100}$$

To bring out the physical significance of the generators and the algebra, it is customary to split them into the momentum three-vector $\mathbf{P}$, the energy H, the angular momentum three-vector $\mathbf{J}$ and the boost three-vector $\mathbf{K}$ According to

$$\mathbf{P} = (P^1, P^2, P^3) \tag{3.101}$$

$$H = P^0 \tag{3.102}$$

$$\mathbf{J} = (J^{23}, J^{31}, J^{12}) \equiv (J_1, J_2, J_3) \tag{3.103}$$

$$\mathbf{K} = (J^{10}, J^{20}, J^{30}) \equiv (K_1, K_2, K_3) \tag{3.104}$$

The Lorentz subalgebra then takes the form

$$[J_i, J_j] = i\epsilon_{ijk}J_k \tag{3.105}$$

$$[J_i, K_j] = i\epsilon_{ijk}K_k \tag{3.106}$$

$$[K_i, K_j] = -i\epsilon_{ijk}J_k \tag{3.107}$$

and the rest of the nonzero commutators are

$$[J_i, P_j] = i\epsilon_{ijk}P_k \tag{3.108}$$

$$[K_i, P_j] = iH\delta_{ij} \tag{3.109}$$

$$[K_i, H] = -iP_i \tag{3.110}$$

Of these commutators, (3.105), (3.106) and (3.108) tell us that **J**, **K** and **P** transform as three-vectors under space rotations.

The form of the Poincaré algebra reviewed here can be thought of as a quantum version of the algebra. A trivial representation is given by

$$P_\mu = p_\mu \quad \text{and} \quad J_{\mu\nu} = x_\mu p_\nu - x_\nu p_\mu \quad \text{with} \quad [x_\mu, p_\nu] = i\eta_{\mu\nu} \tag{3.111}$$

An explicit representation that we will often use is (with the Lorentz generators denoted by L)

$$P_\mu = -i\partial_\mu \quad \text{and} \quad L_{\mu\nu} = -i(x_\mu\partial_\nu - x_\nu\partial_\mu) \tag{3.112}$$

It may be convenient to use a different realization, more classically oriented, in order to remove the occurrence of i in the commutators. With the explicit representation,

$$P_\mu = -\partial_\mu \quad \text{and} \quad M_{\mu\nu} = x_\mu\partial_\nu - x_\nu\partial_\mu \tag{3.113}$$

we get for the nonzero commutators

$$[M_{\mu\nu}, M_{\rho\sigma}] = \eta_{\mu\rho}M_{\sigma\nu} - \eta_{\nu\rho}M_{\sigma\mu} - \eta_{\mu\sigma}M_{\rho\nu} + \eta_{\nu\sigma}M_{\rho\mu} \tag{3.114}$$

$$[P_\rho, M_{\mu\nu}] = \eta_{\mu\rho}P_\nu - \eta_{\nu\rho}P_\mu \tag{3.115}$$

This is sometimes referred to as the *coordinate representation* because it corresponds to a transformation of a scalar field φ: $\delta\varphi = -\xi^\mu\partial\varphi$ with $\xi^\mu = \lambda^\mu{}_\nu x^\nu + \epsilon^\mu$.

In preparation for the discussion of representations of the Lorentz group, it is convenient to do one more rewriting of the Lorentz Lie algebra of equations (3.105)–(3.107). Introduce the non-Hermitian operators

$$M_i = \frac{1}{2}(J_i + iK_i) \quad \text{and} \quad N_i = \frac{1}{2}(J_i - iK_i) \tag{3.116}$$

In terms of these generators, the Lorentz algebra breaks up in two conjugated $\mathfrak{su}(2)$ algebras

$$[M_i, M_j] = i\epsilon_{ijk}M_k \tag{3.117}$$

$$[N_i, N_j] = i\epsilon_{ijk}N_k \tag{3.118}$$

$$[M_i, N_j] = 0 \tag{3.119}$$

3.4.3 Connectedness properties of the Lorentz group

Equation (3.86), which reads $\eta_{\mu\nu}\Lambda^{\mu}{}_{\rho}\Lambda^{\nu}{}_{\sigma} = \eta_{\rho\sigma}$, has two important consequences. First, it follows that $|\det \Lambda^{\mu}{}_{\nu}| = \pm 1$. This implies that the Lorentz transformations split into two disconnected sets. The transformations with $|\det \Lambda^{\mu}{}_{\nu}| = +1$ are called *proper Lorentz transformations* as they are continuously connected to the identity. A further split follows from setting $\rho = \sigma = 0$ in the formula. Then

$$-\Lambda^{0}{}_{0}\Lambda^{0}{}_{0} + \sum_{i=1}^{3}\Lambda^{i}{}_{0}\Lambda^{i}{}_{0} = -1 \tag{3.120}$$

from which follows the either $\Lambda^{0}{}_{0} \geq 1$ or $\Lambda^{0}{}_{0} \leq -1$. Lorentz transformations with $\Lambda^{0}{}_{0} \geq 1$ are referred to as *ortochronous Lorentz transformations* as they do not change the sign of the time coordinate. The Lorentz transformations with both $|\det \Lambda^{\mu}{}_{\nu}| = +1$ and $\Lambda^{0}{}_{0} \geq 1$ are called *proper ortochronous Lorentz transformations*. These are continuously connected to the identity. This is also called the *restricted Lorentz group*. The other three components are disconnected from the proper ortochronous, and from each other.[24]

The full group of Lorentz transformation can be recovered by adjoining the proper ortochronous Lorentz transformations with time reversal $\mathcal{T} : x^0 \to -x^0$, space inversion $\mathcal{P} : x^i \to -x^i$ and the combination $\mathcal{PT} : x^{\mu} \to -x^{\mu}$. In the representation theory of the Lorentz group, the operations of time reversal and space inversion must be treated separately.

Two notions of "connectedness"

There are two notions of connectedness involved here. What we have discussed above is the concept of a set – or space – being *connected*. In simple terms: a set is connected if it consists of "one piece" not being the union of two or more disjoint open sets (where "open sets" are defined by the topology of the set).

Then there is the notion simple – or path – connectedness. A space is *simply connected* – or *path connected* – if every path in the space can be continuously contracted to a point. Pictorially: there are no "holes" in the space.

Consider now the group of proper ortochronous Lorentz transformations, denoted by $L_{+}^{\uparrow} = SO(3,1)^{\uparrow}$. It has a *double covering group*, the spin group Spin(3, 1). Here, we will instead work with the isomorphic group of complex 2×2 matrices of unit determinant $SL(2, \mathbf{C})$. The $2 \to 1$ homomorphism (double cover) from $SL(2, \mathbf{C})$ to $L_{+}^{\uparrow}$ can be made explicit in the following way (see, for instance, [244]).

24 It is not possible to pass continuously from a positive real number to a negative real number without passing zero.

For any real four-vector v^μ, we can compute a unique Hermitian 2×2 matrix V

$$V = v^\mu \sigma_\mu = \begin{pmatrix} v^0 + v^3 & v^1 - iv^2 \\ v^1 + iv^2 & v^0 - v^3 \end{pmatrix} \tag{3.121}$$

where σ_μ are the Pauli matrices of (1.11). Conversely, any 2×2 complex matrix V determines a unique four-vector through

$$v^\mu = \frac{1}{2} \operatorname{Tr}(V\sigma^\mu) \tag{3.122}$$

where matrix multiplication is understood in $V\sigma^\mu$. Since the diagonal elements of an Hermitian matrix must be real, the formula (3.122) indeed yields a real four-vector.

In this language, Lorentz transformations are given by

$$V \to \lambda V \lambda^\dagger \tag{3.123}$$

with λ complex 2×2 matrices with unit determinant. The Hermiticity of the matrices V are clearly preserved by such transformations. The square of the four-vector v^ν can be computed as

$$v_\mu v^\mu = (v^0 + v^3)(v^0 - v^3) - (v^1 - iv^2)(v^1 + iv^2) = -\det v \tag{3.124}$$

This determinant is preserved by the transformations (3.123) provided that $|\det \lambda| = 1$.[25] The correspondence between the real 4×4 matrices Λ and the complex 2×2 matrices can be explicated through

$$\lambda v^\mu \sigma_\mu \lambda^\dagger \equiv (\Lambda^\mu{}_\nu(\lambda) v^\nu)\sigma_\mu \tag{3.125}$$

from which it follows that $\Lambda(\lambda\lambda') = \Lambda(\lambda)\Lambda(\lambda')$ for two matrices λ and λ'.

However, it is clear from (3.123) that two matrices λ whose quotient is a phase factor, produce the same transformation of v. Therefore, consistent with (3.125), one may take $\det \lambda = 1$. This yields the group $SL(2, \mathbf{C})$, the *special linear group of* 2×2 *complex matrices*.

The map (group homomorphism) from $SL(2, \mathbf{C})$ to $L_+^\uparrow$ is $2 \to 1$. Intuitively, if λ is a matrix in $SL(2, \mathbf{C})$, so is $-\lambda$ and they both produce the same Lorentz transformation as can be seen from (3.123) or (3.125). The result can be proved[26] by considering the kernel of the transformation (3.123), that is, those transformations for which $V \to \lambda V \lambda^\dagger = V$. It follows that $\Lambda(\lambda) = \Lambda(\lambda')$ implies $\lambda' = \pm\lambda$. Therefore, $SL(2, \mathbf{C})$ covers $L_+^\uparrow$ twice as we run through all 2×2 complex matrices with unit determinant. In the language to be introduced in Section 3.9, this can be written as $L_+^\uparrow = SL(2, \mathbf{C})/Z_2$ where Z_2 is the invariant subgroup consisting of the two elements I and $-I$. The group $SL(2, \mathbf{C})$ is itself simply connected (as are all groups $SL(n, \mathbf{C})$).

25 This condition brings down the number of parameters of λ from eight to six.

26 See [246], Section 17.2.

Group coverings

---

i The fact that the group homomorphism from $SL(2,\mathbf{C})$ to $L_+^\uparrow$ is $2 \to 1$ is also expressed as $SL(2,\mathbf{C})$ being the two-fold *cover* of $L_+^\uparrow$. The concept of one group covering another is naturally generalized to n-fold coverings when the phenomenon arises. Note that the group Lie algebras, being infinitesimal objects, are exactly the same for the group and the cover group.

A much more complete discussion of the topology of the Lorentz group can be found in [18]. This topology – nonsimple connectedness – explains the existence of the spinor representations.

3.5 Poincaré and Lorentz representations and particle states

In brief outline, we will now review the crucial steps in deriving the representations of the Poincaré algebra on quantum one-particle states. This is the method of *induced representation* of Wigner. We will follow [18]. We focus on the logical steps, leaving out the calculational details. These are not difficult, but the surrounding arguments are somewhat involved, and these we want to focus on.

The underlying logic of the method is the following. The Poincaré group, as well as the Lorentz group, is noncompact. This implies that unitary representations – the ones we need in quantum mechanics for the states – are infinite dimensional. The Wigner method handles this by employing the infinite dimensionality of the translation group: there are finite dimensional unitary representations for each value of the momentum p_μ, precisely the representations of the little group. This works well for massive representations where the little group is compact. In the case of zero mass, there are interesting complications, since the little group is also noncompact. The complications are, as may be guessed, connected to gauge invariance.

The one-particle states are taken as eigenstates of the momentum operator P_μ and are denoted by $\Psi_{p,\sigma}$ where p is the momentum and σ is a collective label for all other distinguishing degrees of freedom such as spin, which is what we are actually focusing on here. We thus have $P_\mu \Psi_{p,\sigma} = p_\mu \Psi_{p,\sigma}$. Under Poincaré transformations, these states are supposed to transform as $\Psi \to U(\Lambda, a)\Psi$. For pure translations, we get

$$U(1,a)\Psi_{p,\sigma} = e^{-ip\cdot a}\Psi_{p,\sigma} \tag{3.126}$$

The remaining work resides in working out the effect of pure Lorentz transformations on the states. Acting on the state $\Psi_{p,\sigma}$ with $U(\Lambda, 0) \equiv U(\Lambda)$ should give a state with momenta Λp. Working this out, using equation (3.97) yields[27]

$$P^\mu U(\Lambda)\Psi_{p,\sigma} = \Lambda^\mu{}_\nu p^\nu U(\Lambda)\Psi_{p,\sigma} \tag{3.127}$$

---

27 Or rather, with an inverse Lorentz matrix, see formula (3.92).

This means that $U(\Lambda)\Psi_{p,\sigma}$ must be a linear combination of states $\Psi_{\Lambda p,\bar{\sigma}}$

$$U(\Lambda)\Psi_{p,\sigma} = \sum_{\bar{\sigma}} C_{\bar{\sigma}\sigma}(\Lambda, p)\Psi_{\Lambda p,\bar{\sigma}} \tag{3.128}$$

In general, it may be that the matrix $C_{\bar{\sigma}\sigma}$ can be broken up into block-diagonal form. Each such block then corresponds to an irreducible representation, and those are the ones that we focus on finding.

The quantities $p{\cdot}p = p^2$ and sign (p^0) are invariant under Lorentz transformations. We now choose $p^0 > 0$ and $p^2 \leq 0$, the last clause which splits up into massive ($p^2 < 0$) and massless ($p^2 = 0$) cases, respectively. Due to the invariance, one can for each value of $p^2 \leq 0$ and positive sign (p^0) choose a *standard momentum* k^μ and write any other momentum within the class[28] as $p^\mu = L^\mu{}_\nu k^\nu$ where $L^\mu{}_\nu$ is a Lorentz transformation that depends on p^μ and also on the choice of k^μ. Having done this, it is natural to define – due to equation (3.127) – the state $\Psi_{p,\sigma}$ as the corresponding Lorentz transformation of the *standard state* $\Psi_{k,\sigma}$

$$\Psi_{p,\sigma} = N(p)U(L(p))\Psi_{k,\sigma} \tag{3.129}$$

where $N(p)$ is a normalization factor. The normalization is discussed in detail in [18]. We will follow this reference and choose

$$N(p) = \sqrt{k^0/p^0} \tag{3.130}$$

Orbits of the Lorentz group

The possible combinations of p^2 and sign (p^0) can be worked out taking the condition $p^2 = -m^2$ into account. We then get the

- *time-like* orbits with $p^2 < 0$ and disconnected branches $p^0 > 0$ or $p^0 < 0$,
- *light-like* orbits with $p^2 = 0$ and branches $p^0 > 0$ and $p^0 < 0$ connected at $p^0 = 0$,
- *space-like* connected orbit with $p^2 > 0$,
- *vacuum* orbit with $p^\mu = 0$.

Time-like orbits correspond to massive particles and light-like orbits to massless particles.

The next step is to figure out the transformation properties of the standard state, and to reduce the transformation properties of the general state to the properties of the standard state. In order to do that, one applies a Lorentz transformation $U(\Lambda)$ to both sides of the definition (3.129), and using the group multiplication law, rewrites the result in the form

$$U(\Lambda)\Psi_{p,\sigma} = N(p)U(L(\Lambda p))U(L^{-1}(\Lambda p)\Lambda L(p))\Psi_{k,\sigma} \tag{3.131}$$

28 Those are indeed equivalence classes.

The second U-factor produces a Lorentz transformation that takes the standard momentum k to $p = L(p)k$, then to Λp and finally back to k. It therefore belongs to the subgroup of Lorentz transformations W that leave the standard momentum invariant, that is,

$$W^{\mu}{}_{\nu}k^{\nu} = k^{\mu} \tag{3.132}$$

These Lorentz transformations define the so-called *little group*. On the standard states, the transformation law (3.128) is realized as[29]

$$U(W)\Psi_{k,\sigma} = \sum_{\bar{\sigma}} D_{\bar{\sigma}\sigma}(W)\Psi_{k,\bar{\sigma}} \tag{3.133}$$

The matrices $D_{\sigma'\sigma}$ are precisely what we want to find.

Denote the particular little group transformation occurring in (3.133) as

$$W(\Lambda, p) = L^{-1}(\Lambda p)\Lambda L(p) \tag{3.134}$$

Using this particular transformation in the transformation formula (3.131) and using the definition (3.129) connecting general states and standard states, the transformation law (3.131) becomes

$$U(\Lambda)\Psi_{p,\sigma} = \frac{N(p)}{N(\Lambda p)} \sum_{\bar{\sigma}} D_{\bar{\sigma}\sigma}(W(\Lambda, p))\Psi_{\Lambda p,\bar{\sigma}} \tag{3.135}$$

Comparing to the transformation (3.128) we see that this is precisely what we aimed for.

Finally, according to the composition rule (3.89) we have $U(\Lambda, a) = U(1, a)U(\Lambda, 0)$, and we can combine the transformation rules (3.126) and (3.135) into

$$U(\Lambda, a)\Psi_{p,\sigma} = e^{-i(\Lambda p)\cdot a} \sqrt{(\Lambda p)^0/p^0} \sum_{\bar{\sigma}} D_{\bar{\sigma}\sigma}(W(\Lambda, p))\Psi_{\Lambda p,\bar{\sigma}} \tag{3.136}$$

where we have used the normalization (3.130). This transformation law generalizes in an obvious way to many-particle states. Note also that the states can just as well be labelled by the three-vector **p** part of p. So far, the formula (3.136) is general. In the following, it will be specialized to the two most important cases: first massive, and then, massless representations.

This method of constructing representations of the Poincaré group from representations of the little group, is called the *method of induced representations*. The formula (3.136) for a general Poincaré transformation is referred to as *orbit completion*.

29 Put $p = k$ and $\Lambda = W$, and k becomes redundant in $C_{\bar{\sigma}\sigma}(W, k)$, then denote by $D_{\bar{\sigma}\sigma}(W)$.

Massive representations

Massive representations are defined by $p^2 = -m^2$ and $p^0 > 0$. The standard momentum is taken as $k^\mu = (0,0,0,m)$.[30] The little group is SO(3), the group of rotations in three spatial dimensions. The irreducible representations can be given by matrices $D^{(j)}_{\bar\sigma\sigma}(R)$ of dimension $2j+1$ for $j = 0, \frac{1}{2}, 1, \ldots$ (the indices σ run $j, j-1, \ldots, -j$). Explicitly, for infinitesimal rotations $R_{ik} = \delta_{ik} + \theta_{ik}$ with $\theta_{ik} = -\theta_{ki}$, we have

$$D^{(j)}_{\bar\sigma\sigma}(1+\theta) = \delta_{\bar\sigma\sigma} + \frac{i}{2}\theta_{ik}\left(J^{(j)}_{ik}\right)_{\bar\sigma\sigma} \tag{3.137}$$

$$\left(J^{(j)}_{23} \pm iJ^{(j)}_{31}\right)_{\bar\sigma\sigma} = \delta_{\bar\sigma,\sigma\pm1}\sqrt{(j\mp\sigma)(j\pm\sigma+1)} \tag{3.138}$$

$$\left(J^{(j)}_{12}\right)_{\bar\sigma\sigma} = \sigma\delta_{\bar\sigma\sigma} \tag{3.139}$$

It can be shown that the little group elements, the *Wigner rotations* $W(\Lambda, p)$, has the following property: when Λ is a three-dimensional rotation R, then $W(R,p) = R$. It does not depend on the momentum of the particle, and the states of a moving particle transform under spatial rotations as the particle at rest. The formulas are the same as in nonrelativistic quantum mechanics.

Example 1 (A reminder on SO(3) in quantum mechanics). We recognize the formulas for massive representations, given in the box above, as the angular momentum operators $\mathbf{J}$ acting on states $|j,m\rangle$. Representing $\mathbf{J}$ as $\{J_3, J_\pm = J_1 \pm iJ_2\}$ then we have, with m running over the values $j, j-1, \ldots, -j$,

$$J_3|j,m\rangle = |j,m\rangle m \tag{3.140}$$

$$J_\pm|j,m\rangle = |j,m\pm1\rangle\sqrt{j(j+1)-m(m\pm1)} \tag{3.141}$$

$$\mathbf{J}^2|j,m\rangle = |j,m\rangle j(j+1) \tag{3.142}$$

The two simplest nontrivial examples are with $j = 1/2$ and $j = 1$. For $j = 1/2$, we get precisely $J_i = \frac{1}{2}\sigma_i$ in terms of the 2×2 Pauli matrices acting on the spin 1/2 states

$$|\tfrac{1}{2},\tfrac{1}{2}\rangle = \begin{pmatrix}1\\0\end{pmatrix} \quad\text{and}\quad |\tfrac{1}{2},-\tfrac{1}{2}\rangle = \begin{pmatrix}0\\1\end{pmatrix} \tag{3.143}$$

with

$$J_3 = \begin{pmatrix}1 & 0\\0 & -1\end{pmatrix} \qquad J_+ = \begin{pmatrix}0 & 1\\0 & 0\end{pmatrix} \qquad J_- = \begin{pmatrix}0 & 0\\1 & 0\end{pmatrix} \tag{3.144}$$

It is easy to convince one-self that it all works as expected and that the algebra of generators, for instance $[J_+, J_-] = 2J_3$, is satisfied. J_+ thus raises the spin and J_- lowers it.

30 Since it is conventional to consider boosts in the 3 direction, it is convenient to have the 0 and 3 directions adjacent in concrete vectors and matrices, therefore, the ordering 1, 2, 3, 0.

In the case $j = 1$, the 3×3 matrices become

$$J_3 = \begin{pmatrix} 1 & 0 & 0 \\ 0 & 0 & 0 \\ 0 & 0 & -1 \end{pmatrix} \qquad J_+ = \begin{pmatrix} 0 & \sqrt{2} & 0 \\ 0 & 0 & \sqrt{2} \\ 0 & 0 & 0 \end{pmatrix} \qquad J_- = \begin{pmatrix} 0 & 0 & 0 \\ \sqrt{2} & 0 & 0 \\ 0 & \sqrt{2} & 0 \end{pmatrix} \tag{3.145}$$

and the basis states upon which they act are (written as row vectors)

$$|1,1\rangle = (1 \quad 0 \quad 0) \qquad |1,0\rangle = (0 \quad 1 \quad 0) \qquad |1,-1\rangle = (0 \quad 0 \quad 1) \tag{3.146}$$

These should not be confused with three-dimensional space-time basis vectors. For a space-time representation, one should instead use matrices $(R_k)_{ij} = -i\epsilon_{ijk}$. These work out to

$$R_1 = -i\begin{pmatrix} 0 & 0 & 0 \\ 0 & 0 & 1 \\ 0 & -1 & 0 \end{pmatrix} \qquad R_2 = -i\begin{pmatrix} 0 & 0 & -1 \\ 0 & 0 & 0 \\ 1 & 0 & 0 \end{pmatrix} \qquad R_3 = -i\begin{pmatrix} 0 & 1 & 0 \\ -1 & 0 & 0 \\ 0 & 0 & 0 \end{pmatrix} \tag{3.147}$$

These matrices generate infinitesimal rotations around the k axes with angles θ_k respectively. This can be seen by multiplying space basis vectors

$$\mathbf{e}_1 = \begin{pmatrix} 1 \\ 0 \\ 0 \end{pmatrix} \qquad \mathbf{e}_2 = \begin{pmatrix} 0 \\ 1 \\ 0 \end{pmatrix} \qquad \mathbf{e}_3 = \begin{pmatrix} 0 \\ 0 \\ 1 \end{pmatrix} \tag{3.148}$$

with matrices $I - i\theta_k R_k$. The relation between these two realizations of the SO(3) rotation group will come into play when we study spin 1 quantum fields. Another property of the matrices that we need is

$$(R^2)^i{}_j = \sum_k (R_k)^i{}_l (R_k)^l{}_j = 2\delta^i{}_j \tag{3.149}$$

Textbook references are for instance [26], Section 27 or [33], Chapter 17. ◂

3.5.1 From state transformations to field transformations

Weinberg, in [18], has a detailed discussion on how to build relativistic quantum fields from relativistic quantum one-particle states. As the logic itself is very interesting from the point of view of understanding the role of the Poincaré symmetry in physics, we will outline the main steps here, again leaving out the computational details.

The question we want to answer is the following. *What is the relation between the little group transformations – and finite dimensional unitary representations – for the states, and the Lorentz group transformations – and finite dimensional nonunitary representations – for the fields? What, in fact, is the relation between the representations of the Lorentz and Poincaré groups?*

The question is actually acute, since the representations for the states are unitary while the representations for the fields are not unitary. This potentially confusing issue is important to clarify. As we saw in the historical chapter, this was indeed a confused issue in the earlier days of quantum field theory.

Weinberg's starting point is the formula for the interactions[31]

$$V(t) = \int d^3x \mathcal{H}(x, y, z, t) \tag{3.150}$$

where the Hamiltonian density $\mathcal{H}(x)$ is built from the creation and annihilation operators corresponding to the particle states of the theory considered.[32] Interactions, modelled as $V(t)$, will produce a Lorentz-invariant S-matrix if $\mathcal{H}(x)$ is a scalar under Poincaré transformations

$$U(\Lambda, a)\mathcal{H}(x)U^{-1}(\Lambda, a) = \mathcal{H}(\Lambda x + a) \tag{3.151}$$

It must also satisfy the requirement

$$[\mathcal{H}(x), \mathcal{H}(x')] = 0 \quad \text{for } (x - x')^2 \geq 0 \tag{3.152}$$

The Poincaré invariance requirement is satisfied by building the Hamiltonian density, not out of creation and annihilation operators directly, but out of certain linear combinations of them, called *quantum fields*

$$\psi_l^+(x) = \sum_\sigma \int d^3p\, u_l(x; \mathbf{p}, \sigma)\alpha(\mathbf{p}, \sigma) \tag{3.153}$$

$$\psi_l^-(x) = \sum_\sigma \int d^3p\, v_l(x; \mathbf{p}, \sigma)\alpha^\dagger(\mathbf{p}, \sigma) \tag{3.154}$$

Here, the $\psi_l^+(x)$ are annihilation fields, and $\psi_l^-(x)$ are creation fields. The intuition is that the oscillators annihilate or create states of definite momentum and spin that are linearly combined with configuration space "wave function" coefficients u_l and v_l. These functions carry Lorentz indices l that will be related to the state labels σ.[33] Denoting the vacuum with Ψ_0, a one-particle state is created out of the vacuum

$$\Psi_{p,\sigma} = \alpha_{p,\sigma}^\dagger \Psi_0 \tag{3.155}$$

---

31 Compare to Section 2.6.1 in the historical chapter.

32 For instance, in QED: operators creating and destroying photons, electrons and positrons, combined to model the fundamental (cubic) interactions of the theory. In general, it is a weighted sum of products of creation and annihilation operators.

33 The + and – notation refers to the signs in the exponentials of $ip \cdot x$ soon to appear. We suppress, for simplicity, any further labeling of particle types in case there are several particles with the same spin.

The Hamiltonian is written as a sum of products of annihilation and creation fields, corresponding to the various basic interaction vertices of the theory, with constant coupling coefficients.[34] Requiring such a Hamiltonian to transform as a scalar, force the fields to transform according to

$$U(\Lambda,a)\psi_l^{\pm}(x)U^{-1}(\Lambda,a)=\sum_{\bar{l}}D_{l\bar{l}}(\Lambda^{-1})\psi_{\bar{l}}^{\pm}(\Lambda x+a) \tag{3.156}$$

where the $D_{l\bar{l}}$ matrices must form a representation of the Lorentz group.[35] What we want to do now is to understand how these nonunitary finite dimensional matrices $D_{l\bar{l}}(\Lambda^{-1})$ are related to the unitary finite dimensional matrices $D_{\sigma\bar{\sigma}}(W)$ of the little group.[36]

From the transformation law (3.136) for the one-particle states, follows the transformation formulas for the annihilation and creation operators

$$U(\Lambda,a)\alpha(\mathbf{p},\sigma)U^{-1}(\Lambda,a)=e^{i(\Lambda p)\cdot a}\sqrt{\frac{(\Lambda p)^0}{p^0}}\sum_{\bar{\sigma}}D^{(j)}_{\sigma\bar{\sigma}}(W^{-1}(\Lambda,p))\alpha(\mathbf{p}_\Lambda,\bar{\sigma}) \tag{3.157}$$

$$U(\Lambda,a)\alpha^{\dagger}(\mathbf{p},\sigma)U^{-1}(\Lambda,a)=e^{-i(\Lambda p)\cdot a}\sqrt{\frac{(\Lambda p)^0}{p^0}}\sum_{\bar{\sigma}}D^{(j)^*}_{\sigma\bar{\sigma}}(W^{-1}(\Lambda,p))\alpha^{\dagger}(\mathbf{p}_\Lambda,\bar{\sigma}) \tag{3.158}$$

Some changes of notation have been introduced here as compared to (3.136). The unitarity of the $D_{\sigma\bar{\sigma}}$ matrices has been used to rewrite $D_{\bar{\sigma}\sigma}(W)=D^*_{\sigma\bar{\sigma}}(W^{-1})$. Furthermore, the $D_{\sigma\bar{\sigma}}$ matrices have been labeled by the spin j. Finally, the momentum label p has been changed to the three momentum $\mathbf{p}$ (since the states are on-shell anyway) and $\mathbf{p}_\Lambda$ stands for the three-vector part of Λp.[37]

Now, the formulas (3.157) and (3.158) are consequences of the analysis of the unitary transformations of the Poincaré group as expressed in the language of operators creating and annihilating quantum states. The formulas (3.153) and (3.154) are definitions, and the formula (3.156) is a requirement. Combining them yield[38]

$$\sum_{\bar{\sigma}}u_{\bar{l}}(\Lambda x+a;\mathbf{p}_\Lambda,\bar{\sigma})D^{(j)}_{\bar{\sigma}\sigma}(W(\Lambda,p))=\sqrt{\frac{p^0}{(\Lambda p)^0}}\sum_{l}D_{\bar{l}l}(\Lambda)\exp(i(\Lambda p)\cdot a)u_l(x;\mathbf{p},\sigma) \tag{3.159}$$

34 The coupling constants should be Lorentz covariant in a sense made clear in [18].

35 Although there could be different matrices for the creation and annihilation fields, it is always possible make choices as to make them equal; see [18].

36 Remember the $l\bar{l}$ are indices for field components while the $\sigma\bar{\sigma}$ label representations.

37 The formulas (3.157) and (3.158) are adjoints of each other. Note that the adjoint in the Hilbert space of states means complex conjugation for the c-number matrices D.

38 There is a certain amount of algebra involved here. Note in particular that there is an implicit transposition of the $\sigma\bar{\sigma}$ indices in order that coefficient functions $u_l(x;\mathbf{p},\sigma)$ and $v_l(x;\mathbf{p},\sigma)$ can be treated as row vectors in σ-space. Treating the coefficient functions as column vectors in l space, the formulas then become matrix equations with σ-labeled columns and l-labeled rows. This is convenient for practical calculation.

$$\sum_{\bar{\sigma}} v_{\bar{l}}(\Lambda x + a; \mathbf{p}_\Lambda, \bar{\sigma}) D^{(j)^*}_{\bar{\sigma}\sigma}(W(\Lambda, p)) = \sqrt{\frac{p^0}{(\Lambda p)^0}} \sum_l D_{\bar{l}l}(\Lambda) \exp(-i(\Lambda p) \cdot a) v_l(x; \mathbf{p}, \sigma) \tag{3.160}$$

First, specializing to translations with $\Lambda = 1$ the formulas show that the fields must be Fourier transforms

$$\psi_l^+(x) = \frac{1}{(2\pi)^{3/2}} \sum_\sigma \int d^3p\, e^{ip\cdot x} u_l(\mathbf{p}, \sigma) \alpha(\mathbf{p}, \sigma) \tag{3.161}$$

$$\psi_l^-(x) = \frac{1}{(2\pi)^{3/2}} \sum_\sigma \int d^3p\, e^{-ip\cdot x} v_l(\mathbf{p}, \sigma) \alpha^\dagger(\mathbf{p}, \sigma) \tag{3.162}$$

Using this, and specializing to Lorentz transformations, we get

$$\sum_{\bar{\sigma}} u_{\bar{l}}(\mathbf{p}_\Lambda, \bar{\sigma}) D^{(j)}_{\bar{\sigma}\sigma}(W(\Lambda, p)) = \sqrt{\frac{p^0}{(\Lambda p)^0}} \sum_l D_{\bar{l}l}(\Lambda) u_l(\mathbf{p}, \sigma) \tag{3.163}$$

$$\sum_{\bar{\sigma}} v_{\bar{l}}(\mathbf{p}_\Lambda, \bar{\sigma}) D^{(j)^*}_{\bar{\sigma}\sigma}(W(\Lambda, p)) = \sqrt{\frac{p^0}{(\Lambda p)^0}} \sum_l D_{\bar{l}l}(\Lambda) v_l(\mathbf{p}, \sigma) \tag{3.164}$$

These formulas can then be further specialized to boost and rotations.

Boost formulas

For *boosts*, choose $\mathbf{p} = 0$ corresponding to a state at rest, and let Λ be the standard boost $L(q)$ that takes the state to the four-momentum q^μ, that is, $q = L(q)p$. Then $L(p) = 1$ and the little group element becomes $W(\Lambda, p) = L^{-1}(\Lambda p)\Lambda L(p) = 1$. The formulas (3.163) and (3.164) become

$$u_{\bar{l}}(\mathbf{q}, \sigma) = \sqrt{\frac{m}{q^0}} \sum_l D_{\bar{l}l}(L(q)) u_l(0, \sigma) \tag{3.165}$$

$$v_{\bar{l}}(\mathbf{q}, \sigma) = \sqrt{\frac{m}{q^0}} \sum_l D_{\bar{l}l}(L(q)) v_l(0, \sigma) \tag{3.166}$$

These formulas relate the functions $u_l(\mathbf{p}, \sigma)$ and $v_l(\mathbf{p}, \sigma)$ at arbitrary three-momentum $\mathbf{p}$ to the rest-frame objects $u_l(0, \sigma)$ and $v_l(0, \sigma)$. As we will see, these are the "germs out of which the fields are grown". So far there are no conditions on $u_l(0, \sigma)$ and $v_l(0, \sigma)$. Such will come when rotations are considered.

Rotation formulas

For *rotations*, again take $\mathbf{p} = 0$ but now choose Λ to be a rotation R. Then $\mathbf{p}_\Lambda = 0$ and the little group element is $W(\Lambda, p) = R$. The formulas (3.163) and (3.164) become

$$\sum_{\bar{\sigma}} u_{\bar{l}}(0, \bar{\sigma}) D^{(j)}_{\bar{\sigma}\sigma}(R) = \sum_l D_{\bar{l}l}(R) u_l(0, \sigma) \tag{3.167}$$

$$\sum_{\bar{\sigma}} v_{\bar{l}}(0,\bar{\sigma}) D^{(j)^*}_{\bar{\sigma}\sigma}(R) = \sum_{l} D_{\bar{l}l}(R) v_l(0,\sigma) \tag{3.168}$$

These are formulas that relate little group representations – which are rotations of the states labeled by σ – with Lorentz group rotations of the functions labeled by "spacetime" indices l. It is convenient to bring this aspect to the fore a little by writing the equations as

$$\sum_{\bar{\sigma}} u_{\bar{l}}(0,\bar{\sigma}) \mathbf{J}^{(j)}_{\bar{\sigma}\sigma} = \sum_{l} \mathcal{R}_{\bar{l}l} u_l(0,\sigma) \tag{3.169}$$

$$\sum_{\bar{\sigma}} v_{\bar{l}}(0,\bar{\sigma}) \mathbf{J}^{(j)*}_{\bar{\sigma}\sigma} = -\sum_{l} \mathcal{R}_{\bar{l}l} v_l(0,\sigma) \tag{3.170}$$

The minus sign appearing in the last equation comes from the complex conjugated matrices of the rotation group; see formulas (3.147). In the next section, we will work out the consequences of these formulas in detail for the interesting case of spin 1.

Bottom line on unitary vs. nonunitary representations

! In quantum theory, wave functions are states, and must therefore transform unitarily. Fields, however, are operators and need not transform unitarily.

In more detail: The c-number wave functions $u_l(\mathbf{p},\sigma)e^{ip\cdot x}$ and $v_l(\mathbf{p},\sigma)e^{-ip\cdot x}$ transform unitarily under the Poincaré group. They are coefficient functions that relate the state creation and annihilation operators $a^\dagger(\mathbf{p},\sigma)$ and $a(\mathbf{p},\sigma)$ to the quantum fields $\psi_l^+(x)$ and $\psi_l^-(x)$. The quantum fields transform as finite dimensional nonunitary representations of the Lorentz group. In short: states transform unitarily, operators need not, c-number wave functions straddle the gap.

3.5.2 The cardinal example: massive spin 1 fields

Let us refer to the pairs of formulas (3.165)–(3.166) and (3.169)–(3.170) relating Poincaré and Lorentz representations as the *boost* and *rotation* formulas, respectively. As a backdrop for spin 1, we start by applying the formulas to the case of a spin 0 scalar field. Then the label σ takes just one value 0 and can be dropped, and all the representation matrices are identity matrices. The rotation formulas trivialize. The constants $u(0)$ and $v(0)$ in the boost formulas can be taken as $(2m)^{-1/2}$. These formulas then give

$$u(\mathbf{p}) = (2p^0)^{-1/2} \quad \text{and} \quad v(\mathbf{p}) = (2p^0)^{-1/2} \tag{3.171}$$

and we immediately get the annihilation and creation fields

$$\phi^+(x) = \frac{1}{(2\pi)^{3/2}} \int d^3p (2p^0)^{-1/2} e^{ip\cdot x} \alpha(\mathbf{p}) \tag{3.172}$$

$$\phi^-(x) = \frac{1}{(2\pi)^{3/2}} \int d^3p (2p^0)^{-1/2} e^{-ip\cdot x} \alpha^\dagger(\mathbf{p}) \tag{3.173}$$

The combination $\phi(x) = \phi^+(x)+\phi^-(x)$ is recognizable as a quantum field for a neutral, that is, charge-less, scalar field, although the normalization is different from what is perhaps more common in textbooks.[39]

For spin 1, we expect the coefficient wave-functions in the quantum fields to be four-vector functions $u^\mu(\mathbf{p},\sigma)$ and $v^\mu(\mathbf{p},\sigma)$. That is, the l indices are taken as space-time vector indices μ, and consequently the Lorentz representation matrices $D_{\bar{l}l}(\Lambda)$ are taken as $D(\Lambda)^\mu{}_\nu = \Lambda^\mu{}_\nu$ as is appropriate for vectors. Therefore, we can write $D_{\bar{l}l}(\Lambda(p)) \to L(p)^\mu{}_\nu$. The boost formulas specialize to

$$u^\mu(\mathbf{p},\sigma) = \sqrt{\frac{m}{p^0}} L(p)^\mu{}_\nu u^\nu(0,\sigma) \tag{3.174}$$

$$v^\mu(\mathbf{p},\sigma) = \sqrt{\frac{m}{p^0}} L(p)^\mu{}_\nu v^\nu(0,\sigma) \tag{3.175}$$

The rotation formulas will provide more interesting information. They now read

$$\sum_{\bar{\sigma}} u^\mu(0,\bar{\sigma})\mathbf{J}^{(j)}_{\bar{\sigma}\sigma} = \mathcal{R}^\mu{}_\nu u^\nu(0,\sigma) \tag{3.176}$$

$$\sum_{\bar{\sigma}} v^\mu(0,\bar{\sigma})\mathbf{J}^{(j)*}_{\bar{\sigma}\sigma} = -\mathcal{R}^\mu{}_\nu v^\nu(0,\sigma) \tag{3.177}$$

where $\mathcal{R}^\mu{}_\nu$ denote the rotation generators in the vector representation. We have kept the spin j arbitrary on the left (little group) side of the formulas since we want to decide the possible values for j. To proceed, we write down the explicit form for $\mathcal{R}^\mu{}_\nu$ which is (see formulas (3.147))

$$(\mathcal{R}_k)^0{}_0 = (\mathcal{R}_k)^0{}_i = (\mathcal{R}_k)^i{}_0 = 0 \quad \text{and} \quad (\mathcal{R}_k)^i{}_j = -i\epsilon_{ijk} \tag{3.178}$$

Let us now focus on the formula (3.176) for the u coefficient. We observe that it is three-vector equation. The idea is to "rotate" the left- and right-hand sides of the equation, for each three-direction k, with $J_k^{(j)}$ and use the formula again for the k-direction to write the right-hand sides as a product of $\mathcal{R}_k$ matrices. That sequence of manipulations result in

$$\sum_{\bar{\sigma}\sigma} u^\mu(0,\bar{\sigma})(J_k^{(j)})_{\bar{\sigma}\sigma}(J_k^{(j)})_{\sigma\sigma'} = (\mathcal{R}_k)^\mu{}_\nu(\mathcal{R}_k)^\nu{}_\rho u_\rho(0,\sigma') \tag{3.179}$$

Next, summing over k and using equations (3.178) yields

$$\sum_{\bar{\sigma}} u^0(0,\bar{\sigma})(\mathbf{J}^{(j)})^2_{\bar{\sigma}\sigma'} = 0 \tag{3.180}$$

---

39 I am following Weinberg in [18] closely.

$$\sum_{\bar{\sigma}} u^i(0,\bar{\sigma})(\mathbf{J}^{(j)})^2_{\bar{\sigma}\sigma'} = 2u^i(0,\sigma') \tag{3.181}$$

where we have also used (3.149). Then since $(\mathbf{J}^{(j)})^2_{\bar{\sigma}\sigma'} = j(j+1)\delta_{\bar{\sigma}\sigma'}$ we get

$$j(j+1)u^0(0,\sigma') = 0 \tag{3.182}$$
$$j(j+1)u^i(0,\sigma') = 2u^i(0,\sigma') \tag{3.183}$$

The only solutions to these equations are

$$j = 0 \quad \text{with} \quad u^0 \neq 0 \quad \text{and} \quad u^i = 0 \tag{3.184}$$
$$j = 1 \quad \text{with} \quad u^0 = 0 \quad \text{and} \quad u^i \neq 0 \tag{3.185}$$

Analyzing the equation (3.177) for the v coefficient yields the same constraints on j and $v^\mu(0,\sigma)$. A vector field can thus describe a spin 0 particle or a spin 1 particle.

Spin zero

For spin 0, we drop the σ label and the boost equations yield

$$u^\mu(\mathbf{p}) = \sqrt{\frac{m}{p^0}} L(p)^\mu{}_0 u^0(0) \tag{3.186}$$

$$v^\mu(\mathbf{p}) = \sqrt{\frac{m}{p^0}} L(p)^\mu{}_0 v^0(0) \tag{3.187}$$

where $L(p)^\mu{}_0$ is a boost that takes the standard momentum $k^\mu = (0,0,0,m)$ to p^μ. This means that $L(p)^\mu{}_0 m = p^\mu$. Then

$$u^\mu(\mathbf{p}) = (mp^0)^{-1/2} p^\mu u^0(0) \tag{3.188}$$
$$v^\mu(\mathbf{p}) = (mp^0)^{-1/2} p^\mu v^0(0) \tag{3.189}$$

These formulas indicate that the field is a derivative of a scalar field. Choosing $u^0(0) = i(m/2)^{1/2}$ and $v^0(0) = -i(m/2)^{1/2}$ yields

$$u^\mu(\mathbf{p}) = i(2p^0)^{-1/2} p^\mu \tag{3.190}$$
$$v^\mu(\mathbf{p}) = -i(2p^0)^{-1/2} p^\mu \tag{3.191}$$

Comparing to the formulas for a scalar field (3.171), we get the identification for a spin zero vector field $\phi^\mu = \partial^\mu \phi$.

Spin one

For spin 1, we have $u^0 = v^0 = 0$ and we want to determine $u^i(0,\sigma)$ and $v^i(0,\sigma)$ for $\sigma = 1, 0, -1$. Working out the consequences of the rotation equation (3.176) for $J_3^{(1)}$, yield the following partial determination of the $u^\mu(0,\sigma)$ coefficients:

$$u^\mu(0,1) = \begin{pmatrix} u^1 \\ iu^1 \\ 0 \\ 0 \end{pmatrix} \qquad u^\mu(0,0) = \begin{pmatrix} 0 \\ 0 \\ u^3 \\ 0 \end{pmatrix} \qquad u^\mu(0,-1) = \begin{pmatrix} u^2 \\ -iu^2 \\ 0 \\ 0 \end{pmatrix} \tag{3.192}$$

where u^1, u^2 and u^3 are undetermined constants. Next, working out the raising and lowering cases of the rotation equation (3.176) result in fixing $u^1 = -u^2 = -2^{-1/2}u^3$. Then u^3 can be normalized to $(2m)^{-1/2}$.

The corresponding calculations can then be done for the $v^\mu(0,\sigma)$ coefficients using the rotation formula (3.177). Doing that leads to $v^\mu(0,1) = -u^\mu(0,-1)$, $v^\mu(0,0) = u^\mu(0,0)$ and $v^\mu(0,-1) = -u^\mu(0,1)$. Therefore, $u^\mu(0,\sigma)^* = v^\mu(0,\sigma)$. It is conventional to introduce *polarization vectors* through the definition $e^\mu(0,\sigma) = \sqrt{2m}u^\mu(0,\sigma)$. These then become

$$e^\mu(0,1) = -\frac{1}{\sqrt{2}}\begin{pmatrix} 1 \\ i \\ 0 \\ 0 \end{pmatrix} \qquad e^\mu(0,0) = \begin{pmatrix} 0 \\ 0 \\ 1 \\ 0 \end{pmatrix} \qquad e^\mu(0,-1) = \frac{1}{\sqrt{2}}\begin{pmatrix} 1 \\ -i \\ 0 \\ 0 \end{pmatrix} \tag{3.193}$$

Finally, using the boost equations (3.174) and (3.175) we can be put everything together into a massive spin 1 quantum field

$$\phi^\mu(x) = \frac{1}{(2\pi)^{3/2}} \int \frac{d^3p}{\sqrt{2p^0}} \left(e^\mu(\mathbf{p},\sigma)\alpha(\mathbf{p})e^{ip\cdot x} + e^{\mu *}(\mathbf{p},\sigma)e^{-ip\cdot x}\alpha^\dagger(\mathbf{p})\right) \tag{3.194}$$

where $e^\mu(\mathbf{p},\sigma) = L(\Lambda)^\mu{}_\nu e^\nu(0,\sigma)$. From the general formulas (3.161) and (3.162) as applied to vector fields, we recognize the identifications

$$u^\mu(\mathbf{p},\sigma) = (2p^0)^{-1/2} e^\mu(\mathbf{p},\sigma) \quad \text{and} \quad v^\mu(\mathbf{p},\sigma) = (2p^0)^{-1/2} e^{\mu *}(\mathbf{p},\sigma) \tag{3.195}$$

This "normalization" will be used also in the massless case, although the polarization vectors will be different in a significant way. Also note that the field $\phi^\mu(x)$ is divergence-free as it should be, since $p_\mu e^\mu(\mathbf{p},\sigma) = k_\mu e^\mu(0,\sigma) = 0$ due to Lorentz invariance.

3.5.3 The little group for zero mass

Massless representations are defined by $p^2 = 0$ and $p^0 > 0$. The standard momentum is taken as $k^\mu = (0,0,1,1)$.[40] The little group in the massless case is ISO(2), the inhomogeneous group of translations and rotations in two space dimensions. This group

40 Remember the concrete index ordering 1, 2, 3, 0.

is also referred to as the *Euclidean group* in two dimensions and denoted by E_2. The general little group element can be written as[41]

$$W(\theta,\alpha,\beta) = S(\alpha,\beta)R(\theta) \tag{3.196}$$

where the matrices S and R are given by

$$S^{\mu}{}_{\nu}(\alpha,\beta) = \begin{pmatrix} 1 & 0 & -\alpha & \alpha \\ 0 & 1 & -\beta & \beta \\ \alpha & \beta & 1-\zeta & \zeta \\ \alpha & \beta & -\zeta & 1+\zeta \end{pmatrix} \tag{3.197}$$

$$R^{\mu}{}_{\nu}(\theta) = \begin{pmatrix} \cos\theta & \sin\theta & 0 & 0 \\ -\sin\theta & \cos\theta & 0 & 0 \\ 0 & 0 & 1 & 0 \\ 0 & 0 & 0 & 1 \end{pmatrix} \tag{3.198}$$

where $\zeta = (\alpha^2+\beta^2)/2$ in $S(\alpha,\beta)$.

The form of the massless little group

? An argument leading to this form of the massless little group Lorentz transformations goes like this. Consider a time-like four-vector $t^{\mu} = (0,0,0,1)$ and a particular little group element S. Using invariance of the scalar product, one finds that $(St)^{\mu}k_{\mu} = (St)^{\mu}(Sk)_{\mu} = t^{\mu}k_{\mu} = -1$ and $(St)^{\mu}(St)_{\mu} = t^{\mu}t_{\mu} = -1$. The first of these equations can be satisfied by $(St)^{\mu} = (\alpha,\beta,\zeta,1+\zeta)$ with undetermined α, β and ζ. The second equation then forces $\zeta = (\alpha^2+\beta^2)/2$.

This argument, however, does not fix W to S completely, since the four-vectors t and k are totally inert to pure rotations and rotations around the third axis, respectively. Indeed, from $Wt = St$ we find that $S^{-1}W$ is a pure rotation and from $Wk = Sk$ find that $S^{-1}W$ is a rotation $R(\theta)$ around the third axis. This yields the equation (3.196) with S and R given by (3.197) and (3.198).

The two subgroups generated by S and R are Abelian, and explicitly we have $S(\alpha',\beta')S(\alpha,\beta) = S(\alpha'+\alpha,\beta'+\beta)$ and $R(\theta')R(\theta) = R(\theta'+\theta)$ as is appropriate for translations and rotations in two dimensions. Furthermore, from

$$R(\theta)S(\alpha,\beta)R^{-1}(\theta) = S(\alpha\cos\theta+\beta\sin\theta, -\alpha\sin\theta+\beta\cos\theta) \tag{3.199}$$

we see that S generates an invariant Abelian subgroup. The little group for massless representations is thus not semisimple (see Section 3.11.1).

The infinitesimal group element is

$$W(\theta,\alpha,\beta)^{\mu}{}_{\nu} = \delta^{\mu}{}_{\nu} + \lambda^{\mu}{}_{\nu} \quad \text{with} \quad \lambda^{\mu}{}_{\nu} = \begin{pmatrix} 0 & \theta & -\alpha & \alpha \\ -\theta & 0 & -\beta & \beta \\ \alpha & \beta & 0 & 0 \\ \alpha & \beta & 0 & 0 \end{pmatrix} \tag{3.200}$$

41 To clarify, this is E_2 represented on four-dimensional space-time vectors.

The corresponding unitary operator on the space of states is

$$U(W(\theta,\alpha,\beta)) = 1 + i\alpha A + i\beta B + i\theta J_3 \quad \text{where} \begin{cases} A = -J^{13} + J^{10} \\ B = -J^{23} + J^{20} \end{cases} \tag{3.201}$$

The Lie algebra is (see Section 3.4.2)

$$[J_3, A] = iB \qquad [J_3, B] = -iA \qquad [A, B] = 0 \tag{3.202}$$

or in terms of non-Hermitian operators $T_+ = A + iB$ and $T_- = A - iB$

$$[J_3, T_\pm] = \pm T_\pm \qquad [T_+, T_-] = 0 \tag{3.203}$$

An interesting phenomenon occurs when one simultaneously diagonalizes T_+ and T_- by states Ψ_{k,t_+,t_-}

$$T_+\Psi_{k,t_+,t_-} = t_+\Psi_{k,t_+,t_-} \quad \text{and} \quad T_-\Psi_{k,t_+,t_-} = t_-\Psi_{k,t_+,t_-} \tag{3.204}$$

From the group law (3.199) follows

$$U(R(\theta))T_\pm U^{-1}(R(\theta)) = T_\pm e^{\pm i\theta} \tag{3.205}$$

It then follows that there is an infinite set of eigenstates Ψ^θ_{k,t_+,t_-} of the translation generators $T_\pm$ with eigenvalues $e^{\pm i\theta}t_\pm$ parametrized by θ. Namely,

$$T_\pm\Psi^\theta_{k,t_+,t_-} = (e^{\pm i\theta}t_\pm)\Psi^\theta_{k,t_+,t_-} \quad \text{with} \quad \Psi^\theta_{k,t_+,t_-} = U^{-1}(R(\theta))\Psi_{k,t_+,t_-} \tag{3.206}$$

Physically, this means that if we have an eigenstate of the translation generators, then all the rotated states are eigenstates, too.

Massless particles in nature, the photon for instance, do not possess any such continuous, dimensionful, eigenvalue. Therefore, for the regular massless representations one chooses eigenvalues $t_\pm = 0$. However, from a theoretical higher spin perspective, the nonzero eigenvalue representations are interesting. These are called *continuous spin representations*, not because the spin is continuous, but because of this parametrization in terms of the rotation angle θ. Spin, or helicity, is still discrete, but runs from zero to infinity in integer or half-integer steps. We will devote a special Section 3.6, to the representation theory of this, from the higher spin perspective, very interesting group.

Massless representations

---

i

Massless representations are defined by $p^2 = 0$ and $p^0 > 0$. The standard momentum is taken as $k^\mu = (0, 0, 1, 1)$, and the little group is ISO(2). The eigenvalues for the translation generators T_+ and T_- are chosen to be zero.

From the form of the general little group element in equation (3.196) and the analysis just preformed, we have the unitary operators $U\big(W(\theta,\alpha,\beta)\big)$ acting on states $\Psi_{k,\sigma}$,

$$U\big(S(\alpha,\beta)R(\theta)\big)\Psi_{k,\sigma} = \exp(i\alpha A + i\beta B)\exp(i\theta J_3)\Psi_{k,\sigma} = \exp(i\theta\sigma)\Psi_{k,\sigma} \tag{3.207}$$

Comparing to the general formula (3.133), we find the little group representation matrices for massless states

$$D_{\bar{\sigma}\sigma}(W) = \exp(i\theta\sigma)\delta_{\bar{\sigma}\sigma} \tag{3.208}$$

Finally, referring back to the general formula (3.136), a Lorentz transformation of massless state of helicity σ is

$$U(\Lambda)\Psi_{p,\sigma} = \sqrt{(\Lambda p)^0/p^0}\,\exp(i\sigma\theta(\Lambda,p))\Psi_{\Lambda p,\sigma} \tag{3.209}$$

where $\theta(\Lambda,p)$ is defined by $W(\theta,\alpha,\beta) = S\big(\alpha(\Lambda,p),\beta(\Lambda,p)\big)R\big(\theta(\Lambda,p)\big)$. It should be clear that even though the S part of the little group is trivially represented on the states, it does not mean that S itself is a unit matrix.

From the transformation rule (3.209), it is clear that helicity is invariant under proper orthochronous Lorentz transformations. But we are used to the fact that, in four dimensions, a massless particle – or field – has two degrees of freedom corresponding to the helicities $\pm\sigma$ where σ takes integer or half-integer values. The restriction to integer or half-integer values for the helicity comes from the topology of the Lorentz group. The connection between positive and negative helicities is mediated by the space inversion piece of the Lorentz group. Space inversion in three space dimensions is connected to left- or right-handedness, or in general dimensions to what we refer to as *chirality* in particle physics.

3.5.4 The Pauli–Lubanski vector, Casimir operators and the little group

Much of the preceding discussion can be streamlined and understood with the help of the *Pauli–Lubanski vector*

$$W_\mu = \frac{1}{2}\epsilon_{\mu\nu\rho\lambda}P^\nu J^{\rho\lambda} \tag{3.210}$$

In quantum field theory, it is common to split up the Lorentz generators as

$$J_{\mu\nu} = L_{\mu\nu} + S_{\mu\nu} \tag{3.211}$$

with $L_{\mu\nu}$ denoting the orbital angular momentum part, explicitly realized as in formula (3.112), and $S_{\mu\nu}$ the spin angular momentum part, explicitly realized from case to case on the indices of quantum fields. Due to the commutativity of partial derivatives, the angular momentum part of $J_{\mu\nu}$ drops out of the Pauli–Lubanski vector, and only the spin part contributes.

The Pauli–Lubanski vector has he following algebraic properties

$$W_\mu P^\mu = 0 \tag{3.212}$$

$$[P_\mu, W_\nu,] = 0 \tag{3.213}$$

$$[J_{\mu\nu}, W_\rho] = i(\eta_{\mu\rho} W_\nu - \eta_{\nu\rho} W_\mu) \tag{3.214}$$

$$[W_\mu, W_\nu] = i\epsilon_{\mu\nu\rho\lambda} W^\rho P^\lambda \tag{3.215}$$

The first two equations follow from the commutativity of momentum operators, the second can be derived from the Poincaré algebra but is also a necessary consequence of W_μ being built from vectors and tensors, it thus transforms as a vector. The fourth equation follows from the second and third. Furthermore, the square of the Pauli–Lubanski vector commutes with the whole Poincaré algebra, it thus serves as a *second Casimir operator* alongside the square of the momentum vector.

Generically, the little group is a subgroup of the Lorentz group that leaves a certain momentum vector k invariant, that is, Lorentz transformations for which

$$W^\mu{}_\nu k^\nu = k^\mu \tag{3.216}$$

This is the same equation as (3.132). The choice of the symbol W here is not arbitrary. When acting on one-particle states, the *first Casimir operator* P^2 gives the squared mass of the particle. Furthermore, due to equation (3.213), when acting on such states, the momentum operator in W_μ can be replaced by its eigenvalue p_μ.

Massive representations

We can now make the connection to little groups. Choosing a standard momentum $k^\mu = (0, 0, 0, m)$ for massive representations, we find $W_0 = 0$, while the interesting components are

$$W_i = -\frac{m}{2}\epsilon_{ijk} J^{jk} \tag{3.217}$$

generating the rotation algebra in three dimensions. The Casimir operator becomes

$$W_\mu W^\mu = W_i W^i = m^2 J_i J^i = m^2 \mathbf{J}^2 \tag{3.218}$$

On a spin j representation, the Pauli–Lubanski vector evaluates to $m^2 j(j+1)$.

Massless representations

Choosing a standard momentum $k^\mu = (0, 0, \omega, \omega)$ for massless representations, we find the components

$$\begin{aligned} W_1 &= -\omega(J_1 + K_2) \\ W_2 &= -\omega(J_2 - K_1) \\ W_3 &= -W_0 = -\omega J_3 \end{aligned} \tag{3.219}$$

generating the algebra of $\mathfrak{iso}(2)$. Here, W_0 i redundant. Identifying with the notation of formulas (3.202), we find $W_1 = \omega B$ and $W_2 = -\omega A$. The Casimir operator becomes

$$W_\mu W^\mu = (W_1)^2 + (W_2)^2 = \omega^2(A^2 + B^2) \tag{3.220}$$

There are two types of massless representations. The regular one-dimensional *helicity representations* are labelled by helicity λ taking positive and negative integer or half-integer values. These representations have eigenvalues zero for the translation generators A and B. The second Casimir operator is zero. The second type of massless representations, the continuous spin representations, are discussed in Section 3.6.

3.5.5 The emergence of gauge invariance

We will now try to construct a quantum field for a massless spin 1 particle, transforming as a vector, along the lines followed in Section 3.5.2 for a massive particle. This will lead to the emergence of gauge invariance. We look for a field of the form of equation (3.194) with polarization vectors, or wave-functions, as in (3.195) but now appropriate for massless particles. But we start more general and consider wave-functions u_l and v_l and representation matrices $D_{\bar{l}l}(\Lambda)$, and write the formulas for Lorentz transformations corresponding to (3.163) and (3.164),

$$u_{\bar{l}}(\mathbf{p}_\Lambda, \sigma) \exp(i\sigma\theta(\Lambda, p)) = \sqrt{\frac{p^0}{(\Lambda p)^0}} \sum_l D_{\bar{l}l}(\Lambda) u_l(\mathbf{p}, \sigma) \tag{3.221}$$

$$v_{\bar{l}}(\mathbf{p}_\Lambda, \sigma) \exp(-i\sigma\theta(\Lambda, p)) = \sqrt{\frac{p^0}{(\Lambda p)^0}} \sum_l D_{\bar{l}l}(\Lambda) v_l(\mathbf{p}, \sigma) \tag{3.222}$$

where we have used (3.208) for the massless little group $D_{\bar{\sigma}\sigma}(W)$ matrices. Note that for massless particles $p^0 = |\mathbf{p}|$.

In analogy to the massive case, we first consider boosts. Then let p^μ be a standard momentum that can be chosen with $p^0 = |\mathbf{p}|$ and $\mathbf{p} = (0, 0, |\mathbf{p}|)$. Also let $\Lambda = L(q)$ be a Lorentz transformation that boosts the particle to momentum $\mathbf{q}$. Then $\theta = 0$ (so that $D_{\bar{\sigma}\sigma}(\Lambda)$ is a unit matrix) and $(\Lambda p)^0 = q^0$ or $p_\Lambda = \Lambda p = q$. The equations then become[42]

$$u_{\bar{l}}(\mathbf{q}, \sigma) = \sqrt{\frac{|\mathbf{p}|}{q^0}} \sum_l D_{\bar{l}l}(L(q)) u_l(\mathbf{p}, \sigma) \tag{3.223}$$

42 Note the subtle difference as compared to the massive case. There is no rest-frame here, but the equations still relate wave functions at general momentum $\mathbf{q}$ to wave functions at the standard momentum $\mathbf{p}$.

$$v_{\bar{l}}(\mathbf{q},\sigma) = \sqrt{\frac{|\mathbf{p}|}{q^0}} \sum_l D_{\bar{l}l}(L(q))v_l(\mathbf{p},\sigma) \tag{3.224}$$

Next, in the massive case, we considered SO(3) rotations R in space. Here, we must consider little group ISO(2, 1) transformations W of the type discussed in Section 3.5.3 above. The standard momentum is still given by $p^0 = |\mathbf{p}|$ and $\mathbf{p} = (0, 0, |\mathbf{p}|)$ but the Lorentz transformation Λ is a little group transformation W. These leave the standard momentum invariant.The equations then become

$$u_{\bar{l}}(\mathbf{p},\sigma)\exp(i\sigma\theta(W,p)) = \sum_l D_{\bar{l}l}(W)u_l(\mathbf{p},\sigma) \tag{3.225}$$

$$v_{\bar{l}}(\mathbf{p},\sigma)\exp(-i\sigma\theta(W,p)) = \sum_l D_{\bar{l}l}(W)v_l(\mathbf{p},\sigma) \tag{3.226}$$

The equations can now be specialized to rotations $R(\theta)$ around the 3-axis and boosts $S(\alpha,\beta)$ in the 1-2 plane. Using equations (3.198) and(3.197), we can write for rotations

$$u_{\bar{l}}(\mathbf{p},\sigma)\exp(i\sigma\theta) = \sum_l D_{\bar{l}l}(R(\theta))u_l(\mathbf{p},\sigma) \tag{3.227}$$

$$v_{\bar{l}}(\mathbf{p},\sigma)\exp(-i\sigma\theta) = \sum_l D_{\bar{l}l}(R(\theta))v_l(\mathbf{p},\sigma) \tag{3.228}$$

and for boosts

$$u_{\bar{l}}(\mathbf{p},\sigma) = \sum_l D_{\bar{l}l}(S(\alpha,\beta))u_l(\mathbf{p},\sigma) \tag{3.229}$$

$$v_{\bar{l}}(\mathbf{p},\sigma) = \sum_l D_{\bar{l}l}(S(\alpha,\beta))v_l(\mathbf{p},\sigma) \tag{3.230}$$

In the sequel, it is enough to discuss the u wave-functions since the equations are complex conjugate of each other and we can choose $v = u^*$.[43]

Let us now be more specific and see if we can construct a four-vector field ϕ_μ corresponding to the Lorentz group representation $(\frac{1}{2}, \frac{1}{2})$. As in the massive case, we then have $D(\Lambda)^\mu{}_\nu = \Lambda^\mu{}_\nu$. We also work in terms polarization vectors $e^\mu(\mathbf{p},\sigma)$ related to the wave-functions in the same way as in the massive case; see formula (3.195). The equation (3.223), that encodes boosting a standard momentum wave-function to arbitrary momentum, then reads

$$e^\mu(\mathbf{q},\sigma) = L^\mu{}_\nu(\mathbf{q})e^\nu(\mathbf{p},\sigma) \tag{3.231}$$

Equation (3.227) becomes

$$e^\mu(\mathbf{p},\sigma)\exp(i\sigma\theta) = R^\mu{}_\nu(\theta)e^\nu(\mathbf{p},\sigma) \tag{3.232}$$

43 This amounts to normalization. Details are given in [18], Chapter 5.

This equation requires $e^\mu(\mathbf{p},\sigma)$ to have zero components in the 3 and 0 directions and forces $\sigma = \pm 1$. With a suitable normalization, we get the familiar solutions

$$e^\mu(\mathbf{p},\pm 1) = \frac{1}{\sqrt{2}}(1,\pm i,0,0) \tag{3.233}$$

Next, equation (3.229) becomes

$$e^\mu(\mathbf{p},\sigma) = S^\mu{}_\nu(\alpha,\beta)e^\nu(\mathbf{p},\sigma) \tag{3.234}$$

When the solution (3.233) is inserted into this equation with $\sigma = \pm 1$ it becomes

$$\frac{1}{\sqrt{2}}(1,\pm i,0,0) = \frac{1}{\sqrt{2}}(1,\pm i,\alpha \pm i\beta,\alpha \pm i\beta) \tag{3.235}$$

This requires $\alpha \pm i\beta = 0$ which would mean that $S^\mu{}_\nu(\alpha,\beta)$ becomes a unit matrix and the little group collapses into rotations around the third axis.

The interpretation of this, according to Weinberg, is that it is impossible to use the creation and annihilation operators for massless helicity ± 1 particle states to build a vector quantum field.

It could perhaps be objected that we started from a little group representation with zero eigenvalues for the translation generators of the little group, and that therefore one could put α and β to zero with good conscience. However, that would entail conflating Poincaré group representations on states, with Lorentz group representations on quantum fields, the very distinction that this whole discussion aims to clarify. The fields must still transform covariantly under the full Lorentz group and, therefore, under the full little group, not just under the rotation factor.

While the $R^\mu{}_\nu$ factor of the little group transforms the polarization vectors covariantly, the $S^\mu{}_\nu$ factor transforms the polarizations according to

$$S^\mu{}_\nu(\alpha,\beta)e^\nu(\mathbf{p},\pm 1) = e^\nu(\mathbf{p},\pm 1) + \frac{1}{\sqrt{2}}(\alpha \pm i\beta)\frac{p^\mu}{|\mathbf{p}|} \tag{3.236}$$

with an inhomogeneous term. The full little group transformation law for the polarization vectors become

$$D^\mu{}_\nu(W(\theta,\alpha,\beta))e^\nu(\mathbf{p},\pm 1) = \exp(\pm i\theta)\left(e^\nu(\mathbf{p},\pm 1) + \frac{1}{\sqrt{2}}(\alpha \pm i\beta)\frac{p^\mu}{|\mathbf{p}|}\right) \tag{3.237}$$

Towards gauge invariant vector fields

?

Still following [18], one could go on and consider a vector field $a_\mu(x)$ constructed out of these polarization vectors, even though they do not transform covariantly. Doing that, by boosting the polarization vectors to arbitrary momentum, leads to the conditions $a^0 = 0$ and $\nabla \cdot \mathbf{a} = 0$ on the vector field components, apart from the Klein–Gordon equation $\Box a_\mu(x) = 0$. From formula (3.237) follows that under a general Lorentz transformation we have

$$U(\Lambda)a_\mu(x)U^{-1}(\lambda) = \Lambda^\nu{}_\mu a_\nu(\Lambda x) + \partial_\mu \Omega(x,\Lambda) \tag{3.238}$$

where the gauge parameter itself is a linear combination of creation and annihilation operators.

This formula can then be abstracted to a general vector A_μ field transforming covariantly under the Lorentz group. In order for such a field to describe massless states it must be subject to gauge transformations $\delta A_\mu(x) = \partial_\mu \xi(x)$ with an arbitrary gauge parameter $\xi(x)$. Its gauge invariant, free field equation is $\Box A_\mu - \partial_\mu \partial \cdot A = 0$. When gauge-fixing such a field, equation (3.238) returns in the guise of the need to perform a compensating gauge transformation when a Lorentz transformation takes the field "out of the gauge". We will see a concrete example when gauge-fixing to the light-cone gauge (see Section 6.1.4).

The problems encountered here do not appear if one instead attempts to construct an antisymmetric tensor field $f_{\mu\nu}$ for massless spin 1 states. Such a field transforms as $(1,0) \oplus (0,1)$ under the Lorentz group, and it is of course related to the vector field through the familiar formula $f_{\mu\nu} = \partial_\mu a_\nu - \partial_\nu a_\mu$.

3.5.6 Finite dimensional representations of the Lorentz group

The representation theory of the Lorentz group is quite complicated, but since the mid 1940s it is well understood (see Section 2.4.2). The group is noncompact, and all its unitary representations are infinite-dimensional. For field theory, at least finite component field theory, it is however the nonunitary finite dimensional representations that are interesting. As we have already discussed, there is no paradox in this fact. For the sake of completeness, we will here very briefly just state the facts regarding the representation theory of the restricted Lorentz group.

We will only review the finite-dimensional nonunitary representations. As we saw at the end of Section 3.4.2, the Lorentz algebra can be written in terms of two independent, but conjugated, $\mathfrak{so}(3)$ Lie algebras with generators M_i and N_i (see formulas (3.117)–(3.119)). Each of these two algebras has matrix representations in terms of angular momentum matrices as in formulas (3.137)–(3.139). The combined algebras can be represented as a direct sum of the components. We therefore only have to introduce a convenient notation for the present context. To that end, let the indices m and n run over the values $-M, -M+1, \dots, M$ and $-N, -N+1, \dots, N$, respectively, where M and N may be integer or half-integer. The combined representation matrices are taken as

$$\mathbf{M}_{\bar{m}\bar{n},mn} = \delta_{\bar{n}n}\mathbf{L}^{(M)}_{\bar{m}m} \tag{3.239}$$

$$\mathbf{N}_{\bar{m}\bar{n},mn} = \delta_{\bar{m}m}\mathbf{L}^{(N)}_{\bar{n}n} \tag{3.240}$$

where the $\mathbf{L}^{(M)}_{\bar{m}m}$ matrices are given by the formulas

$$(\mathbf{L}^{(M)}_1 \pm i\mathbf{L}^{(M)}_2)_{\bar{m}m} = \delta_{\bar{m},m\pm1}\sqrt{(M \mp m)(M \pm m + 1)} \tag{3.241}$$

$$(\mathbf{L}_3^{(M)})_{\bar{m}m} = m\delta_{\bar{m}m} \tag{3.242}$$

and correspondingly for the $\mathbf{L}_{\bar{n}n}^{(N)}$ matrices. Both sets of matrices are unitary. The dimension of the representation is $(2M + 1)(2N + 1)$.

The boost part of the Lorentz group, with generators K_i, is represented non-unitarily through the combination $K_i = -i(M_i - N_i)$, while the rotation part with generators J_i, is represented unitarily through the combination $J_i = M_i + N_i$. This can then be used to classify how the fields transform according to the $D(M,N)$ representations by using the rules of angular momentum addition. A field in the $D(M,N)$ representation will rotate like states with spin j where

$$j = M + N, M + N - 1, \dots, |M - N| + 1, |M - N| \tag{3.243}$$

and where for each j the m_j quantum number runs over $-j, -j + 1, \dots, j - 1, j$ as usual.

General quantum fields transforming in the $D(M,N)$ representation are constructed in [18]. The procedure is quite straightforward, but results in a somewhat unwieldy formalism. We will not need the particulars in our work. However, we will take the opportunity to derive the restriction, that we reviewed in Section 2.6.1, on the possibilities to represent massless states in terms of fields.

Weinberg's restriction on massless field realizations

? Remember the equations relating momentum space wave functions to representations. For rotations, we have (3.227) and (3.228) and for boosts (3.229) and (3.230). In order to have wave functions transforming in the $D(M,N)$ representation, we label them as $u_{\bar{m}\bar{n}}(\mathbf{p},\sigma)$. Following Weinberg, infinitesimal generators for rotations and boosts in the space of wave functions are given by

$$(\mathcal{J}_{ij})_{\bar{m}\bar{n},mn} = \epsilon_{ijk}\left[(M_k)_{\bar{m}\bar{n},mn} + (N_k)_{\bar{m}\bar{n},mn}\right] \tag{3.244}$$

$$(\mathcal{K}_i)_{\bar{m}\bar{n},mn} = (\mathcal{J}_{i0})_{\bar{m}\bar{n},mn} = -i\left[(M_k)_{\bar{m}\bar{n},mn} - (N_k)_{\bar{m}\bar{n},mn}\right] \tag{3.245}$$

in terms of the matrices (3.239) and (3.240). Considering first rotations, for an infinitesimal θ we have $D(R(\theta)) = 1 + i\theta\mathcal{J}_{23}$. The rotation equation (3.227) gives

$$u_{\bar{m}\bar{n}}(\mathbf{p},\sigma)(1 + i\sigma\theta) = \left[\delta_{\bar{m}m}\delta_{\bar{n}n} + i\theta(L_3^{(M)})_{\bar{m}m}\delta_{\bar{n}n} + i\theta(L_3^{(N)})_{\bar{n}n}\delta_{\bar{m}m}\right]u_{mn}(\mathbf{p},\sigma) \tag{3.246}$$

Using (3.242), we get $\sigma = \bar{m} + \bar{n}$. The corresponding calculation for the $v_{\bar{m}\bar{n}}$ wave function yields $-\sigma = \bar{m} + \bar{n}$. Considering next a boost with infinitesimal α and $\beta = 0$, we get (see formula (3.201))

$$\begin{aligned}
0 &= i\alpha(\mathcal{J}_{31} + \mathcal{J}_{01})_{\bar{m}\bar{n},mn}u_{mn} \\
&= i\alpha\left[(L_2^{(M)})_{\bar{m}m}\delta_{\bar{n}n} + (L_2^{(N)})_{\bar{n}n}\delta_{\bar{m}m} + i(L_1^{(M)})_{\bar{m}m}\delta_{\bar{n}n} - i(L_1^{(N)})_{\bar{n}n}\delta_{\bar{m}m}\right]u_{mn}(\mathbf{p},\sigma) \\
&= i\alpha\left[(L_2^{(M)} + iL_1^{(M)})_{\bar{m}m}u_{m\bar{n}}(\mathbf{p},\sigma) + (L_2^{(N)} - iL_1^{(N)})_{\bar{n}n}u_{\bar{m}n}(\mathbf{p},\sigma)\right]
\end{aligned} \tag{3.247}$$

The two terms must be zero separately, so we have

$$(L_1^{(M)} - iL_2^{(M)})_{\bar{m}m}u_{m\bar{n}}(\mathbf{p},\sigma) = 0 \tag{3.248}$$

$$(L_1^{(N)} + iL_2^{(N)})_{\bar{n}n} u_{\bar{m}n}(\mathbf{p},\sigma) = 0 \tag{3.249}$$

A boost with $\alpha = 0$ and infinitesimal β, yields precisely the same two equations. Furthermore, the computations for the $v_{\bar{m}\bar{n}}$ wave function gives the same equations. Now, since the first of these equations is a lowering equation, it implies $\bar{m} = -M$. Likewise, the second equation, being a raising equation, implies $\bar{n} = N$. In conclusion, we get the restrictions on possible wave function representations

$$\text{for } u_{\bar{m}\bar{n}}: \quad \sigma = M - N \tag{3.250}$$

$$\text{for } v_{\bar{m}\bar{n}}: \quad \sigma = N - M \tag{3.251}$$

For instance, this shows way it is not possible to represent a particle-antiparticle pair with helicities $(1, -1)$ with a vector field A_μ representation $D(1/2, 1/2)$, while a $D(1,0) \oplus D(0,1)$ representation corresponding to an antisymmetric tensor $F_{\mu\nu}$ is possible.

As noted in [96] (see our Section 2.6.1), we can now see that it is precisely the structure of the massless little group that brings this restriction about. More exactly, it is its non-semisimplicity, the fact that the internal translation generators commute.[44] However, as we have also seen in Section 3.5.5, the little group then makes a second entrance, and saves the day, by introducing vector field gauge transformations corresponding to the internal translations.

3.6 Representations of the two-dimensional Euclidean group

The two-dimensional Euclidean group turns up as the little group for massless representations of the Poincaré group. This group is therefore not surprisingly quite interesting from a higher spin perspective, and we will study its representation theory in some detail.[45]

The group consists of translations $T(a_1, a_2) = T(\mathbf{a})$ and rotations $R(\theta)$ in two-dimensional space $\mathbf{R}^2$, together constituting group elements $g(\theta, \mathbf{a})$. As the notation indicates, translations are parametrized by a vector (a_1, a_2) and rotations by an angle θ. The effect on the space coordinates is given by

$$\begin{pmatrix} x_1' \\ x_2' \end{pmatrix} = \begin{pmatrix} \cos\theta & -\sin\theta \\ \sin\theta & \cos\theta \end{pmatrix} \begin{pmatrix} x_1 \\ x_2 \end{pmatrix} + \begin{pmatrix} a_1 \\ a_2 \end{pmatrix} \quad \text{or} \quad \mathbf{x}' = R(\theta)\cdot\mathbf{x} + \mathbf{a} \tag{3.252}$$

This is the defining formula for the group. The group law then follows as

$$g(\theta_2, \mathbf{a}_2) g(\theta_1, \mathbf{a}_1) = g(\theta_2 + \theta_1, R(\theta_2)\cdot\mathbf{a}_1 + \mathbf{a}_2) \tag{3.253}$$

where we have used the relation $R(\theta_2)R(\theta_1) = R(\theta_2 + \theta_1)$ satisfied by the rotation subgroup. There are indeed two readily recognizable subgroups; the subgroup of rotations

---

44 See Wigner's discussion reviewed in our Section 2.3.4.

45 We are following [245], Chapter 9.

$g(\theta, 0) = R(\theta)$ and the subgroup of translations $g(0, \mathbf{a}) = T(\mathbf{a})$. The subgroup of translations is Abelian and can be realized in the standard way as a unitary operator

$$U(T(\mathbf{a})) = e^{-i\mathbf{a}\cdot\mathbf{P}} \tag{3.254}$$

in terms of a momentum operator $\mathbf{P}$. Likewise, the subgroup of rotations can be realized as

$$U(R(\theta)) = e^{-i\theta J} \tag{3.255}$$

in terms of the rotation angular momentum operator J. The group law (3.253) applied to $g(\theta, \mathbf{a})g(-\theta, 0)$ implies that a general group element can be written as

$$g(\theta, \mathbf{a}) = T(\mathbf{a})R(\theta) \tag{3.256}$$

We have seen this formula before in (3.196) where E_2 is represented as Lorentz transformations on four-dimensional Minkowski space-time. The Lie algebra that we have already encountered in (3.202) is

$$[P_1, P_2] = 0 \tag{3.257}$$

$$[J, P_k] = i\epsilon_{kl}P_l \tag{3.258}$$

This way of writing the algebra makes it clear that P_k transforms as a vector under rotations

$$e^{-i\theta J}\mathbf{P}e^{i\theta J} = R(-\theta)\cdot\mathbf{P} \tag{3.259}$$

It is also clear that the group has one Casimir operator, namely $\mathbf{P}^2$. Furthermore, a short calculation using (3.256) and the group law, shows that $g(\theta, \mathbf{b})T(\mathbf{a})g(\theta, \mathbf{b})^{-1} = T(R(\theta)\mathbf{a})$. This means that the translation subgroup is an invariant subgroup. Since it is Abelian, this shows that E_2 is not semisimple. The group is also noncompact, since the parameters for translations (a_1, a_2) have infinite ranges.

The representation theory of the group can be approached in two ways, corresponding to the two natural sets of operators to diagonalize; either the momentum operators $\mathbf{P}$, or the angular momentum operator J. We will look for unitary representations, but in order to simplify notation we will write the unitary operators plainly as $U(\theta)$ and $U(\mathbf{a})$ instead of $U(R(\theta))$ and $U(T(\mathbf{a}))$.

3.6.1 Induced plane wave representations

The method of induced representations that we used for the Poincaré group in Section 3.5, can be used for the little group itself, precisely because it contains an invariant Abelian subgroup.[46]

46 Apart from dimension and signature of space, the groups are of the same type: inhomogeneous $SO(p, q)$ groups; see Section 3.11.2.

Reminder of where we left off continuous spin representations

---

Let us return to the discussion about the little group representations in the massless case where we left it off at the end of Section 3.5.3. There we considered states that are eigenstates of the translation generators A and B. With nonzero eigenvalues a and b of A and B, the second Casimir operator evaluates to $\rho^2(a^2 + b^2)$ in the standard momentum $(0, 0, \rho, \rho)$ state. In two-dimensional transverse space, $(\rho a, \rho b)$ is thus a vector of constant length ρ and can therefore be parametrized by an angle φ, and we can write the states as $\Psi(\varphi)$ suppressing all other quantum numbers. Working with the generators A and B, we write the eigenvalues as $a = \cos\varphi$ and $b = \sin\varphi$. Thus

$$A\Psi(\varphi) = \cos\varphi\Psi(\varphi) \quad \text{and} \quad B\Psi(\varphi) = \sin\varphi\Psi(\varphi) \tag{3.260}$$

or in terms of $T_\pm = A \pm iB$

$$T_\pm\Psi(\varphi) = \exp(\pm i\varphi)\Psi(\varphi) \tag{3.261}$$

Intuitively, the action of J_3 is to rotate these states by an angle.

Based on the above reminder, we denote the states – using Dirac notation – with any of the expressions $|\mathbf{p}\rangle = |p\cos\varphi, p\sin\varphi\rangle = |p, \varphi\rangle$. Choosing a standard two-momentum $\mathbf{k} = (k, 0)$, the corresponding state is $|k, 0\rangle$. Then we have

$$P_1|k,0\rangle = k|k,0\rangle \qquad P_2|k,0\rangle = 0 \qquad \mathbf{P}^2|k,0\rangle = k^2|k,0\rangle \tag{3.262}$$

The only rotation leaving this state invariant is $U(\theta = 0)$. The "little group of the little group" is trivial. Acting on the state with $U(\theta = \varphi)$ should give a rotated state. Let us check this

$$\mathbf{P}U(\varphi)|k,0\rangle = U(\varphi)[U^{-1}(\varphi)\mathbf{P}U(\varphi)]|k,0\rangle \tag{3.263}$$

$$= U(\varphi)R(-\varphi)\cdot\mathbf{P}|k,0\rangle = U(\varphi)|k,0\rangle R(-\varphi)\cdot\mathbf{k} \tag{3.264}$$

We find that $U(\varphi)|k,0\rangle$ is a new eigenstate $|\mathbf{p}\rangle$ rotated by the angle φ

$$|\mathbf{p}\rangle \equiv U(\varphi)|k,0\rangle = |k,\varphi\rangle \tag{3.265}$$

of momentum $(k\cos\varphi, k\sin\varphi)$. Such eigenstates, considered as continuous set of vectors, are closed under the group transformations

$$U(\mathbf{b})|\mathbf{p}\rangle = |\mathbf{p}\rangle e^{-i\mathbf{b}\cdot\mathbf{p}} \tag{3.266}$$

$$U(\theta)|\mathbf{p}\rangle = |\mathbf{q}\rangle = |p, \varphi + \theta\rangle \tag{3.267}$$

Note that the length of the vectors are fixed so that $p = |\mathbf{p}| = k = |\mathbf{k}| = q = |\mathbf{q}|$. The states are distinguished by the angle $0 \le \theta < 2\pi$. The basis $|\mathbf{p}\rangle = |p,\varphi\rangle$ vectors are eigenstates of the translation operators

$$P_1|p,\varphi\rangle = p\cos\varphi|p,\varphi\rangle \qquad P_2|p,\varphi\rangle = p\sin\varphi|p,\varphi\rangle \tag{3.268}$$

while for the rotation generator we have $J = i\partial/\partial_\varphi$ consistent with (3.255) and (3.267). The full group transformation is

$$U(\theta, \mathbf{b})|p, \varphi\rangle = U(\mathbf{b})U(\theta)|p, \varphi\rangle = e^{-ip(b_1\cos(\varphi+\theta)+b_2\sin(\varphi+\theta))}|p, \varphi + \theta\rangle \tag{3.269}$$

The states are subject to the orthonormality conditions

$$\langle \mathbf{p}'|\mathbf{p}\rangle = \langle p, \theta'|p, \theta\rangle = 2\pi\delta(\theta' - \theta) \tag{3.270}$$

3.6.2 Angular momentum representation

In the angular momentum basis, one chooses to diagonalize the rotation generator J and considers states $|p, m\rangle$ where again p^2 is the value of the Casimir operator $\mathbf{P}^2$, thus we have

$$\mathbf{P}^2|p, m\rangle = p^2|p, m\rangle \tag{3.271}$$

$$J|p, m\rangle = m|p, m\rangle \tag{3.272}$$

The operators $P_\pm = P_1 \pm iP_2$ are step operators, and the unitary representation space is given by the direct sum of the SO(2) subspaces parametrized by $m = 0, \pm1, \pm2, \ldots$.

We saw in Section 3.5.3 that the regular helicity representations for massless Poincaré states corresponds to representations of the little group with zero eigenvalues for the translation operators. The nonzero eigenvalue representations, on the other hand, are the continuous spin representation. Let us now study this in the angular momentum basis. Normalizing the states so that $\langle p, m|p, m\rangle = 1$, one gets

$$\big|P_\pm|p, m\rangle\big| = \langle p, m|P_\mp P_\pm|p, m\rangle = \langle p, m|\mathbf{P}^2|p, m\rangle = p^2 \tag{3.273}$$

Taking $p^2 = 0$ implies $P_\pm|p, m\rangle = 0$ and the representations consist of single states $|0, m\rangle$ with $J|0, m\rangle = m|0, m\rangle$ and

$$U(\mathbf{b})|0, m\rangle = |0, m\rangle \tag{3.274}$$

$$U(\theta)|0, m\rangle = e^{-im\theta}|0, m\rangle \tag{3.275}$$

In the Poincaré context, m is interpreted as helicity.

Multi-valued representations

The group SO(2) has multivalued representations with $U(\theta) = e^{-im\theta/n}$. The group SO(3), however, has only two-valued representations corresponding to integer and half-integer spin for massive states. Even for massless states, only one- or two-valued representations have to be considered [247]. The "valuedness" of representations are determined by the global, topological, property of path connectedness of the group manifold.

The situation for $p^2 > 0$ is more interesting. Given any initial reference state $|p, m_0\rangle$, repeated application of the raising and lowering operators yields an infinite representations space of states $\{|p, m\rangle : m = 0, \pm 1, \pm 2, \dots\}$. We define

$$P_\pm |p, m\rangle = \mp ip|p, m \pm 1\rangle \tag{3.276}$$

The phase factors $\mp i$, allowed by normalization, will be motivated below. These states are eigenstates of rotation, but we must work out their properties under translations since translations mix the states (but note that p is constant characterizing the representation). We do this by calculating matrix elements of $U(\theta)$ and $U(\mathbf{b})$ between states of different m and m'. In a notation conforming to the one used for the Poincaré group, we want to compute

$$\begin{aligned} D^{(p)}_{\bar{m}m}(\theta, \mathbf{b}) = \langle p, \bar{m}|U(\theta, \mathbf{b})|p, m\rangle &= \sum_{m'} \langle p, \bar{m}|U(\mathbf{b})|p, m'\rangle \langle p, m'|U(\theta)|p, m\rangle \\ &= \sum_{m'} \langle p, \bar{m}|U(\mathbf{b})|p, m'\rangle e^{-i\theta m} \delta_{m'm} = \langle p, \bar{m}|U(\mathbf{b})|p, m\rangle e^{-i\theta m} \end{aligned} \tag{3.277}$$

To compute the $U(\mathbf{b})$ matrix element, we write the translation vector in polar coordinates $\mathbf{b} = (b, \varphi)$, and then refer it back to a translation along the 1 direction through $U((b, \varphi)) = U(\varphi)U((b, 0))U(\varphi)^{-1}$. This gives

$$\langle p, \bar{m}|U(\mathbf{b})|p, m\rangle = e^{i(m-\bar{m})\varphi} \langle p, \bar{m}|U((b, 0))|p, m\rangle \tag{3.278}$$

where

$$U((b, 0)) = e^{-ibP_1} = e^{-ib(P_+ + P_-)/2} = \sum_{k=0}^{\infty} \sum_{l=0}^{\infty} \frac{(b/2)^{k+l}}{k!l!} (-iP_+)^k (-iP_-)^l \tag{3.279}$$

The matrix elements are nonzero when $\bar{m} = m - l + k$. Each factor $(-iP_+)^k(-iP_-)^l$ will yield a factor $(-1)^k(i(-i)p)^k(-i \cdot ip)^l = (-1)^k p^{k+l}$ when using (3.276) so that the matrix elements become

$$\langle p, \bar{m}|U((b, 0))|p, m\rangle = \sum_{k,l} (-1)^k \frac{(pb/2)^{k+l}}{k!l!} \tag{3.280}$$

Since $k - l$ is fixed to $\bar{m} - m$ the sum can be rearranged into a single sum that turns out to precisely yield the Bessel function $J_{m-\bar{m}}(pb)$. Putting all together, we get the representations matrices

$$D^{(p)}_{\bar{m}m}(\theta, (b, \varphi)) = e^{i(m-\bar{m})\varphi} J_{m-\bar{m}}(pb) e^{-i\theta m} \tag{3.281}$$

Series for Bessel functions

Put $\bar{m} - m = c$. With a new summation index $n = k + l$, we get $k = (n + c)/2$ and $l = (n - c)/2$. Then $k \geq 0$ and $l \geq 0$ implies $n \geq |c|$. In case $c \leq 0$, shift the summation index to $n' = (n + c)/2$ and recognize the sum formula for $J_c(pb)$. In case $c \geq 0$ shift the summation index to $n' = (n - c)/2$ and recognize the sum formula for $J_{-c}(pb)$.

Relation between the representations

It should be clear that the angular momentum eigenstates $|p, m\rangle$ and the plane wave eigenstates $|p, \theta\rangle$ are related through Fourier analysis. The exact relation involves a phase factor. We just quote the formulas [245]

$$|p, \theta\rangle = \sum_{n=-\infty}^{n=\infty} |p, n\rangle e^{-in(\theta+\pi/2)} \tag{3.282}$$

$$|p, n\rangle = \frac{1}{2\pi} \int_0^{2\pi} |p, \theta\rangle e^{in(\theta+\pi/2)} \tag{3.283}$$

3.6.3 Continuous spin representations

The two types of E_2 representations that we considered in the preceding sections play the same role for continuous spin representations as the SO(3) representations do for massive particles and the zero translational eigenvalue representation of E_2 do for regular massless helicity representation. Since the faithful representations of E_2 are infinite dimensional, we expect the corresponding quantum fields to depend on some continuous variable rather than being finite component fields. We will follow Wigner [63] and introduce an auxiliary four-vector coordinate ξ. Alternatively, one could work with complex two-spinor variables ζ^a, $\bar{\zeta}^{\dot{a}}$ as done in [248].

First, we must connect the Poincaré group little group notation with the ISO(2) notation developed above. The second Casimir operator for the Poincaré group evaluates to ρ^2 for a massless state of standard momentum $k^\mu = (0, 0, \rho, \rho)$. This means that in all our formulas for ISO(2) representations derived above, we just put $p = \rho$. The induced plane wave representations that we have denoted by $|p, \varphi\rangle$ are the ones related to the states Ψ^θ_{k,t_+,t_-} discussed at the end of Section 3.5.3. The notation is a bit redundant and we can write just Ψ^θ_k. Likewise, corresponding to the rotation eigenvalue states $|p, m\rangle$ we write Ψ^m_k. The translation vector $\mathbf{b}$ corresponds to the vector (α, β).

Thus for continuous spin representations we can write the little group transformations in two ways. In the *plane wave basis*, also denoted the *angle basis*, we have

$$W(\theta, \alpha, \beta)\Psi^\varphi_k = e^{-i\rho(\alpha\cos(\varphi+\theta)+\beta\sin(\varphi+\theta))}\Psi^{\varphi+\theta}_k = \int_0^{2\pi} d\bar{\varphi} D_{\bar{\varphi}\varphi}(\theta, \alpha, \beta)\Psi^{\bar{\varphi}}_k \tag{3.284}$$

with

$$D_{\bar{\varphi}\varphi}(\theta, \alpha, \beta) = \delta(\bar{\varphi} - \varphi - \theta)e^{-i\rho(\alpha \cos\bar{\varphi} + \beta \sin\bar{\varphi})} \tag{3.285}$$

In the *angular momentum basis*, also denoted the *spin basis*, we have

$$W(\theta, \alpha, \beta)\Psi_k^m = e^{i(m-\bar{m})\varphi} J_{m-\bar{m}}(\rho\sqrt{2\zeta})e^{-i\theta m}\Psi_k^{\bar{m}} = D^{(p)}_{\bar{m}m}(\theta, (\sqrt{2\zeta}, \varphi))\Psi_k^{\bar{m}} \tag{3.286}$$

where we sum over $\bar{m}$.

Physical intuition on Poincaré little groups

Having come so far, the reader may suspect that the massless little group does not have any obvious relation to physical space-time. This is indeed so, as pointed out by Wigner in [247]. This is in contrast to the massive little group SO(3).

Consider a massive spin 1 particle. It has a rest frame that we can imagine boosting ourselves to. There we can study the geometry of the polarization vector **e** of the particle. It behaves under rotations as any other three-dimensional spatial vector. There are no translations in the group, so we don't have to worry about such.

A massless spin 1 particle, on the other hand, has no rest frame. There is no way to "catch up" with the particle to study the geometry of its polarization vector **e**, which by the way lives in just two dimensions. What we can do is to observe the behavior of the polarization vector in the plane of polarization perpendicular to the momentum of the particle. But the translational part of the massless little group has no obvious interpretation in ordinary physical space. As we have seen, however, the translations manifest themselves in the form of gauge transformations. One cannot escape the feeling that there is more to understand here.

3.6.4 Two-component formalism

The two-component formalism is often convenient to use. It is based on the accidental[47] $2 \to 1$ homomorphism from SL(2, **C**) to $L_+^\uparrow$ that we discussed in Section 3.4.3. It provides a concrete realization of the finite-dimensional representations of the Lorentz group.

We have already met this formalism in connection with the Dirac and Fierz higher spin formalism in the historical chapter.[48] There are a lot of variations regarding the basic concrete conventions, but as stressed by Corson, all relations can be derived abstractly from two "axioms", one regarding the complex conjugation of the σ matrices

47 That is, not generalizing to "bigger" groups or algebras.

48 The two-component spinor formalism was introduced by B. L. van der Waerden [43] and further developed by O. Laporte and G. E. Uhlenbeck [44]. Laporte and Uhlenbeck point out that the formalism was implicit in Weyl's book *Gruppentheorie und Quantenmechanik* [249] as well as in a paper by V. Fock referenced as Zeits. f. Physik 57, 261 (1929).

and the other their matrix product.[49] We will however follow tradition and jump right in with a concrete realization. Corresponding to the two conjugated $\mathfrak{su}(2)$ Lie subalgebras there are two carrier linear spinor spaces indexed by *undotted* and *dotted* indices. For raising and lowering indices, we choose the following convention:

$$\psi^\alpha = \epsilon^{\alpha\beta}\psi_\beta \quad \text{and} \quad \psi_\alpha = \psi^\beta\epsilon_{\beta\alpha} \tag{3.287}$$

and likewise for dotted spinors. This convention is consistent with

$$\epsilon_{\alpha\beta} = \epsilon^{\alpha\beta} = \begin{pmatrix} 0 & 1 \\ -1 & 0 \end{pmatrix} \tag{3.288}$$

and likewise for dotted indices. Note also the following consequences of these definitions:

$$\epsilon^{\alpha\gamma}\epsilon_{\gamma\beta} = -\delta^\alpha_\beta \quad \text{and} \quad \phi_\alpha\psi^\alpha = -\phi^\beta\psi_\beta \tag{3.289}$$

and likewise for dotted indices.

The complex conjugate of a spinor with undotted indices is a spinor with dotted indices, and it is quite convenient to enhance the notation with a bar (in particular when indices are not shown), that is,

$$(\psi_\alpha)^* = \bar{\psi}_{\dot\alpha} \quad \Leftrightarrow \quad (\bar{\psi}_{\dot\alpha})^* = \psi_\alpha \tag{3.290}$$

The components of spinors are complex. The notation for complex conjugation is consistent with taking $(\epsilon_{\alpha\beta})^* = \epsilon_{\dot\alpha\dot\beta}$ and likewise for upper indices.

Group theoretically, a spinor with an undotted index down corresponds to the $D(1/2, 0)$ representation. It is also denoted *chiral* or *left-handed.* A spinor with a dotted index up corresponds to the $D(0, 1/2)$ representation. It is also denoted *antichiral* or *right-handed.*

The formulas (3.290) generalize to multispinors

$$(\chi_{\alpha_1\ldots\alpha_m\,\dot\beta_1\ldots\dot\beta_n})^* = \bar{\chi}_{\dot\alpha_1\ldots\dot\alpha_m\,\beta_1\ldots\beta_n} \quad \Leftrightarrow \quad (\bar{\chi}_{\dot\alpha_1\ldots\dot\alpha_m\,\beta_1\ldots\beta_n})^* = \chi_{\alpha_1\ldots\alpha_m\,\dot\beta_1\ldots\dot\beta_n} \tag{3.291}$$

Roughly speaking, a multispinor with m undotted indices and n dotted, corresponds to the $D(m/2, n/2)$ representation.

We now return to the $1 \leftrightarrow 1$ correspondence between Hermitian 2×2 σ-matrices and real four-vectors of Section 3.4.3 explicated by the formulae (3.121) and (3.122).[50] In

49 See [23], Chapter II, Section 9. The large number of differing two-spinor conventions in the literature is bewildering. Grassmann spinors may add further confusion. One should be wary that this is an area where notation tends to clash with itself (see comments in [250]).

50 The vector spaces are indeed isomorphic, it is the transformation groups that are $2 \to 1$ homomorphic.

the present context, it is natural to index the Hermitian 2×2 matrices by an undotted and a dotted index, concretely

$$\sigma^{\mu}{}_{\alpha\dot{\beta}} = (\sigma^0, \sigma^i) \tag{3.292}$$

A four-vector V_μ can then be represented as

$$V_{\alpha\dot{\beta}} = V_\mu \sigma^{\mu}{}_{\alpha\dot{\beta}} \tag{3.293}$$

The group theoretical basis for this indexing is that a vector corresponds to the $D(1/2, 1/2)$ representation.

Raising the indices on the matrices $\sigma^{\mu}{}_{\alpha\dot{\beta}}$ results in a matrix that is conventionally denoted by $\bar{\sigma}^{\mu\dot{\alpha}\alpha}$

$$\bar{\sigma}^{\mu\dot{\alpha}\alpha} = \epsilon^{\dot{\alpha}\dot{\beta}} \epsilon^{\alpha\beta} \sigma^{\mu}{}_{\beta\dot{\beta}} = (\sigma^0, -\sigma^i) \tag{3.294}$$

The bar notation here has, so far, no obvious relation to complex conjugation, it signifies the minus sign in front of the space components σ^i when the spinor indices are raised.[51]

By explicit calculation, one can show the following useful relations for the matrices:

$$\begin{aligned}
\sigma^{\mu}{}_{\alpha\dot{\alpha}} \sigma_{\mu\beta\dot{\beta}} &= -2\epsilon_{\alpha\beta}\epsilon_{\dot{\alpha}\dot{\beta}} && (3.295)\\
(\sigma^{\mu}\bar{\sigma}^{\nu} + \sigma^{\nu}\bar{\sigma}^{\mu})_{\alpha}{}^{\beta} &= -2\eta^{\mu\nu}\delta_{\alpha}{}^{\beta} && (3.296)\\
(\bar{\sigma}^{\mu}\sigma^{\nu} + \bar{\sigma}^{\nu}\sigma^{\mu})^{\dot{\beta}}{}_{\dot{\alpha}} &= -2\eta^{\mu\nu}\delta^{\dot{\beta}}{}_{\dot{\alpha}} && (3.297)\\
\mathrm{Tr}(\sigma^{\mu}\bar{\sigma}^{\nu}) = \mathrm{Tr}(\bar{\sigma}^{\mu}\sigma^{\nu}) &= -2\eta^{\mu\nu} && (3.298)
\end{aligned}$$

In the last three formulas, there are implicit summations over indices. Due to the ϵ metric in spinor space, we have $\psi_\alpha\phi^\alpha = -\psi^\alpha\phi_\alpha$, so one must pay attention to the placement of the indices. We adopt the convention: A suppressed pair of undotted indices are contracted as ${}^{\gamma}{}_{\gamma}$, and a suppressed pair of dotted indices are contracted as ${}_{\dot{\gamma}}{}^{\dot{\gamma}}$.[52]

The first equation (3.295) can be written as

$$\eta_{\mu\nu}\sigma^{\mu}{}_{\alpha\dot{\alpha}}\sigma^{\nu}_{\beta\dot{\beta}} = -2\epsilon_{\alpha\beta}\epsilon_{\dot{\alpha}\dot{\beta}} \tag{3.299}$$

offering a representation of the Minkowski metric in spinor space. The σ-matrices can be interpreted as a kind of "vierbeins" connecting the metrics in tensor space

---

51 This will be further commented on in the box below.

52 The differing conventions are natural if one looks at the placement of the indices on the matrices σ and $\bar{\sigma}$ in the formulas (3.296) and (3.297). There are different conventions of this type in the literature. Ours is the same as in [240].

and spinor space. In fact, the σ-matrices are Clebsch–Gordan coefficients relating the $D(1/2, 1/2)$ representation of $SL(2, C)$ to the vector representation of $SO(3, 1)$.[53] The awkward factor of -2 in the formula is convention dependent. The minus sign is related to our choice of a mostly plus Minkowski metric in relation to our raising and lowering conventions for undotted and dotted spinor indices, as well as choice of undotted and dotted ϵ-matrices. The factor of 2 can be defined away by redefining the σ-matrices with a factor of $1/\sqrt{2}$. For the relation between two-component and four-component spinor formalism, see Section 1.4.

We can now translate between tensor indices and two-component spinor indices through the formulas

$$T^{\dot\beta_1\dots\dot\beta_n}_{\alpha_1\dots\alpha_n} = T_{\mu_1\dots\mu_n}\prod_{i=1}^{n}\sigma_{\mu_i}{}^{\dot\beta_i}{}_{\alpha_i} \tag{3.300}$$

$$T_{\mu_1\dots\mu_n} = T^{\dot\beta_1\dots\dot\beta_n}_{\alpha_1\dots\alpha_n}\prod_{i=1}^{n}\left(\frac{1}{2}\sigma_{\mu_i}{}^{\alpha_i}{}_{\dot\beta_i}\right) \tag{3.301}$$

which are generalizations of (3.121) and (3.122).

Let us explicate the pair of formulas (3.300) and (3.301). Since they work index by index it is enough to check them for a vector. In that case, $V_\mu = \sigma_\mu{}^\alpha{}_{\dot\beta}V^{\dot\beta}_\alpha$ and $V^{\dot\beta}_\alpha = V^\nu\sigma_\nu{}^{\dot\beta}{}_\alpha$. We get the calculation

$$\begin{aligned} V_\mu &= \frac{1}{2}\sigma_\mu{}^\alpha{}_{\dot\beta}(V^\nu\sigma_\nu{}^{\dot\beta}{}_\alpha) = -\frac{1}{2}\sigma_{\mu\alpha\dot\beta}(V^\nu\bar\sigma_\nu{}^{\dot\beta\alpha}) \\ &= -\frac{1}{4}V^\nu(\sigma_{\mu\alpha\dot\beta}\bar\sigma_\nu{}^{\dot\beta\alpha} + \sigma_{\nu\alpha\dot\beta}\bar\sigma_\mu{}^{\dot\beta\alpha}) = -\frac{1}{4}\sum_{\alpha=1}^{2}(-2\eta_{\mu\nu}\delta_\alpha{}^\alpha) = V_\mu \end{aligned}$$

In the third equality, we are using the fact that we are actually computing a matrix trace. The first equality can be written $V_\mu = \frac{1}{2}\operatorname{Tr}(\sigma_\mu V)$ in accordance with the index summation convention mentioned above.

Some peculiarites of two-component formalisms

The relative order between dotted and undotted indices is of no consequence, although it is conventional to keep a certain order as in the matrices σ and $\bar\sigma$. The over-bar notation for the σ-matrix with raised indices can then be understood as Hermitian conjugation in the sense

$$(\sigma_\mu^{\alpha\dot\beta})^\dagger = \bar\sigma_\mu^{\dot\beta\alpha} \tag{3.302}$$

where the matrix elements of σ_μ are first complex conjugated and then transposed. The over-bar notation then signifies complex conjugated matrix elements. Now, since the matrices σ_μ are in fact Hermitian, we get

$$\bar\sigma_\mu^{\dot\beta\alpha} = \sigma_\mu^{\alpha\dot\beta} \tag{3.303}$$

53 These relations are elaborated in Penrose and Rindler [250].

As the reader may discern, there is a notational clash here in the use of the over-bar notation as compared to its use for multispinors in formula (3.291). Here, the over-bar does not signify switching dotted and undotted indices, as that would produce $(\sigma^{\alpha\dot{\beta}})^* = \bar{\sigma}^{\dot{\alpha}\beta}$.[54]

It is however essential to keep track of the order among the dotted indices themselves when they are raised or lowered. The same holds for undotted indices. The indices must be "staggered" in the sense that each lower index must have an "empty" upper position to which it can be raised, and vice versa.[55] In this connection, one may consider not using the Kronecker symbol, and instead use $\epsilon_\alpha{}^\beta$ and $\epsilon^\beta{}_\alpha$ with staggered indices.

Introduce provisionally the symbol $\mathbf{1}_\alpha^\beta$ to denote what we normally would mean by δ_α^β. That is: $\mathbf{1}_1^1 = \mathbf{1}_2^2 = 1$, $\mathbf{1}_1^2 = \mathbf{1}_2^1 = 0$ and do not consider it a part of the spinor algebra. Then everywhere where one would write a δ_α^β, instead write with staggered indices

$$\epsilon_\alpha{}^\beta = -\epsilon^\beta{}_\alpha \text{ numerically equal to } \mathbf{1}_\alpha^\beta \tag{3.304}$$

Instead of the first formula of (3.289), we now write $\epsilon^{\alpha\gamma}\epsilon_{\gamma\beta} = \epsilon^\alpha{}_\beta$.

Another bonus comes when one considers derivatives. The antisymmetric metric in spinor space can play tricks if one is not careful. Start by defining, as is natural

$$\partial_\alpha x^\beta \equiv \frac{\partial}{\partial x^\alpha} x^\beta = \epsilon_\alpha{}^\beta \tag{3.305}$$

Then, curiously enough, the two "equally natural" formulas

$$\partial^\alpha x_\beta = \frac{\partial}{\partial x_\alpha} x_\beta = \mathbf{1}_\beta^\alpha \quad \text{and} \quad \partial^\alpha x_\beta = \epsilon^\alpha{}_\beta$$

produce conflicting results. It is the first formula that has to be given up, because the second is consistent with (3.305) using the raising and lowering conventions. For the rest of the derivatives, we take

$$\partial^\alpha x_\beta = \epsilon^\alpha{}_\beta \qquad \partial_\alpha x_\beta = \epsilon_{\alpha\beta} \qquad \partial^\alpha x^\beta = \epsilon^{\alpha\beta} \tag{3.306}$$

Example 2 (Translating massive wave equations). The translation between the Dirac–Fierz–Pauli spinor formulation of the massive field equations, quoted in the formulas (2.19) and (2.20) of Chapter 2, and the more common tensor formulation of equations (2.140), (2.141) and (2.142) can now be performed.

As already noted, the Klein–Gordon equation for the multispinors follows from inserting one linear spinor equation into the other. Then the Klein–Gordon equation for the tensor field follows immediately from the correspondence (3.301) using equations (3.296) and (3.297) to show

$$p^{\dot{\alpha}\gamma} p_{\gamma\dot{\beta}} = -p^2 \delta^{\dot{\alpha}}{}_{\dot{\beta}} \quad \text{and} \quad p_{\alpha\dot{\gamma}} p^{\dot{\gamma}\beta} = -p^2 \delta_\alpha{}^\beta \tag{3.307}$$

54 I know of no set of conventions that avoids this clash in one way or another. Most authors seem to use the over-bar notation for both complex and Hermitian conjugation, letting context determine when is what. A resolution of the clash would most likely need a more elaborate formalism. The only reference I know of, which note and discuss the clash is Penrose and Rindler [250] on pages 123–124. Indexing σ-matrices by $\alpha\dot{\alpha}$ one may escape noting it at all (ibid. p. 114).

55 See [250], Section 2.5.

Symmetry in the spinor indices leads to the divergence condition on the multispinor. It can be exemplified for a spin 3/2 field where the field equations read

$$p^{\dot{\alpha}\beta}A^{\dot{\gamma}}_{\beta\gamma} = -mB^{\dot{\alpha}\dot{\gamma}}_{\gamma} \quad \text{and} \quad p_{\alpha\dot{\beta}}B^{\dot{\beta}\dot{\gamma}}_{\gamma} = -mA^{\dot{\gamma}}_{\alpha\gamma} \tag{3.308}$$

Contracting, for instance, the first equation with $\epsilon_{\dot{\alpha}\dot{\gamma}}$ and using symmetry of $B^{\dot{\alpha}\dot{\gamma}}_{\gamma}$ in $\dot{\alpha}\dot{\gamma}$, yields $p_{\dot{\gamma}}^{\ \beta}A^{\dot{\gamma}}_{\ \beta\gamma} = 0$, and similarly for the B field.

Tracelessness of the tensor fields follow from the translation formula (3.301) using formula (3.295) and symmetry in all dotted and undotted spinor indices. Thus, the DFP spinor field equations for massive integer spin fields, contain exactly the same information as the field equations in terms of traceless, divergence-free tensors. ◂

3.7 Basic algebraic structures

Much of abstract algebra is pivoted around the concept of vector spaces and it is also central to field theory in general and higher spin theory in particular. Here, we will collect basic definitions and formulas, pertaining to algebra, trying to motivate them with what will follow. The text can be read as a collection of "reminders" and it is quite informal. Detailed expositions of the material can be found in many places, for instance, in [251–254] .

3.7.1 Morphisms

Morphisms are structure-preserving mappings between mathematical objects of the same "kind". For instance, mappings between sets are ordinary functions, mappings between vector spaces are linear transformations and mappings between topological spaces are continuous functions. In the case of sets, there is no structure to preserve. For vector spaces, the sum and product with scalars, must be preserved. In Table 3.1, we list various types of general morphisms and their properties.

Table 3.1: Vocabulary of morphisms.

Type of morphism	Meaning
mono-morphism	injective (one-to-one)
epi-morphism	surjective (onto)
iso-morphism	injective and surjective (bijective)
endo-morphism	morphism to the same set
auto-morphism	isomorphism to the same set

3.7.2 Groups

Remember that a *group* is a set endowed with an internal binary operation that is associative, has a left and right identity and where every element has an inverse. Denoting the set with X and its elements by $x, y, \ldots$ and so on, the unit element by ι and inverses by x^{-1} we have a map $X \times X \to X : (x, y) \mapsto xy$ with properties for all elements x, y, z,

$$(xy)z = x(yz) \tag{3.309}$$

$$x\iota = \iota x = x \tag{3.310}$$

$$x^{-1}x = xx^{-1} = \iota \tag{3.311}$$

Morphisms between groups are called *homomorphisms*. They map the group operation in one group to the group operation in another, thus preserving the group structure.

From the simple concept of a group – in itself very rich – we can construct new useful algebraic structures that play various roles in particle and field theory. The groups are generically transformation groups, acting on the states of systems which are modeled on vector spaces, further on promoted to normed vector spaces or Hilbert spaces depending on context. We then talk about (linear) representations of the groups. In much of theoretical physics, groups, algebras, vector spaces and Hilbert spaces are the most commonly used structures. But occasionally (and some authors prefer generality) a more elaborate set of concepts may be needed. Thus the following list of structures. The basic definitions will be collected here, pointing out the roles they play. Deeper properties of these structures will be reviewed as need arise. First, we have rings and fields which are abstractions of the usual number systems.

3.7.3 Rings and fields

A *ring* is a set X endowed with two internal binary operations $(x, y) \mapsto xy$ and $(x, y) \mapsto x + y$ (multiplication and addition) such that X is an Abelian group under addition and multiplication is associative and distributive over addition. Thus for all x, y, z

$$(xy)z = x(yz) \tag{3.312}$$

$$x(y + z) = xy + xz \tag{3.313}$$

$$(y + z)x = yx + zx \tag{3.314}$$

A ring is Abelian if multiplication is Abelian. A ring may have a multiplicative unit element ι and it may be the case that every element (except the additive neutral element 0) has a multiplicative inverse. In that case, the ring is called a *division ring*. If such a ring is also Abelian, then it is a *field*.

Fields are abstractions of our ordinary number systems **Q**, **R** and **C**.[56] The role model for a ring with a multiplicative unit but without multiplicative inverses is the integers **Z**. Another such example is the ring of polynomials in some indeterminate x with coefficients again in some ring. Note also that the set of real valued functions of a real variable on some given subset of the real numbers is a ring where addition and multiplication is defined pointwise. An example of a noncommutative ring is given by the set of $n \times n$ matrices with real entries. The main point is: *rings do not have division.*

One use of rings and fields are to form linear combinations of states of systems. From now on, to avoid unnecessary confusion with the "fields" of physics, we will write *number systems* for the "fields" of mathematics. Anyway, this leads to the concepts of modules and vectors spaces.

3.7.4 Modules

A *module* over a ring R is an Abelian group X (with the operation denoted by $+$) together with an external operation (scalar multiplication) $R \times X \to X : (\alpha, x) \mapsto \alpha x$ with properties

$$\alpha(x + y) = \alpha x + \alpha y \tag{3.315}$$

$$(\alpha + \beta)x = \alpha x + \beta x \tag{3.316}$$

$$(\alpha\beta)x + \alpha(\beta x) \tag{3.317}$$

for all $\alpha, \beta \in R$ and $x, y \in X$. If the ring has an identity, then $\iota x = x$.

Modules are the simplest sets with enough structure to serve as representation spaces for groups. The point is that the elements of a module can be linearly combined, as is clear from the properties above. However, modules are not well behaved with respect to the concept of bases.[57] There is a risk of confusion here since in representation theory a G-module is also defined as vector space upon which a group G is linearly represented.[58]

56 Hamilton's quaternions is a noncommutative division ring, closely related to $D = 4$ special relativity. The eight-dimensional octonions is a division ring where the associativity of multiplication is not required.

57 The reason for this, related to the ring not having multiplicative inverses, is quite deep. See [255], Article III.81.

58 Groups can also be *realized* on sets that are not modules or vector spaces. However, as vector spaces are well understood due to their linear structure and existence of bases, they are often preferable. It seems that in the literature, the word "module" is often used simply as a synonym to "linear representation space".

3.7.5 Vector spaces

A *vector space* is a module for which the ring is a field. In practice, we almost always work with vector spaces over the real numbers or the complex numbers. The role models for vector spaces are the ordinary m-dimensional Euclidean spaces $\mathbf{R}^m$. Vector spaces are also called *linear spaces*.

Recall that in an m-dimensional vector space V we can set up a basis consisting of m linearly independent basis vectors e_j such that any vector v can be expressed as a linear combination $v = v^1 e_1 + v^2 e_2 + \cdots + v^m e_m$ with unique components v^i. Furthermore, the *dual vector space* V^* is introduced by considering linear functions f (with $f(0) = 0$) defined on the space V. From the linearity, it then follows that $f(v) = v^1 f(e_1) + v^2 f(e_2) + \cdots + v^m f(e_m)$. It is therefore enough to know $f(e_i)$ for all i in order to compute $f(v)$ for any vector v. This makes the set of linear functions a vector space itself, namely the dual vector space. The basis vectors in V^* is conveniently denoted by e^{*i}. Thinking about these as linear functions, they are completely specified by giving their values $e^{*i}(e_j)$ for all j. A particularly simple choice is the *dual basis* given by $e^{*i}(e_j) = \delta^i{}_j$.

Using this machinery, we can now compute the *inner product* between a dual vector $f = f_i e^{*i}$ and a vector $v = v^j e_j$ as $f(v) = f_i v^i$. It is convenient to have a notation for the inner product, such as $\langle\ ,\ \rangle : V^* \times V \to \mathbf{R}$. Then we have

$$\langle e^{*i}, e_j \rangle = \delta^i{}_j \tag{3.318}$$

Using this notation, we can write

$$f(v) = \langle f, v \rangle = \langle f_i e^{*i}, v^j e_j \rangle = f_i v^j \langle e^{*i}, e_j \rangle = f_i v^j \delta^i{}_j = f_i v^i \tag{3.319}$$

This is an abstraction of the ordinary scalar product between row and column vectors in elementary linear algebra. To bring this out more clearly, we use the fact that since the dual vector space has the same dimension as the vector space, they are actually isomorphic. Let g be such an isomorphism $V \to V^* : v_i = g_{ij} v^j$.

Now we can compute the inner product between two vectors u and v in the vector space itself as $\langle u, v \rangle = \langle g(u), v \rangle = g_{ij} u^j v^i$. It is very natural to interpret g_{ij} as a metric on the vector space and to think of it as a matrix and its action on a vector v^i as lowering indices. Therefore, we may require g_{ij} to be a positive definite as well as symmetric. Then $\langle v, v \rangle$ becomes a squared norm of the vector v and we have effectively turned our vector space into a linear metric space.

Of course, in relativity, we have metrics that are not positive definite, and we then get vectors that may classified as space-like, null, or time-like.

There is clearly a certain "symmetry" between vectors and dual vectors. Therefore, just as we can think of the inner product $\langle f, v \rangle$ as a function $f(v)$, we can think of it as a function $v(f)$. This is useful when introducing tensors and tensor spaces.

All these operations have a natural generalization to complex vector spaces over the complex numbers $\mathbf{C}$. Furthermore, they work just as well in the tangent and cotangent vector spaces of manifolds (see Section 3.10).

3.7.6 Banach and Hilbert spaces

We are now very close to the spaces of quantum mechanics. Let us be very brief and just list the basic definitions. A *Banach space* is a complete normed linear (vector) space. That the space is normed means that there is a real number $\|v\|$ for every vector v with the properties

$$\begin{aligned} &\|v\| \geq 0 \quad \text{and} \quad \|v\| = 0 \Leftrightarrow v = 0 \\ &\|u + v\| \leq \|u\| + \|v\| \quad \text{(triangle inequality)} \\ &\|av\| = |a|\|v\| \end{aligned} \tag{3.320}$$

That the space is complete means that notions of analysis such as limits and derivatives can be defined using the norm. More precisely, every Cauchy sequence of vectors is convergent.[59] Intuitively, in the finite dimensional case, Banach spaces are abstractions of Euclidean spaces $\mathbf{R}^n$. Then, one further generalization is to differential manifolds. Another application of Banach spaces are spaces of functions between two sets.

Finally, a *Hilbert space* is a complex Banach space whose norm derives from an inner product. This is precisely what we need for quantum mechanics. This is indeed very nice. The arena of quantum mechanics is Hilbert spaces – finite dimensional or infinite dimensional – which are *complex linear metric spaces where one can use the methods of analysis*! If one is not so concerned about rigor, the methods of calculus are sufficient.

3.7.7 Algebras

Intuitively, an algebra is a vector space where one has defined a multiplication. More exactly, it is a vector space equipped with a binary operation $\diamond : V \times V \to V$ that is *bilinear*

$$(u + v) \diamond w = u \diamond w + v \diamond w \quad \text{and} \quad u \diamond (v + w) = u \diamond v + u \diamond w \tag{3.321}$$

$$(\alpha u) \diamond (\beta v) = (\alpha\beta) u \diamond v \tag{3.322}$$

for all elements u, v, w in the algebra, and α, β in the number system.[60] An algebra is *associative* if the following formula holds:

$$(u \diamond v) \diamond w = u \diamond (v \diamond w) \tag{3.323}$$

59 See, for instance, [253].

60 As for vector spaces, some physicists follow mathematical customs, writing "algebra over **K**" where **K** is the chosen number system. We will do that when it is needed for clarity, otherwise letting it be implicit by the context.

for all elements u, v, w in the algebra. Algebras that have a unit element are called *unital*.

Important examples of associative unital algebras are the matrix algebras $\mathfrak{gl}(n)$ of $n \times n$ arbitrary matrices and their many subalgebras. The axioms for unital, associative algebras are quite strong, and one can prove that any such algebra of finite dimension is actually isomorphic to a subalgebra of the algebra of $n \times n$ matrices over the same number system.

The following concepts are useful for algebras in general. A *subalgebra* H of an algebra G is a subspace $H \subseteq G$ which is itself an algebra. This requirement is often written as $H \diamond H \subseteq H$ (any $h_1, h_2 \in H$ implies $h_1 \diamond h_2 \in H$) where now $\diamond$ signifies all operations in the algebra.

A specific kind of subalgebras are the *invariant subalgebras*, often called *ideals*.[61] The intuition is that multiplying elements in the invariant subalgebra H with any element in G still gives an element in H. Since the algebra may not be commutative, one actually needs to define left and right invariant ideals. The defining property for a *right ideal* can be written $H \diamond G \subseteq H$ (any $h \in H$ and $g \in G$ implies $h \diamond g \in H$). Left ideals are defined analogously. If H is both a right and left ideal, then it is called an invariant subalgebra or *two-sided ideal.*

When an algebra A possessess an invariant subalgebra I, one can define the quotient algebra A/I of equivalence classes. The quotient is with respect to the multiplication in the algebra. An invariant subalgebra must first of all be a vector subspace of the underlying vector space. Then multiplication by any element in the algebra (right or left) produces an element in the ideal. Further properties for algebras will be defined in the context of Lie-algebras (see Section 3.11).

3.7.8 Gradings and derivations

For many algebraic structures, one can define the concepts of gradings and derivations, and they play important roles in higher spin theory. These concepts are often most usefully defined when needed and relevant, but the underlying intuition can be captured in general terms. A graded vector space is a vector space that can be decomposed into a direct sum of vector spaces. The grading is often over natural numbers or the integers, but other "index" sets are possible. If V_n denotes the individual, homogeneous, vector spaces, then one writes for the full graded vector space

$$V = \bigoplus_{n=0}^{\infty} V_n \tag{3.324}$$

or something analogous for other kinds of index sets.

61 There are, however, many concepts of "ideals" in mathematics, prompting caution.

A role model is given by ordinary polynomials in an indeterminate variable x over the real numbers. Then the homogeneous elements V_n are given by the monomials. Supplying this vector space with ordinary multiplication of monomials, we get a *graded algebra* with $V_n \diamond V_m \subset V_{n+m}$. This can be generalized, but that may just as well be done in the proper contexts.

Furthermore, it may be possible to define a *derivation* d with the property $dV_n \subset V_{n-1}$ and satisfying a generalized Leibniz type rule. Also here, the exact definitions are most usefully given as needed and relevant. In the example of polynomials in a variable x, the derivation d can be taken as the ordinary derivative d/dx.

3.7.9 Tensor products and tensors

Tensor products are extremely useful mathematical constructions that play important roles in many areas of theoretical physics; for instance in classical tensor calculus itself, in quantum mechanics and in higher spin theory. Since the concept is not without its subtleties, let us approach it from a few different angles.

In Section 3.7.5 when discussing vectors and dual vectors, we saw that it was natural to write vectors with upper indices and dual vectors with lower indices.[62] In preparation for manifold theory, we now call vectors (upper indices) *contravariant vectors* and dual vectors (lower indices) *covariant vectors*. This is also in conformity to the usage within the special theory of relativity as we reviewed it in Section 3.4.1. This now generalizes to tensors, indeed just as in special relativity.

A (p, q) tensor is an object $T^{i_1 \ldots i_q}_{j_1 \ldots j_p}$ with p covariant and q contravariant indices. In analogy to how a contravariant vector – a $(0, 1)$ tensor – maps covariant vectors to real numbers, and how a covariant vector – a $(1, 0)$ tensor – maps contravariant vectors to real numbers, we can think of a (p, q)-tensor as mapping q covariant vectors and p contravariant vectors to real numbers.

It can be thought of as a multi-linear map (linear in each "index slot") from the Cartesian product of q copies of the vector space V and p copies of the dual vector space V^* to the real numbers, that is a map T^q_p

$$V^{\times q} \times (V^*)^{\times p} \xrightarrow{T^q_p} \mathbf{R} \tag{3.325}$$

explicitly given by

$$T^q_p(u_1, \ldots, u_q; v_1, \ldots, v_p) = T^{i_1 \ldots i_q}_{j_1 \ldots j_p} u_{1,i_1} \ldots u_{q,i_q} v^{i_1}_1 \ldots v^{i_p}_p \tag{3.326}$$

62 Which is which, is a convention.

Tensors of the same type can be multiplied by numbers and added, therefore, the space of (p, q) tensors is a itself vector space. As noted in Section 3.4.1, tensors can be multiplied. This offers the possibility to turn the set of all tensors into an algebra.

Indeed, denoting the vector space of (p, q) tensors by Υ_p^q, the *tensor product*

$$\Upsilon_p^q \times \Upsilon_m^n \xrightarrow{\otimes} \Upsilon_{p+m}^{q+n} \subset \Upsilon_p^q \otimes \Upsilon_m^n : \quad T_p^q \otimes T_m^n = T_{p+m}^{q+n} \tag{3.327}$$

This, somewhat heavy formula, says that multiplying a (p, q)-tensor from the vector space of (p, q)-tensors and a (m, n)-tensor from the vector space of (m, n)-tensors, produces a $(p+m, q+n)$-tensor in the vector space of $(p+m, q+n)$-tensors, this vector space being a subspace of the *tensor product of the vector spaces* of (p, q)-tensors and (m, n)-tensors.

It is the "subspace" that is the interesting and somewhat subtle point here. Namely, not all tensors in the tensor product space $\Upsilon_p^q \otimes \Upsilon_m^n$ can be factored as products of tensors $T_p^q \otimes T_m^n$ (similar symbol notwithstanding). The concept can be distilled a little bit by considering two vector spaces V and U with generic vectors v and u. The product of vectors $v \otimes u$ then lives in the tensor product $V \otimes U$, this space, however, contains tensors that cannot be factored. This will be explained in the box below.

3.7.10 Tensor algebra

The algebra aspect of the tensor product can be brought forth if one considers the following list of successive tensor products of a single vector space V, of covariant vectors say, with itself

$$\mathbf{R}, V, V \otimes V, V \otimes V \otimes V, \ldots \tag{3.328}$$

where $\mathbf{R}$ represents the scalars. The elements of this list are covariant tensors of rank $n \in \mathbf{N}$. Clearly, any tensor can be multiplied by a scalar. Any two tensors of the same rank can be linearly combined. This is an example of an *graded vector space*.[63] Furthermore, any two tensors can be multiplied using the product $\otimes$. This gives us a example of a *graded algebra*. We will use the notation $\Upsilon(V)$ for this *tensor algebra* and $\Upsilon_n(V)$ for the tensors of rank n.

Even though a tensor cannot in general be factored as a product of vectors, it can always be written as a sum of products of vectors. This means that the tensor algebra $\Upsilon(V)$ is generated by the vector space V in the sense of the list (3.328). More exactly, mathematicians say that it is *freely generated*, or that it is *free* graded algebra over V, "free" meaning that it is in a certain sense the most general such algebra. These notions will be further elaborated in Section 3.7.12. Let us now compare the tensor product to the direct product, or direct sum.

---

63 Adding tensors of different rank cannot be done, nor is it needed.

3.7.11 Direct sums (or products)

Somewhat confusingly, a *direct sum* and a *direct product* of vector spaces, is actually the same thing.[64] Intuition can be gained by considering ordinary Euclidean $\mathbf{R}^2$ vector space. The underlying set is the Cartesian product of two copies of the real line $\mathbf{R}_x \times \mathbf{R}_y$ with elements represented as ordered pairs (x, y). Vector addition and scalar multiplication is lifted from the two real lines in the following standard way:

$$(x_1, y_1) + (x_2, y_2) = (x_1 + x_2, y_1 + y_2) \tag{3.329}$$

$$a(x, y) = (ax, ay) \tag{3.330}$$

These formulae generalize to the case of a direct product of two vector spaces X and Y of dimension m and n, respectively, just by interpreting $x_1 + x_2$ as vector addition in X and analogously for the rest of the indicated operations. From this, it is clear that the dimension of the product vector space is $m + n$. For notation, the product of the vector spaces may be denoted by the same symbol $\times$ as the Cartesian product of the underlying sets, or the symbol $\oplus$, in which case one writes $X \oplus Y$. The reason for this latter choice of notation may be that $\otimes$ is already used for tensor products.[65] As we saw above, the tensor product of the two vector spaces are of dimension $m \cdot n$, so the concepts are fundamentally different. We can now write for the graded vector space (and algebra) based on (3.328) discussed in Section 3.7.9,

$$\Upsilon(V) = \bigoplus_{i=0}^{\infty} \Upsilon_i(V) \tag{3.331}$$

A further comment on tensor versus Cartesian products

! Working abstractly one might still get momentarily confused over the difference between Cartesian products $\times$ and tensor products $\otimes$. The difference is most easily understood for vector spaces.

Consider a two-dimensional vector space V with basis vectors $\mathbf{e}_1$ and $\mathbf{e}_2$. Then consider two arbitrary vectors $\mathbf{u} = u^1\mathbf{e}_1 + u^2\mathbf{e}_2$ and $\mathbf{v} = v^1\mathbf{e}_1 + v^2\mathbf{e}_2$. Their "product" can be written by formally multiplying the vectors and collecting all terms $u^1v^1\mathbf{e}_1\mathbf{e}_1 + u^1v^2\mathbf{e}_1\mathbf{e}_2 + u^2v^1\mathbf{e}_2\mathbf{e}_1 + u^2v^2\mathbf{e}_2\mathbf{e}_2$. Having done that one would like to consider $\mathbf{e}_1\mathbf{e}_1$, $\mathbf{e}_1\mathbf{e}_2$, $\mathbf{e}_2\mathbf{e}_1$ and $\mathbf{e}_2\mathbf{e}_2$ as basis elements in a new vector space and consider for example vectors of the form $w^{11}\mathbf{e}_1\mathbf{e}_1 + w^{22}\mathbf{e}_2\mathbf{e}_2$. This is precisely what the tensor product allows us to do. Thinking in terms not of a product of the vectors *per se*, but rather of a product of the vector space V with itself (indeed the tensor product $V \otimes V$) allows us to consider tensors such as $w^{11}\mathbf{e}_1\mathbf{e}_1 + w^{22}\mathbf{e}_2\mathbf{e}_2$

---

64 For other algebraic structures, direct sums and direct products may differ in some respects. In my opinion, the terminology is not very well chosen.

65 Some further insight into this notational conundrum can be gleaned from the concept of a product of two groups G and H, where the "product" in the product group $G \times H$ is defined as $(g_1, h_1) \cdot (g_2, h_2) = (g_1g_2, h_1h_2)$. For an illuminating discussion, see [255] Section I3.

that cannot be written as products of vectors. It can, however, be written as a linear combination of products of vectors.[66]

So what is the Cartesian product of two vector spaces? Perhaps it is best to think of it as just the Cartesian product of the underlying sets. The *Cartesian product* is a set theoretic notion, the corresponding concept for vector spaces is the direct sum, or direct product, denoted by $\oplus$. The connotation of "direct sum" is in relation to the "adding of vector spaces" as explicated by the formula (3.329). The connotation of "direct product" is in relation to the underlying Cartesian product of the sets.

3.7.12 Free algebras

The concept of a *free algebra* is natural and very useful. Consider a set of n "indeterminates" (variables of some unspecified kind) $x_1, x_2, \ldots, x_n$. The x_i's can also be viewed as letters of an alphabet. From these, one can form strings (or words) of any length by concatenating them. This can be considered as formal products where the order is important. There are no restrictions, conditions or equations relating products of the indeterminates x_i, so no simplifications can be done. This is indeed the meaning of the characterization "free". Then consider formal linear combinations of such words with coefficients chosen from some number system

$$S(x_i) = C + \sum_{k=1} C^{i_1 \ldots i_k} x_{i_1} \ldots x_{i_k} \tag{3.332}$$

where the first term C signifies the empty word. Each index i_k run from 1 to n, and the sum is a sum over index sets $\{i_1, \ldots, i_k\}$ for each word length k. A little thought convinces one-self that the $S(x_i)$ span an algebra by concatenation (product) and sum. Indeed, the countable set of words $\{x_{i_1} \ldots x_{i_k}\}$ with $k = 0, 1, 2, \ldots$, form a basis for the algebra. One can also think of the $S(x_i)$ as noncommutative polynomials in the variables x_i. The algebra is however clearly associative (since word concatenation is associative and concatenation distributes over the summation).

If one chooses as indeterminates the basis vectors e_i of a n-dimensional vector space V, it turns out that the free algebra of the $S(e_i)$ is isomorphic to the tensor algebra $\Upsilon(V)$ of (3.331). This isomorphism will become important when we consider universal enveloping algebras in Volume 2. The isomorphism hinges on the one-to-one correspondence between the coefficients $C^{i_1 \ldots i_k}$ and rank k tensors.[67]

---

66 The quantum mechanical phenomena of entanglement is captured mathematically by these concepts. Some states – the entangled ones – of a system built from two systems cannot be realized as a product of states of the individual systems.

67 It should be quite clear that the graded tensor algebra of (3.331) is indeed a free algebra, once the concept as such is grasped.

3.8 Exterior algebra and differential forms

The exterior algebra of antisymmetric covariant tensors – or differential forms – play an important role in field theory. Here, we will see how such an algebra can be constructed[68] out of the tensor algebra $\Upsilon(V)$. The subject will then be returned to in Section 3.10.2 where we discuss differential forms on manifolds.

Consider a vector space V of covariant vectors and the tensor algebra $\Upsilon(V)$ built upon it according to Section 3.7.10. For antisymmetric tensors, we need an antisymmetric product, to be denoted by $\wedge$ and called a *wedge product*. The product $\otimes$ of general tensors have no particular symmetry apart form it being associative but not commutative.[69] Thus we require of the product $\wedge$ that for any tensors s and t: $s \wedge t = -t \wedge s$. This can be achieved by mapping the product $t \otimes t$ (and all multiplies of it) to zero. This results in a quotient algebra, which turn out to be precisely the exterior algebra of antisymmetric tensors. Since for any two tensors s and t, we have $s \wedge s = 0$ and $t \wedge t = 0$ we can do the following short calculation:

$$(s+t) \wedge (s+t) = 0 \quad \Rightarrow \quad s \wedge t + t \wedge s = 0 \tag{3.333}$$

thus ensuring antisymmetry of the new product. This new algebra will be denoted by $\Omega(V)$ and called an *exterior algebra*. It has certain special properties. It is graded just as the tensor algebra $\Upsilon(V)$ and the components will be analogously denoted by $\Omega_p(V)$. The tensor type p is called *form degree*. However, due to the antisymmetry, the form degree cannot be higher that the dimension n of the underlying vector space V, and we now have

$$\Omega(V) = \bigoplus_{i=0}^{n} \Omega_i(V) \tag{3.334}$$

The reason for the name *differential forms* has to do with the fact that the coordinate differentials dx^μ provide a natural basis for covariant vectors. Somewhat deeper, differential forms are objects that are naturally integrated.

A basis for the vector space of p-forms $\Omega_p(V)$ can be constructed from a basis of the vector space V. Let θ_i with $1 \le \theta \le n$ be a basis for V. Consider two vectors

$$u = u_i \theta^i \quad \text{and} \quad v = v_i \theta^i \tag{3.335}$$

It is easy to see by direct computation that the antisymmetry $u \wedge v = -v \wedge u$ requires for the basis vectors

$$\theta^i \wedge \theta^j = -\theta^j \wedge \theta^i \tag{3.336}$$

68 Inspired by [256], Chapter VII.

69 On a lower level, this is what one have for word concatenation: stringing one m letter word to an n letter word produces an $m+n$ letter word. This is clearly an associative but not commutative operation.

This also leads to an explicit expression for the wedge product of the two 1-forms

$$u \wedge v = \sum_{i<j} (u_i v_j - u_j v_i) \theta^i \wedge \theta^j \tag{3.337}$$

Then any antisymmetric 2 index covariant tensor t, that is, a 2-form, can be expanded as

$$t = \sum_{i<j} t_{ij} \theta^i \wedge \theta^j \tag{3.338}$$

Therefore, the set $\{\theta^i \wedge \theta^j : i < j\}$ is a basis for the vector space $\Omega_2(V)$ of 2-forms. Continuing this argument for higher order wedge products of vectors shows that a basis for $\Omega_p(V)$ is the set $\{\theta^{i_1} \wedge \cdots \wedge \theta^{i_p} : i_1 < i_2 < \ldots < i_p$. To formalize this, we can introduce the index sets I_p for every p

$$I_p = \{i_1, i_2, \ldots, i_p\} \quad \text{where } i_1 < i_2 < \ldots < i_p \tag{3.339}$$

and the independent basis vectors

$$\theta^{I_p} = \theta^{i_1} \wedge \theta^{i_2} \wedge \ldots \theta^{i_p} \tag{3.340}$$

The dimension of the vector space spanned by this basis is $\binom{n}{p}$ since there are that many different index sets I_p. Using this notation, a p-form α can formally be written as

$$\alpha = \alpha_{I_p} \theta^{I_p} \tag{3.341}$$

where the sum over the abstract index I_p runs over the $\binom{n}{p}$ independent index combinations of (3.339).

The exterior product of any two p- and q-forms can now be defined and consistently computed according to

$$\Omega_p(V) \times \Omega_q(V) \to \Omega_{p+q}(V) : (\alpha_p, \beta_q) \mapsto \alpha_p \wedge \beta_q \tag{3.342}$$

with α_p and β_q expanded in the basis (3.340).

The linearly independent basis (3.340) is convenient in explicit calculations. One can also use an overcomplete basis and express a p-form as

$$\alpha = \frac{1}{p!} \alpha_{i_1 i_2 \ldots i_p} \theta^{i_1} \wedge \theta^{i_2} \wedge \cdots \wedge \theta^{i_p} \tag{3.343}$$

where all indices are summed over all values. The combinatorial factor $1/p!$ ensures equality to (3.341). This expansion is useful when writing general formulas.

The formalism developed so far has been entirely algebraic with no reference to an underlying space supporting the vector space V or the p-forms built upon it. Alternatively, one could think of everything so far as applying to a specific point x in a flat space E. Moving around with this point, we easily generalize to vector fields and p-form fields on E. In particular, we could identify E and V (as is often the case in elementary applications). For the next step in the development – introducing the exterior derivative – such a generalization is needed.

3.8.1 Exterior derivative

To be specific, consider a n-dimensional Euclidean vector space E (identified with V). As in elementary vector analysis, the coordinate differentials dx^i form a natural basis of covariant vectors (as does the partial derivatives ∂_i for contravariant vectors). Everything from the previous section can now be carried over with the coordinate differentials dx^i playing the role of the basis vectors θ^i. Furthermore, this holds at every point x in E so we can think about fields defined on E. This opens up the possibility to compute partial derivatives.

Indeed, since the derivatives $\partial/\partial x^i = \partial_i$ can be viewed as covariant vectors, we can consider the 1-form

$$dx^i\partial/\partial x^i = dx^i\partial_i \equiv d \tag{3.344}$$

Acting with this operator on a p-form will produce a $(p+1)$-form according to the following calculation:

$$d \wedge \omega_p = dx^i \frac{\partial}{\partial x^i} \wedge \omega_p = dx^i \frac{\partial}{\partial x^i} \wedge \omega_I dx^I = \frac{\partial}{\partial x^i}\omega_I dx^i \wedge dx^I \tag{3.345}$$

This derivative is called the *exterior derivative*. It is a linear operator that satisfies the following rules:

$$d(\alpha_p \wedge \beta_q) = d\alpha_p \wedge \beta_q + (-1)^p \alpha_p \wedge d\beta_q \tag{3.346}$$

$$d(d\alpha) = 0 \tag{3.347}$$

The first equation resembles the Leibniz rule for derivatives. The last equation follows from the equality of mixed partial derivatives. Alternatively, one can take the properties of the exterior derivatives as axioms defining it.

Example 3 (Forms in three dimensions). In three-dimensional Euclidean space, we have the following forms:

$$\begin{aligned} &\text{0-form:} && f \\ &\text{1-form:} && \alpha = a_x dx + a_y dy + a_z dz \\ &\text{2-form:} && \beta = b_z dxdy + b_y dzdx + b_x dydz \\ &\text{3-form:} && \gamma = g dxdydz \end{aligned} \tag{3.348}$$

where, as is usual when no confusion can arise, the wedge symbol is suppressed.[70] Then letting d act on these forms, we get

---

70 The choice of the order $dzdx$ in the second term for the 2-form is dictated by getting the conventional sign in the second term for $d\beta$ below.

$$
\begin{aligned}
df &= \frac{\partial f}{\partial x}dx + \frac{\partial f}{\partial y}dy + \frac{\partial f}{\partial z}dz \\
d\alpha &= \left(\frac{\partial a_y}{\partial x} - \frac{\partial a_x}{\partial y}\right)dxdy + \left(\frac{\partial a_x}{\partial z} - \frac{\partial a_z}{\partial x}\right)dzdx + \left(\frac{\partial a_z}{\partial y} - \frac{\partial a_y}{\partial z}\right)dydz \\
d\beta &= \left(\frac{\partial b_x}{\partial x} + \frac{\partial b_y}{\partial y} + \frac{\partial b_z}{\partial z}\right)dxdydz \\
d\gamma &= 0
\end{aligned}
\tag{3.349}
$$

Here, we recognize the usual derivative operators **grad**, **curl** and **div** of three-dimensional vector analysis. Furthermore, forms are natural objects to integrate over. ◂

!

So far, we have not made clear in what space the forms are valued. When working on the base manifold $\mathbf{R}^n$, we can always think, mathematically, of the components of vectors and tensors as being real valued and consequently, vectors being valued in $\mathbf{R}^n$. However, this $\mathbf{R}^n$ is only isomorphic to the base space $\mathbf{R}^n$ and not identical. As physicists, we only have to think of electromagnetic fields, to realize that there is a difference – electric fields do not really "point" in geometric space even though we often depict it that way. The solution to the riddle is, of course, that the electric force on a charged test particle can be thought of as balanced by a mechanical force (from a spring dynamo-meter for instance) that can be represented in geometrical space. Alternatively, one can think of the acceleration that the electric force would impart if the particle are free to move.[71] In principle though, the forms describing physical fields are valued in vector spaces appropriate to the phenomena to be described. Mathematically, we can consider the tensor vector space as built on the tangent and cotangent spaces of the manifold.

3.8.2 Integration

One source of importance of differential forms is that they are natural objects to integrate. This is important since in field theory we want to integrate Lagrangian densities to get actions. Furthermore, integrals are global objects, and as such they are sensitive to the topological properties of the underlying space.

The three-dimensional example of the previous section is enough to bring out the idea. 1-forms leads to line integrals, 2-forms to surface integrals and so on. In particular, in d dimensions, d-forms are natural objects to integrate over – pieces of, or all of – the underlying space.

3.8.3 de Rahm cohomology

The introduction of the exterior derivative turns the graded algebra $\Omega(V)$ into a *differential graded algebra*. The operator d maps the vector space of p-forms into the vector

71 What we observe are the effects of electromagnetic fields on charged particles mapped via the Lorentz force, into geometrical space.

space of $(p+1)$-forms. One gets a sequence of maps

$$\Omega_0 \xrightarrow{d} \Omega_1 \xrightarrow{d} \Omega_2 \xrightarrow{d} \cdots \xrightarrow{d} \Omega_p \xrightarrow{d} 0 \tag{3.350}$$

More specifically, this structure is called the *de Rahm complex* on $\mathbf{R}^n$. Due to the property $d^2 = 0$ of the map d, one can define two interesting types of forms.

In general, the *kernel* of a map m between two spaces X and Y, consists of those elements in X that are mapped to zero in Y. Here, forms in any of the spaces Ω_i that are mapped to zero in the space Ω_{i+1} by the derivation d, are called *closed*. That is, the kernel of d are the *closed forms*. On the other hand, the *image* of the derivation d are called *exact* or *exact forms*. The exact forms are automatically closed.

Why is this interesting? One may gain intuition from the simple example of differential forms on $\mathbf{R}^2$. Consider 1-forms $fdx + gdy$. To find the closed 1-forms, one must solve the differential equation $\partial g/\partial x - \partial f/\partial y = 0$. Among the solutions are the trivial "uninteresting" exact forms that are closed automatically. One would like "divide out" the exact forms in order to "measure" the size of the interesting solution space. This, admittedly vague idea, may serve as a motivation for the definition of the qth *de Rahm co-homology* of $\mathbf{R}^n$ as the vector spaces

$$\begin{aligned} H^q_{dR}(\mathbf{R}^n) &= \{\text{closed } q - \text{forms}\}/\{\text{exact } q - \text{forms}\} \\ &= \ker(d)|_{\Omega_q} / \operatorname{im}(d)|_{\Omega_q} \end{aligned} \tag{3.351}$$

The "dividing out" – a modulo construction – can be made mathematically sound (since the spaces involved are vector spaces and, therefore, Abelian groups as well) using the concepts reviewed in Section 3.9.

The *de Rahm co-homology* is the cohomology of the full de Rahm complex (3.350). It is meant to capture topological properties of the space $\mathbf{R}^n$. Now, this space is obviously trivial topologically, and this is reflected in the so-called *Poincaré lemma* which states[72]

$$H^q_{dR}(\mathbf{R}^n) = \begin{cases} \mathbf{R} & \text{for } q = 0 \\ 0 & \text{for } 1 \leq q \leq n \end{cases} \tag{3.352}$$

However, the de Rahm cohomology can be generalized to differential manifolds with more complicated topology. Since differential forms are natural objects to integrate, and integrals depend on global properties of the manifold, one may suspect that we here have a tool to analyze the topology of manifolds. Furthermore, the fairly concrete constructions that we have mentioned here, can be generalized by carefully extracting the generic properties of co-homology and turning them into axioms. In that way yielding more abstract differential graded algebras and cohomology theories. This is

72 For more discussion and a proof, see [257].

a huge area of mathematics. We will introduce examples as they may be relevant for higher spin gauge theory.

Example 4 (Poincaré lemma for the real line). For the real line $\mathbf{R}$, the Poincaré lemma can be proved using only elementary calculus. The sequence of spaces (3.350) is very short: $\Omega_0 \xrightarrow{d} \Omega_1 \xrightarrow{d} 0$. On the space of functions $\Omega_0(\mathbf{R})$, the kernel of d are the constant functions. The image of d is empty, so $H^0_{dR}(\mathbf{R}) = \mathbf{R}$. On the space of 1-forms $\Omega_1(\mathbf{R})$, the kernel of d is all the 1-forms. It remains to show that every 1-form is exact. If $\omega = g(x)dx$ is a 1-form, we can integrate it to $f(x) = \int_0^x g(t)dt$. Then $df(x) = g(x)dx$ and we see that every 1-form is exact. Therefore, $H^1_{dR}(\mathbf{R}) = 0$. ◂

3.8.4 The Hodge dual

The dimensions of the vector spaces Ω_p and Ω_{n-p} are the same. This makes it possible to define a duality operation that map p-forms into $(n-p)$-forms and vice versa

$$\star(dx^{i_1} \wedge \cdots \wedge dx^{i_p}) = \frac{1}{(n-p)!}\epsilon^{i_1\ldots i_p}{}_{i_{p+1}\ldots i_n} dx^{i_{p+1}} \wedge \cdots \wedge dx^{i_n} \tag{3.353}$$

with $\epsilon_{i_1\ldots i_p i_{p+1}\ldots i_n}$ the totally antisymmetric tensor in n dimensions. This operation is called the *Hodge duality* transformation. It can be used to define an inner product between two p-forms α_p and β_p as the integral

$$\langle \alpha_p, \beta_p \rangle = \int_M \alpha_p \wedge (\star\beta_p) \tag{3.354}$$

This is handy when writing actions for field theories formulated in the form language (see example 6). In terms of the coefficient functions of the forms one gets

$$\langle \alpha_p, \beta_p \rangle = p! \int_M \alpha_{i_1\ldots i_p}\beta_{i_1\ldots i_p} dx^1 \wedge \cdots \wedge dx_n \tag{3.355}$$

This shows that $\langle \alpha_p, \beta_p \rangle = \langle \beta_p, \alpha_p \rangle$ from which we also conclude

$$\alpha_p \wedge (\star\beta_p) = \beta_p \wedge (\star\alpha_p) \tag{3.356}$$

Applying the Hodge operation twice yields $\star\star\,\omega_p = (-1)^{p(n-p)}\omega_p$.

3.9 Transformation groups

Groups can be studied in the abstract, but in physics they are most often defined and studied as transformation groups acting on sets or spaces. When a group acts on a set (or a space) it "moves" the elements (or points) around, resulting in a permutation.[73] When the set that the group acts on has structure of its own (which is the common case in physics), then the group action should preserve that structure.

Intuitively, a permutation of the elements of any set X, with a finite of infinite number of elements, can be thought of as a rearrangement of the elements. Mathematically, a *permutation* is a bijection in X. As an example, any linear function is a permutation of the real numbers. It is quite clear that the set of permutations constitutes group in itself: two consecutive permutations is a permutation, permutations can be inverted (undone) and "no" permutation is the unit element. Furthermore, according to Cayley's theorem:[74]

> Any group G is isomorphic to a subgroup of the permutations $\sigma : X \to X$ of a suitable set X. The group $\sigma(X)$ is called the *symmetric group* of X.

Cayley's theorem offers the possibility to study realizations of a group G in the group of permutations $\sigma(X)$ of some appropriate set X. To bring this forth, we must introduce some terminology and notation.

First, remember that a homomorphism between two groups G and H is a map $\gamma : G \to H$ that preserves the group law so that $\gamma(g_1 g_2) = \gamma(g_1)\gamma(g_2)$ for all g_1 and g_2 in G. Note that such a homomorphism need not be injective or surjective.

Next, consider homomorphisms between a group G and the permutations σ of a set X: A group G is defined to be represented by the permutations of a set X, if there is a homomorphism $\gamma : G \to \sigma(X)$.

By a small shift in perspective, this can be thought of as transformations on the set X produced by the group acting on the set. Writing the permutation $\sigma(g) : X \to X$ induced by the group element g as L_g, the preservation of the group structure becomes

$$L_{g_2} \circ L_{g_1} = L_{g_2 g_1} \quad \text{for all} \quad g_1, g_2 \in G \tag{3.357}$$

We have already seen an instance of this general formula, for the Poincaré group, in (3.89).

A simplified notation is often used. The action of a group element g on the point x gives a new point $\gamma(g)(x) = L_g(x)$. This new point, and the action of g is often just written, with a bit of "syntactic sugar"[75] as gx, saying that the *group acts on the left*.

The group may also be thought of as *acting on the right*, which would be written as xg. The difference between left and right actions appears when one considers con-

---

73 We here adopt the active point of view on transformations.

74 For a proof, see [252].

75 The term "syntactic sugar" was coined by the computer scientist Peter Landin (of Queen Mary College, London). I prefer it over the phrase "abuse of notation".

secutive actions of the group. Two left actions with g_1 and g_2 yields g_2g_1x, while two right actions yields xg_1g_2. The difference shows up in the order of group elements. Left and right actions can be traded for each other using the formula $(g_1g_2)^{-1} = g_1^{-1}g_2^{-1}$, but it is often convenient to work with both. For right actions R_g, the formula (3.357) is modified accordingly: $R_{g_2} \circ R_{g_1} = R_{g_1g_2}$.

3.9.1 Transitive, effective and free actions

Three important types of group actions on a set that may be desirable in various contexts are: transitive, effective and free.

The group action is *transitive* if any two elements x_1 and x_2 in X can be related by a group element g, i. e., if $x_2 = gx_1$ for some g. This can be phrased in space terminology: any point y can be reached from any point x by a transformation $T : y = Tx$.

The group action is *effective* (or *faithful*) if different group elements g_1 and g_2 induce different permutations in the set X. This can be phrased as: for any two different g_1 and g_2, there exist at least one x in X such that $g_1x \neq g_2x$. Taking one group element as the unit e this means that for any other group element $g \neq e$ we have $gx \neq x$ for at least one x. In space terminology, every transformation T except the identity transformation, moves at least one point x of the space.

The occurrence of transformations that fail to move points, prompts the following two concepts. The *kernel* of a group action is defined as the set of group elements g for which $gx = x$ ("no action"). The kernel is obviously a subgroup. The group action is effective precisely when the kernel is trivial, that is, only contains the unit element of the group.

On the other hand, a *fixed point* of the group action is a point x_g for which $gx_g = x_g$ for some group element $g \in G$. A group action without any fixed points is said to be *free*. Free actions are effective, but an action may be effective without being free. One may very well have $gx_g = x_g$ for a fixed-point x_g and still have $gx \neq x$ for some other point in X. On the other hand, if there are no fixed points at all, then one can never have $gx = x$ for any g and any x.

3.9.2 Orbits, stabilizers and cosets

Orbits, stabilizers and cosets are three concepts that aim to capture the structure of group actions.

An *orbit* of the group action through a point x is the set of all points gx traversed as g varies throughout the group. In some more detail, the *orbit* of G through $x \in X$ is defined as the set[76]

$$\mathcal{O}_G|_x = \{gx : g \in G\} \tag{3.358}$$

It should be clear that the orbits through two different points x_1 and x_2 must either coincide or be disjoint. Indeed, choosing orbits through two distinct points x_1 and x_2 and assuming that their orbits coincide at some point, we must have $g_1x_1 = g_2x_2$ for some group elements g_1 and g_2. But then $x_1 = g_1^{-1}g_2x_2$ and x_1 belongs to the orbit of x_2.

So in general, the action of a group on a set, partitions the set into disjoint orbits. These orbits are indeed equivalence classes under the group action. In the special case that the group action is transitive, there is just one orbit (the complete set X) and of course just one equivalence class.

Referring back to the concept of fixed points, one may wish to consider subsets of group actions that do not move certain points. More exactly, fix some point x_0 and consider the subset of G,

$$G_{x_0} = \{g \in G : gx_0 = x_0\} \tag{3.359}$$

This is in fact a subgroup of G called the *stabilizer* or the *stability* subgroup of G. In various contexts, it is also called the *isotropy group* or the *little group*. We have already seen examples of this when discussing representations of the Poincaré group.

An important variant of orbits occurs when a proper subgroup H of a group G acts on the group itself. Consider right actions $R_h(g) = gh$ where we think of g as a particular element in G (but not in H) and h as ranging over the subgroup H. This action is not transitive. Take an element h_1 in H and a "point" g_1 in the complement of H in G. If the action were transitive, then there would exist a $h \in H$ such $g_1 = h_1h$, in conflict with the proper subgroup assumption. Therefore, the group G is partitioned into a set of disjoint orbits of H. In analogy to formula (3.358), we can consider sets

$$l\mathcal{C}_H|_g = \{gh : h \in H\} \tag{3.360}$$

These sets are called *left cosets* of H in G. A strongly sugared notation is used: the set $l\mathcal{C}_H|_g$ is denoted gH and the set of all left cosets is denoted by G/H.

All this can be repeated for left actions $L_h(g)$ of H on G. The set of *right cosets*

$$r\mathcal{C}_H|_g = \{hg : h \in H\} \tag{3.361}$$

is then denoted by $H\backslash G$. The reason why left cosets are defined through right action of H, and vice versa, will become clear in the next section.[77]

76 The orbit is here defined with the group acting on the left.

77 Mnemonics: G/H reads "G partitioned by H" and $H\backslash G$ reads "H partitioning G".

3.9.3 Quotient spaces

One can now ask what happens if the cosets are acted upon by elements of the group G? It may perhaps be suspected that the cosets gets permuted, and that is indeed the case. Consider the left action of an element g_2 of G on a left coset g_1H. This yields the left coset g_2g_1H. The action is transitive since two disjoint cosets g_1H and g_2H can be related by the action of $g_1(g_2)^{-1}$.

This means that the set G/H carries a transitive left action of G. It is therefore quite natural to consider it as a "space". As we will show in example 5, this will, for instance, allow to us to view Minkowski space-time as $\mathrm{ISO}(3,1)/\mathrm{SO}(3,1)$. But in order to do that in an unambiguous way, we need the converse to the above argument:

> If X is a space upon which the group G acts transitively, then X can be realized as G/H for some suitable subgroup H of G.

The proof is an application of the concepts discussed so far. It can be found in [252]. Spaces that are constructed in this way are called *quotient spaces*.

Example 5 (Minkowski space-time as a quotient space). For a relevant example, consider the Poincaré group acting on Minkowski space-time in the standard way. First, note that the Poincaré group acts transitively on Minkowski space-time, but the Lorentz subgroup does not. We want to realize Minkowski space-time as the quotient G/H with $G = \mathrm{ISO}(3,1)$ and $H = \mathrm{SO}(3,1)$. The group law is

$$g(\Lambda_2, a_2)g(\Lambda_1, a_1) = g(\Lambda_2\Lambda_1, \Lambda_2 a_1 + a_2) \tag{3.362}$$

Choose an arbitrary element $g_0 = g(\Lambda_0, a_0)$ in G and consider the right action of elements $h = g(\Lambda, 0)$ in H on g_0. The result is $g_0H = g(\Lambda_0\Lambda, a_0)$. These are the orbits, one for each value of a_0, parametrized by Λ.

Next, consider right action of an arbitrary element $g' = g(\Lambda', a')$ in G on an orbit. The result is $g'(g_0H) = g(\Lambda'\Lambda_0\Lambda, \Lambda' a_0 + a') = (g'g_0)H$ producing a new left coset. The action is clearly transitive.

The intuition is that picking an arbitrary point (Λ_0, a_0) in Poincaré group space and considering all actions of the Lorentz subgroup as equivalent, one is left with the translations a_0 that are isomorphic to Minkowski space-time. ◂

3.9.4 Normal subgroups and the group of cosets

Finally, let us return to the set of left cosets G/H and note that, although a space, it is in general not a group by itself, unless an extra requirement is added: H must be a so-called *normal subgroup* of G. To understand what needs to be done, consider the following reasoning.

The points in the left coset space G/H are the left cosets gH. One may attempt to define a group multiplication through the tentative formula

$$(g_1H)(g_2H) \stackrel{?}{=} g_1g_2H$$

But since we are trying to multiply equivalence classes, we must make sure that the result does not depend on the choice of representatives, in this case the choice of g_1 and g_2.

Now g_1 and g_1hH represents the same coset in G/H for any h in H. Then the tentative formula implies

$$(g_1hH)(g_2H) \stackrel{?}{=} g_1hg_2H \stackrel{?}{=} g_1g_2H$$

The second tentative equality may not be true. It would however become true if one could find an element h' in H such that $g_1hg_2 = g_1g_2h'$ because then we will get the true equality $g_1hg_2H = g_1g_2h'H = g_1g_2H$ of equivalence classes.

Multiplying the equation $g_1hg_2 = g_1g_2h'$ with g_1^{-1} on the right yields $hg_2 = g_2h'$. Thus we have to require that for any h in H and any g in G there should exist a h' in H such that $gh = h'g$. This is the same thing as requiring the left and right actions of H on G to coincide: $gH = Hg$. This means that the left and right cosets are the same. Subgroups that have this property are called *normal*.

3.10 Differential manifolds

The flat space of Euclid – perhaps even the flat space-time of Minkowski – is immediately given to the modern student. And we have no problem of thinking about a scalar field, for instance temperature, varying from point to point. Even vector fields such as the velocity of a particle **v** or the electric field **E**, should cause no great strain on the imagination. Concepts such as differentiation that require considering nearby points can also be thought about quite easily.

However, if the space is not flat, several related problems appear. The velocity **v** or the electric field **E**, “where are they pointing”, so to speak? When differentiating, how do we compare neighboring points if the space is not flat? How do we compare vectors at neighboring points? The pointing problem is in fact acute even in flat space since the electric field, being electrical, is clearly not pointing in geometrical space. This puzzle is solved, or glossed over, in elementary physics by defining the electric field in terms of the force on a test charge; this force in its turn manifested in the geometrical acceleration of the particle.

The established solution to these conundrums lies in the theory of differential manifolds and in the accompanying theory of fiber bundles (see Section 3.12).

Intuitively, a *manifold* is characterized by being locally, around every point, possible to map to a flat space. Such a map may not be possible to define throughout the whole space and, therefore, one has to think about different maps defined in different regions of the space. Where the regions overlap (as they have to do) there must be transition functions relating the maps. These are the coordinate transformations of theoretical physics. Let us draw the conventional picture (figure 3.2) illustrating the situation.

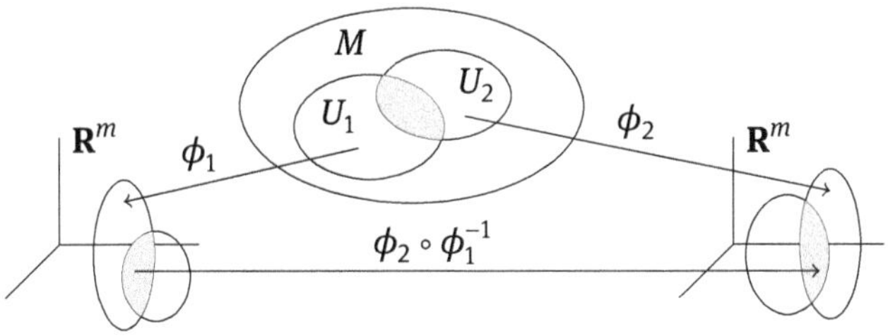

Figure 3.2: Change of coordinate systems.

The manifold M is thought of as covered by open sets U_i. In each U_i, points p are mapped to flat space $\mathbf{R}^m$ by homeomorphic[78] *coordinate maps* ϕ_i. A pair (U_i, ϕ_i) is called a *chart*. Where any two charts (numbered by i and j) overlap, there must be infinitely differentiable *transition functions* $\phi_i \circ \phi_j^{-1}$ and $\phi_j \circ \phi_i^{-1}$.

Coordinate functions

The coordinate functions ϕ_i map points p in the manifold to coordinates x^μ in flat space. This can be formalized as $x^\mu(p) = (u^\mu \circ \phi)(p)$ using *slot functions* $u^\mu : \mathbf{R}^m \to \mathbf{R}$. For physics, this is often too heavy a formalism, and we use x^μ generically for the coordinates. The coordinates in an open set U_i can be thought of as a parametric representation of the manifold in that open set. Although the x^μ are flat space coordinates, in physics we customarily call the indices μ *curved indices* or *world indices*. For further comments regarding tangent spaces, see Section 4.5.1.

The elaborate language of differential manifold theory is seldom used in theoretical physics, where the mathematical infrastructure is taken for granted.[79]

In this context, it is perhaps important to keep in mind that differential geometry is a more general mathematical theory than Einstein General Relativity which can be formulated in a differential geometrical language – with varying degrees of sophistication – through a sequence of physically motivated choices. For instance, so far we have made no choices as to connections or metrics. As we will see, such choices are closely connected to the gauge theory approach to general relativity, a subject that will

78 That is, one-to-one continuous functions with continuous inverses.

79 We will however return to it in Volume 2 when we will introduce jet-spaces.

be treated in Sections 4.5 and 4.6. For mathematically elaborate formulations of differential geometry see, for instance, [258, 259]. For physics informed presentations, the books [260, 127, 261] are helpful.

3.10.1 Tangent space and cotangent space

Due to its (still not entirely understood) significance for generalizing the theory of lower spin fields to in higher spin fields, we will review in some detail the construction of tangent vectors and tangent space. The *coordinate basis* in tangent space can be thought of as spanned by the partial derivatives ∂_μ and a general vector can be written as $V^\mu\partial_\mu$. How does this come about?

The definition of a tangent vector must be[80] intrinsic to the manifold, without reference to any embedding space where the vector can "directed". For the purpose of defining tangent vectors at a point p, one considers parametrized curves $c(\tau)$ mapping an open real interval about $\tau = 0$ into the manifold with $\phi(c(0)) = \phi(p)$. Then one takes a function f from M to $\mathbf{R}$ and defines a tangent vector as the directional derivative of the function along the curve, that is,

$$\left.\frac{df(c(\tau))}{d\tau}\right|_{\tau=0} = \frac{\partial f}{\partial x^\mu}\left.\frac{dx^\mu(c(\tau))}{d\tau}\right|_{\tau=0} \tag{3.363}$$

where, using the chain rule, the derivative has been expressed in terms of the coordinate functions x^μ. In this construction, the function f plays no deep role – it is just a place holder – whereas the choice of curve c corresponds to a particular tangent vector $X(c)$ (among the infinitely many). It is then natural to think of the partial derivatives as spanning a vector space and write an arbitrary tangent vector X as a differential operator

$$X = X^\mu \frac{\partial}{\partial x^\mu} = X^\mu\partial_\mu \tag{3.364}$$

with components X^μ that upon acting on a function produces the directional derivative at the point p through

$$\left.\frac{df(c(\tau))}{d\tau}\right|_{\tau=0} = X^\mu\frac{\partial f}{\partial x^\mu} \equiv X[f] \tag{3.365}$$

To sum up, $X = X^\mu\partial_\mu$ defines a vector, tangent to the manifold at the point $p = c(0)$ along the direction given by the curve $c(\tau)$. In these formulas, $\frac{\partial f}{\partial x^\mu}$ is syntactic sugar for the more correct expression $\frac{\partial(f\circ\phi^{-1}(x))}{\partial x^\mu}$.

80 At least, this is the conventional wisdom, and we will go along with it. For a discussion of various approaches to the definition of tangent space, see [260].

Having so motivated the use of the partial derivatives as a basis in tangent space, we can discard of the curves c and functions f and just think instead of arbitrary vectors X^μ. Before that however, to clinch the construction, one should really consider equivalence classes of curves, where two curves c_1 and c_2 subject to

$$c_1(0) = c_2(0) \tag{3.366}$$

$$\left.\frac{dx^\mu(c_1(\tau))}{d\tau}\right|_{\tau=0} = \left.\frac{dx^\mu(c_2(\tau))}{d\tau}\right|_{\tau=0} \tag{3.367}$$

are considered equivalent. A tangent vector (at a point) is then identified with an equivalence class of curves, rather than with a particular curve.

The coordinate derivative

As we will see, this fact that the coordinate derivatives serve as a natural basis in tangent space, plays a significant role in gauge treatments of gravity and is source of both similarities and differences between spin-1 (Yang–Mills) theory and spin-2 (gravity) theory. It is also basic to many attempts of constructing interacting higher spin theories as generalizations of the lower spin theories. As such, it is a root of both successes and failures. The construction of vectors as tangent vectors to the manifold is indeed elegant, but it has certain weaknesses. !

The *tangent space* at the point x – denoted by T_xM – is then the vector space of all the tangent vectors (all the equivalence classes of curves at the point).[81] The basis vectors $e_\mu = \partial_\mu$ just defined are called the *coordinate basis*. Clearly, the dimension of the tangent space is equal to the dimension of the manifold. The collection of all tangent spaces at all points x is the *tangent bundle* $TM = \bigcup_{x\in M} T_xM$. The word "bundle" here has a special meaning that we will discuss in Section 3.12.

One by-product of the definition of tangent vectors in terms of directional derivatives is that the left-hand side of the formula (3.363) shows that the tangent vector exists independently of any particular coordinate system. Using this, we can derive the transformation properties under a change of coordinates of the components of a vector. Consider two coordinate systems symbolized as x^μ and x'^μ and a vector V. Then we have

$$V^\mu(x)\frac{\partial}{\partial x^\mu} = V'^\mu(x')\frac{\partial}{\partial x'^\mu} \tag{3.368}$$

A simple application of the chain rule yields the transformation formula

$$V'^\mu(x') = V^\nu(x)\frac{\partial x'^\mu}{\partial x^\nu} \tag{3.369}$$

---

81 The notation for tangent spaces and the various other kinds of spaces to be considered is not standardized. But the variation is bounded and it is often possible to recognize the objects. Furthermore, I will be using the notation for the coordinates x as designating the points p of the manifold, not always, but when convenient.

Cotangent space T_p^* (at the point p) is the vector space dual to tangent space T_p. A basis in cotangent space is given by the differentials dx^μ. The intuition here is to consider the differential df of a function f

$$df = \frac{\partial f}{\partial x^\mu} dx^\mu \tag{3.370}$$

A *cotangent vector* ω – or a *one-form* – can now be expanded as

$$\omega = \omega_\mu dx^\mu \tag{3.371}$$

In analogy with general vector space theory, we can now write

$$\langle dx^\mu, \partial_\nu \rangle = \delta^\mu_\nu \tag{3.372}$$

Note that this have nothing to do with the existence or not of a metric on the manifold. As discussed in Section 3.7.5, a metric can be supplied in order to raise and lower indices. Such a metric will then in general not be constant. We will return to this in Section 4.5.1. Tensors in differential geometry are defined on the tangent and cotangent spaces of the manifold. The theory runs just as in the flat space of special relativity.

3.10.2 Differential forms on manifolds

The apparatus of differential forms that we developed in Section 3.8 can be taken over with minor modifications to differential manifolds. The theory of differential forms draws its strength from the interesting properties of totally antisymmetric tensors.[82]

Generalizing the concept of a 1-form, which is a covariant vector, a p-form, or a *differential form* of degree p, is a totally antisymmetric covariant tensor, generically denoted by $\omega_{\mu_1 \dots \mu_p}$. Such tensors form a subspace of the vector space of all type $(p, 0)$ tensors. A short notation for a p-form is ω_p with

$$\omega = \frac{1}{p!} \omega_{\mu_1 \mu_2 \dots \mu_p} dx^{\mu_1} \wedge dx^{\mu_2} \wedge \dots \wedge dx^{\mu_p} \tag{3.373}$$

The vector space of p-forms will be denoted by $\Omega_p(M)|_x$. It is a subspace of the space $(T_x^*)^{\otimes p}$, the tensor product of p copies of the cotangent space. As the point x varies over the manifold, we have a p-form field.[83]

Due to the antisymmetry of the wedge product, the expansion (3.373), while perfectly valid, is redundant. Indeed, the set $\{dx^{\mu_1} \wedge \dots \wedge dx^{\mu_p}\}$ is over-complete as a vector

82 See [243], Section 4.11 which has been an inspiration for the present section.

83 An alternative notation is $\wedge^p T_x^*$ which supports the intuition of an algebra with product $\wedge$ defined on T_x^*.

space basis. Since any particular value for the indices μ_i can only occur once, we can choose to represent a component of a p-form as $\omega_{\mu_1\mu_2\ldots\mu_p}$ with the indices ordered as $\mu_1 < \mu_2 < \cdots < \mu_p$. An independent basis is thus the set $\{dx^{\mu_1} \wedge \cdots \wedge dx^{\mu_p}\}$ with ordered indices.[84] This explains the convenience of the factor $1/p!$ in formula (3.373).

One would like to turn the vector space of all p-forms (for $0 \le p \le m$) into an algebra by defining a product between forms. Let $\alpha_{\mu_1\ldots\mu_p}$ be a p-form and $\beta_{\mu_1\ldots\mu_q}$ be a q-form. The direct product $\alpha\beta$, although a tensor, is not antisymmetric. An antisymmetric product can be defined as follows:

$$(\alpha \wedge \beta)_{\mu_1\cdots\mu_{p+q}} = N(p,q) \sum_{\mathcal{P}} \operatorname{sign}\left((\alpha\beta)_{\mu_{\mathcal{P}(1)}\cdots\mu_{\mathcal{P}(p+q)}}\right) \tag{3.374}$$

The sum is over all permutations $\mathcal{P}$ of the indices needed to make the product fully antisymmetric. The factor $N(p,q)$ determines the "weight" of the anti-symmetrization. There are two often made choices, either "averaging" with $N(p,q) = 1/(p+q)!$ or $N(p,q) = 1/p!q!$. In higher spin theory, it is more common practice to symmetrize and antisymmetrize with unit weight, that is, $N(p,q) = 1$.

The Hodge duality formula (3.353) gets modified

$$^*(dx^{i_1} \wedge \cdots \wedge dx^{i_p}) = \frac{\sqrt{g}}{(n-p)!} \epsilon^{i_1\ldots i_p}{}_{i_{p+1}\ldots i_n} dx^{i_{p+1}} \wedge \cdots \wedge dx^{i_n} \tag{3.375}$$

The factor $\sqrt{g}$ provides the correct factor to make the volume element in the inner product between two p-forms α_p and β_p invariant

$$\langle \alpha_p, \beta_p \rangle = \int_M \alpha_p \wedge^* \beta_p \tag{3.376}$$

Indeed, the curved space Hogde dual is defined in order to achieve just this.

Example 6 (Maxwell theory). The formalism can be usefully calibrated on electrodynamics in four dimensions. The condensed notation $F = dA = d \wedge A$ for the field strength $F_{\mu\nu}$ in terms of the gauge potential A_μ, expands into $\frac{1}{2}F_{\mu\nu}dx^\mu \wedge dx^\nu$. This in its turn rewrites to $\partial_\mu A_\nu dx^\mu \wedge dx^\nu$. A gauge transformation $\delta A_\mu = \partial_\mu \xi$ can be written as $\delta A = d\xi$ from which $\delta F = 0$ follows without calculation.

In order to write the action for electromagnetism in form language, one must use the Hodge-dual of the field strength as the following calculation shows:

$$\begin{aligned} ^*F \wedge F &= \frac{1}{2} F_{\mu\nu} F_{\alpha\beta} \epsilon^{\mu\nu}{}_{\rho\sigma} dx^\rho \wedge dx^\sigma \wedge dx^\alpha \wedge dx^\beta \\ &= \frac{1}{2} F_{\mu\nu} F_{\alpha\beta} \epsilon^{\mu\nu}{}_{\rho\sigma} \epsilon^{\rho\sigma\alpha\beta} dx^1 \wedge dx^2 \wedge dx^3 \wedge dx^4 \\ &= F_{\mu\nu} F_{\alpha\beta} (\eta^{\mu\alpha}\eta^{\nu\beta} - \eta^{\nu\alpha}\eta^{\mu\beta}) dx^1 \wedge dx^2 \wedge dx^3 \wedge dx^4 = 2F_{\mu\nu}F^{\mu\nu} d^4x \end{aligned} \tag{3.377}$$

84 See, for instance, [257], Chapter 1 or [251] Chapter IV.

Thus the action is

$$S_{EM} = -\frac{1}{4g^2}\int F_{\mu\nu}F^{\mu\nu}d^4x = -\frac{1}{8g^2}\int_M {}^*F \wedge F \qquad (3.378)$$

Had we tried $F \wedge F$ we would have found $\frac{1}{2}{}^*F_{\mu\nu}F^{\mu\nu}$ instead, which yields a total derivative. We have employed the useful formula $dx^\alpha dx^\beta dx^\gamma dx^\delta = \epsilon^{\alpha\beta\gamma\delta}dx^0dx^1dx^2dx^3$. This is a purely combinatorial formula, so there is no factor of $\sqrt{g}$. ◄

The epsilon tensor in four dimensions

i Formulas for contractions over pairs of indices of the product of two epsilon tensors are needed now and then. From $\epsilon^{\alpha\beta\gamma\delta}\epsilon_{\alpha\beta\gamma\delta} = 4!$ it is clear that we must have $\epsilon_\mu{}^{\alpha\beta\gamma}\epsilon_{\nu\alpha\beta\gamma} = 3!\eta_{\mu\nu}$. This is its turn requires that we have $\epsilon_{\mu\nu}{}^{\alpha\beta}\epsilon_{\rho\sigma\alpha\beta} = 2(\eta_{\mu\rho}\eta_{\nu\sigma} - \eta_{\nu\rho}\eta_{\mu\sigma})$. Continuing in this way, one can infer equations for epsilon tensors contracted over one index pair and no index pair in terms of sums of products of η tensors (with the correct antisymmetry).

3.11 Lie groups and Lie algebras

A Lie group G is a group which is also a manifold – the *group manifold* – in such a way that the group structure and the manifold structure are compatible. This means that the group multiplication and inversion, viewed as maps $G \times G \to G$ and $G \to G$, are smooth maps.

The compatibility of the group and manifold structures are quite restrictive, and for many purposes it is enough to study the group in the vicinity of the group unit. This is where the Lie algebra resides. The Lie algebra is spanned by generators T which also span a vector space, indeed the tangent space located at the origin (the identity element) in the group manifold. All this can be made exact, and we refer the reader to [262] for details. Here, we will review what we will eventually need from Lie algebra theory itself with only cursory remarks on the corresponding Lie groups. Lie algebra theory is a huge subject in itself, a thorough treatment can be found in [263]. The concept of a Lie algebra can be defined independently of the Lie group and it is often convenient to do so. The association of a Lie algebra to a Lie group can then be made separately.[85]

3.11.1 Lie algebras

A *Lie algebra* $\mathfrak{g}$ is a vector space upon which an internal bilinear operation, denoted by a bracket $[\ ,\ ] : \mathfrak{g} \times \mathfrak{g} \to \mathfrak{g}$, is defined with the following properties:

85 See, for instance, [251], Section III D.

$$[x,y] = -[y,x] \quad \text{for all} \quad x,y \in \mathfrak{g} \tag{3.379}$$

$$[[x,y],z] + [[y,z],x] + [[z,x],y] = 0 \quad \text{for all} \quad x,y,z \in \mathfrak{g} \tag{3.380}$$

The first equation records the antisymmetry (or skew-symmetry) of the bracket, the second the *Jacobi identity*. Given an associative algebra with product $\diamond$, one can always define a Lie algebra with the bilinear operation given by the *commutator*

$$[u,v] = u \diamond v - v \diamond u \tag{3.381}$$

Then the defining properties of Lie algebras are immediately satisfied, and the Jacobi identity is really an identity.[86] The converse operation, constructing an associative product from a Lie product is nontrivial and leads to the very interesting concept of universal covering algebras, a topic that will be treated in Volume 2. For Lie algebras that do not derive from an associative product, the Jacobi identities are nontrivial and an essential part of the definition.

The dimension n of a Lie algebra is the same as for the underlying vector space, and it is spanned linearly by a set of basis elements, or *generators* $\{T^a : a = 1,2,\dots n\}$ (the dimension may be countable infinity).[87] Due to the bilinearity, the structure of the Lie algebra, is fully captured by the commutators between the basis elements

$$[T^a, T^b] = f^{ab}{}_c T^c \tag{3.382}$$

in terms of the *structure constants* $f^{ab}{}_c$. The indicated summation convention in play here may trigger the question about upper and lower indices and metrics on the underlying vector space. Let us pause this question, and just note for now that (3.382) implies that the structure constants are antisymmetric in their upper indices. It should also be clear that the values of structure constants depend on the basis chosen. In all cases, however, using (3.382), the Jacobi identity translates into an equation for the structure constants

$$f^{ab}{}_c f^{cd}{}_e + f^{da}{}_c f^{cb}{}_e + f^{bd}{}_c f^{ca}{}_e = 0 \tag{3.383}$$

A concept that is prominent in gauge field theory is *semisimplicity* of algebras and groups. It is related to the often desirable nonoccurrence of invariant sub-algebras. The point is that if there is an invariant subalgebra $\mathfrak{h}$ then we have $[\mathfrak{h},\mathfrak{g}] \subset \mathfrak{h}$ and we get stuck in the subalgebra if we happen to work with elements in the subalgebra. This motivates the definition of *simple* Lie algebras as non-Abelian Lie algebras that have no

---

86 It may be useful sometimes to think of the algebra elements as acting on a space of functions, then the commutator is written $[u,v]f = (u \diamond v - v \diamond u)f$.

87 Note that the word "generator" is also used in related but different meanings as "generator of symmetry" and "generating element" allowing nonlinear combinations of algebra elements.

invariant subalgebras. Even worse is the occurrence of Abelian invariant subalgebras, because since then $[\mathfrak{h}, \mathfrak{h}] = \emptyset$ and the corresponding basis elements do not show up in the structure constants of the algebra.[88] However, accepting the lesser evil of non-Abelian invariant subalgebras, but not Abelian, leads to the concept of *semisimple algebras*: no Abelian invariant subalgebras are allowed (but non-Abelian are allowed). These concepts are analogous to the corresponding concepts for groups. A group is simple if it does not contain any invariant subgroup. A group is semisimple if it does not contain any Abelian invariant subgroup.[89]

Semisimple algebras can be characterized in several equivalent ways, for instance, as direct sums of simple algebras or as algebras where all elements may be written as commutators of other elements. A further refinement in terminology is *reductive algebras* where Abelian terms are allowed in the direct sum.

The concept of direct sums of algebras needs to be clarified. Consider two Lie algebras $\mathfrak{g}_1$ and $\mathfrak{g}_2$. It should be clear what is meant by the direct sum $V_1 \oplus V_2$ of the underlying vector spaces. The direct sum of the Lie algebras is denoted in the same way as $\mathfrak{g}_1 \oplus \mathfrak{g}_2$ but with the condition that $[\mathfrak{g}_1, \mathfrak{g}_2] = 0$. The concept generalizes easily to more than two algebras. The Lie algebra $\mathfrak{g}$ is a direct sum of Lie algebras $\mathfrak{g} = \mathfrak{g}_1 + \cdots + \mathfrak{g}_n$ if each $\mathfrak{g}_i$ is an ideal of $\mathfrak{g}$.

Direct sum and semidirect sums of Lie algebras

To get this straight and define a natural direct sum of Lie algebras, let $\{\mathfrak{g}_i; i = 1, 2, \ldots n\}$ be a set of Lie algebras. Denote by $\mathfrak{g}$ the vector space direct sum

$$\mathfrak{g} = \bigoplus_{i=0}^{n} \mathfrak{g}_i \tag{3.384}$$

In order to be a direct sum in the Lie algebra sense, we require two properties: First, for any two elements x and y in a certain $\mathfrak{g}_i$, the product is $[x, y] = [x, y]_i$. That is, the product is computed within the Lie algebra to which the elements belong. Second, $[\mathfrak{g}_i, \mathfrak{g}_j] = 0$. That is, elements from different components commute.

The concept of direct sums of algebras can be weakened somewhat. In the case of two Lie subalgebras $\mathfrak{g}_1$ and $\mathfrak{g}_2$ of $\mathfrak{g}$ where $[\mathfrak{g}_1, \mathfrak{g}_2] \subseteq \mathfrak{g}_1$, one says that $\mathfrak{g}$ is a semidirect sum of $\mathfrak{g}_1$ and $\mathfrak{g}_2$ (the order here is essential).

An example is the Poincaré algebra where we take as $\mathfrak{g}_1$ the translation generators P_μ and for $\mathfrak{g}_2$ the Lorentz generators $J_{\mu\nu}$. The algebra of equations (3.98)–(3.100) then falls precisely into the pattern of a semidirect sum of the translations and the Lorentz transformations.

Somewhat ironically – if one can say so – the perhaps most prominent of all algebras in theoretical physics, the Poincaré algebra is not semisimple. It fails semisimplicity since it contains an Abelian invariant subalgebra, namely the translations. The

88 As so well phrased by H. Georgi: "Particularly annoying are Abelian invariant sub-algebras." [264].

89 When we write subalgebra and subgroup here, we mean nontrivial subalgebras and sub-groups.

Poincaré algebra furthermore fails to be a direct sum of algebras since the Lorentz subalgebra and the translation subalgebra do not commute. Thus it is not even a reductive algebra (it does not help to allow Abelian terms since it is not anyway a direct sum). However, it is a semidirect sum of the translations and homogeneous Lorentz algebra.[90]

3.11.2 Structure of Lie algebras and the classical Lie algebras

Just as for groups, abstract Lie algebras may conveniently be represented on vector spaces. As always, the main focus is on irreducible representations, as such representations cannot, by definition, be broken up into "smaller" representations. They are therefore the ones that one may seek to classify. However, vector spaces may be defined over the real numbers or over the complex numbers, and this makes a significant difference for the representation theory. This can be understood in various ways. A simple example will be given below based on the angular momentum Lie algebra.

Complex versus real Lie algebras illustrated: a small algebra redux

---

Consider the rotation part of the Lorentz algebra of (3.105), or alternatively the two conjugated algebras of equations (3.117) and (3.118) Let us write any one of these equivalent algebras in terms of generic Hermitian generators L_i

$$[L_i, L_j] = i\epsilon_{ijk} L_k \tag{3.385}$$

Using complex coefficients, these generators may be linearly recombined into

$$L_\pm = \frac{1}{\sqrt{2}}(L_1 \pm iL_2) \quad \text{with} \quad L_+^\dagger = L_- \tag{3.386}$$

Then the algebra reads

$$[L_3, L_\pm] = \pm L_\pm \quad \text{and} \quad [L_+, L_-] = L_3 \tag{3.387}$$

The question arises as to the equivalence or not of these two algebras. This may be understood by first remembering that a Lie algebra is also a vector space with the generators as basis vectors. Then it is clear that considered as complex vector spaces, the two vector spaces are equivalent. Then they are also equivalent as complex Lie algebras.

To investigate if they are equivalent also as real Lie algebras, that is, Lie algebras where only real coefficients are allowed in linear combinations of the basis generators, one may return to the general three-dimensional orthogonal group Lie algebra, which we can take from formula (3.98) with an as

---

90 This has profound consequences for our subject matter, and may very well be one of the roots of the problems of higher spin field theory.

yet unspecified diagonal metric $g_{\mu\nu}$ (since we want to defer the choice of signature) and with generic generators L_{ij}. We also have the standard transcription $L_{12} = L_3$, $L_{23} = L_1$, $L_{31} = L_2$. Then we get[91]

$$[L_{12}, L_{31}] = -ig_{11}L_{23} \qquad [L_{31}, L_{23}] = -ig_{33}L_{12} \qquad [L_{23}, L_{12}] = -ig_{22}L_{31} \tag{3.388}$$

We may now try to diagonalize one of the generators, say L_{12}, by linearly recombining the others into $aL_{23} \pm bL_{31}$. Then the requirement

$$[L_{12}, aL_{23} \pm bL_{31}] = \pm(aL_{23} \pm bL_{31}) \tag{3.389}$$

can be met, provided that $b = iag_{22}$ and $a = -ibg_{11}$ which implies $g_{11}g_{22} = 1$. If the requirement cannot be met, one continues and tries another generator to diagonalize. Now suppose we are successful with $g_{11}g_{22} = 1$. Then we can take $g_{11} = g_{22} = 1$, implying $b = ia$. Normalizing $a = 1/\sqrt{2}$ we now get

$$[L_{12}, L_\pm] = \pm L_\pm \quad \text{and} \quad [L_+, L_-] = g_{33}L_{12} \tag{3.390}$$

If we choose $g_{33} = 1$, then the metric signature is $(+\,+\,+)$ and we have the compact three-dimensional rotation Lie algebra $\mathfrak{so}(3) \sim \mathfrak{su}(2)$. This is precisely what we had above. However, we may choose $g_{33} = -1$. Then the metric signature is $(+\,+\,-)$ and we get the noncompact Lie algebras $\mathfrak{sl}(2, \mathbf{R}) \sim \mathfrak{su}(1, 1)$.

Suppose now, having these two Lie algebras, we restrict ourselves to real coefficients in linear combinations of the basis generators. We then talk of *real forms of the Lie algebra* defined above in equation (3.385).

A little bit of linear algebra shows that there is no way to linearly recombine the Lie algebra with $g_{33} = -1$ into the one with $g_{33} = 1$ using real coefficients, therefore, we have two different real forms of the same complex Lie algebra (3.385). However, the one with $g_{33} = 1$ may be turned into the one with $g_{33} = -1$ using complex coefficients. Choose for instance $L_\pm \to iL_\pm$. Then the new generators are anti-Hermitian, but that is as it should be, since a noncompact algebra has no finite-dimensional unitary representations.

Finally, returning to our initial question as to whether the algebra (3.387) is equivalent as a real Lie algebra to (3.385), the answer can now be given. It is very easy to be fooled by sloppy notation here. In order to answer the question, without producing to much confusion, let us return to fundamentals and write the algebra (3.385) in a notation that does not assume any one definite metric

$$[L_i, L_j] = i\epsilon_{ij}{}^k L_k \tag{3.391}$$

Along with this, the transcription to "two-index generators" must be written $L_{ij} = \epsilon_{ij}{}^k L_k$. Now we can compare the two ways of writing the algebra: as in (3.388) and as in (3.391) without missing out on any hidden signs. The result of such a comparison is

$$\epsilon_{12}{}^3\epsilon_{31}{}^2 = g_{11} \qquad \epsilon_{23}{}^1\epsilon_{12}{}^3 = g_{22} \qquad \epsilon_{31}{}^2\epsilon_{23}{}^1 = g_{33} \tag{3.392}$$

From here, it follows that lowering the indices using the metric, the structure constants ϵ_{ijk} with lower indices are the same no matter the signature of the metric. This shows that no generality is lost in writing the algebra as we did to begin with.

Now we can proceed with confidence and answer the question. Of course, the algebras are different as real Lie algebras! After all, they are related by a complex linear combination. Choosing different

91 We are following A. O. Barut in [265], recommended for all who relish small algebras!

linear combinations, one can get to either of structures $[L_3, L_\pm] = \pm L_\pm$ with $[L_+, L_-] = \pm L_3$ which are – as already noted – different as real algebras.

More generally, in working out the representation theory, one will have to solve characteristic equations for eigenvalue equations, and the fact that **C** is algebraically closed while **R** is not, becomes crucial and, therefore, the abstract representation theory is done over the complex numbers.

Let us denote by V a generic n-dimensional vector space over **C**. The set of all linear mappings – endomorphisms – of V to itself is again a vector space. Since the composition of linear mappings associate, we get an associative algebra for free. Composing the associative product as in (3.381), we get a Lie algebra called the *general linear algebra* $\mathfrak{gl}(V)$ of V. Then, choosing a basis in V, the linear mappings may be represented by matrices acting on the vectors and the composition of mappings by matrix multiplication. The corresponding Lie algebra is denoted by $\mathfrak{gl}(n)$. This is then the backdrop toward which the classical matrix Lie algebras can be investigated. We will not review any details, but rather just state the facts, point out a few items of importance and then refer the reader to the specialized literature.

The Lie algebra $\mathfrak{gl}(n)$ contains an invariant Abelian subalgebra, consisting of all multiples of the unit matrix. Imposing tracelessness of the matrices, removes this ideal. The resulting complex Lie algebra is denoted $\mathfrak{sl}(n)$ and designated as *special linear*. There is a corresponding Lie group which is the *special linear group* SL(n) of $n \times n$ complex matrices with unit determinant. In the abstract classification scheme of semisimple Lie algebras, this algebra is denoted by A_{n-1}.[92] As n runs from 2 to ∞, we get a denumerable infinite series of Lie algebras. The Cartan classification scheme results in three further series of complex Lie algebras, plainly denoted by B_n, C_n and D_n. Following the comprehensive text book [263] and the succinct summary in [139], we just list the algebras for easy reference, with further explanations, below. The basis matrices A of the algebra corresponding group matrices M are defined near the group identity according to $M = 1 + iA$.

Cartan classification of complex finite dimensional semisimple Lie algebras

---

i

A_n $\mathfrak{sl}(n+1)$ as defined in the paragraph above. Real compact form: special unitary algebra $\mathfrak{su}(n+1)$ with traceless Hermitian matrices as basis.

B_n $\mathfrak{so}(2n+1)$ of matrices $M \in \mathfrak{gl}(2n+1)$ obeying the equation $M^T K' + K' M = 0$. Real compact form: $\mathfrak{so}(2n+1)$ with corresponding group O$(2n+1)$ of unitary and orthogonal (real) matrices. The basis matrices B satisfy $B^* = B^T = -B$. Such matrices are a also traceless.

---

92 The first nontrivial algebra is for $n = 2$. The case $n = 1$ would correspond to the relatively trivial one-dimensional Abelian algebra: think $\mathfrak{u}(1)$.

C_n $\mathfrak{sp}(2n)$ of matrices $M \in \mathfrak{gl}(2n)$ obeying the equation $M^T = JMJ$. Real compact form: $\mathfrak{usp}(2n)$ with corresponding group Usp$(2n)$ of unitary symplectic matrices. The basis matrices C are Hermitian and satisfies $C^T J + JC = 0$.

D_n $\mathfrak{so}(2n)$ of matrices $M \in \mathfrak{gl}(2n)$ obeying the equation $M^T K + KM = 0$. Real compact form: $\mathfrak{so}(2n)$ with corresponding group O$(2n)$ of unitary and orthogonal (real) matrices. The basis matrices D satisfy $D^* = D^T = -D$. Such matrices are a also traceless.

The matrices occurring in the definitions are defined as follows in terms of n-dimensional zero and unit matrices 0_n and 1_n, respectively.

$$K' = \begin{pmatrix} 1 & 0 & 0 \\ 0 & 0_n & 1_n \\ 0 & 1_n & 0_n \end{pmatrix} \qquad J = \begin{pmatrix} 0_n & 1_n \\ -1_n & 0_n \end{pmatrix} \qquad K = \begin{pmatrix} 0_n & 1_n \\ 1_n & 0_n \end{pmatrix} \tag{3.393}$$

In addition to these series of algebras, there are five exceptional algebras : G_2, F_4, E_6, E_7 and E_8.

All these Lie algebras find their applications in theoretical physics: the $\mathfrak{su}(n)$'s as Yang–Mills algebras, the $\mathfrak{so}(n)$'s as infinitesimal rotation algebras in n-dimensional spaces and the $\mathfrak{sp}(2n)$'s as algebras of infinitesimal canonical transformations in classical phase spaces.

One very important fact that comes out of the general analysis of semisimple Lie algebras $\mathfrak{g}$ is that they have a vector space direct sum structure (called *triangular decomposition* in [263])

$$\mathfrak{g}_- \oplus \mathfrak{g}_0 \oplus \mathfrak{g}_+ \tag{3.394}$$

with

$$[\mathfrak{g}_0, \mathfrak{g}_0] = 0 \qquad [\mathfrak{g}_+, \mathfrak{g}_-] \subseteq \mathfrak{g}_0 \qquad [\mathfrak{g}_\pm, \mathfrak{g}_0 \oplus \mathfrak{g}_\pm] \subseteq \mathfrak{g}_\pm \tag{3.395}$$

The notation is intended to show that the subalgebras $\mathfrak{g}_+$ and $\mathfrak{g}_-$ consists of raising and lowering operators respectively, while the subalgebra $\mathfrak{g}_0$ is Abelian.

3.11.3 Bottom line on real Lie algebras

The Lie algebras most often used in physics are real Lie algebras, meaning that linear combinations over the basis generators are taken with real coefficients. This is so even if the basis generators are represented with matrices with complex entries and/or structure constants that are imaginary. It also leads to the distinction between compact and noncompact Lie algebras. It becomes quite complex (!).

One way to think straight about it, is to take stock of the following facts, starting from the compact/noncompact distinction. This distinction only makes sense if there is a metric on the underlying vector space that may be of definite or nondefinite signature. Such a metric can be defined for real Lie algebras, but not for complex Lie algebras. The metric is based on a bilinear inner product $\kappa(x, y)$ (that exists for the complex Lie algebras considered here) called the *Killing form*. A complex linear combination such as $x \to ix$, $y \to iy$ leads to $\kappa(x, y) \to -\kappa(x, y)$ which upsets the signature of the metric (as we saw also in the box above).

Furthermore, for any simple real Lie algebra, κ is nondegenerate and there is an orthonormal basis in which κ takes the form

$$\kappa = \begin{pmatrix} -1_p & 0 \\ 0 & 1_{n-p} \end{pmatrix} \tag{3.396}$$

Such a metric can be used to raise the indices for the structure constants of the algebra. The three upper indices then become totally antisymmetric. When the metric is of a definite sign, that is, when $p = 0$, and $\kappa^{ab} = \delta_{ab}$, then the real Lie algebra is said to be *compact*. This Lie algebra is actually unique up to isomorphism. However, since κ can in general be of various signatures, there are several real forms corresponding to one and the same complex Lie algebra. A list can be found in [263], Section 8.4. There we can also find information on the isomorphism between low dimensional Lie algebra. For instance, the three Lie algebras A_1, B_1 and C_1 are isomorphic. This leads to the following isomorphisms (among others) for their compact forms: $\mathfrak{su}(2) \cong \mathfrak{so}(3)$ and for their noncompact forms: $\mathfrak{su}(1,1) \cong \mathfrak{sl}(2, \mathbf{R}) \cong \mathfrak{so}(2,1)$.

3.12 Fiber bundle theory

Fiber bundles are generalizations and abstractions of the basic theoretical physics concept of fields carrying degrees of freedom valued in "internal" spaces. Since this is so central to field theory in general and higher spin theory in particular, we will discuss it at some length here. "Solving" the higher spin problem may very well involve some deep rethinking at precisely this point. We will recite the mathematics terminology and try to explain it in reference to physics concepts, but leave out proofs.

3.12.1 Basic intuition

Mathematically, the fields of physics are functions $y = f(x)$ from one space X to another Y. A simple example is a scalar temperature field $T(x)$ and for many purposes one need not think of it any more deeply than that. It is simply a function $T : \mathbf{R}^3 \to \mathbf{R}_+$. Another example is the electromagnetic potential $\mathbf{A}(x)$ which is a three-dimensional vector field. Now, one may start to worry in what space the vector is pointing? It is certainly not geometrical space, even though it can be illustrated in that way. We can resort to view it is a function $\mathbf{A} : \mathbf{R}^3 \to \mathbf{R}^3$ where the points in the range carry physical dimension of Vsm^{-1}. And, of course, the points in the range of the temperature field carry physical dimension of K. Still another example is the Yang–Mills field A_μ^a which is four-vector and carries a color index a. In this context, where one normally use units where $\hbar = c = 1$, the dimension of the field is the same as momentum and mass.

However, one may also think of a function $f : X \to Y$ in terms of its *graph* which is the set of points $(x, f(x))$ in the Cartesian product $X \times Y$.[93] The graph is itself a function gr_f and we can write $\text{gr}_f : X \to X \times Y$, where for any point x in the domain X, the graph evaluates according to $\text{gr}_f(x) = (x, f(x))$.

Anticipating the terminology to be introduced, the set X is called the *base space* and the set $X \times Y$ the *total space*. Incidentally, the reason to call these sets *spaces* is that in physics (and in the corresponding mathematics) these sets maintain additional structure. It is also convenient to introduce projections from $X \times Y$ to X and Y. Then $\text{pr}_1(X \times Y) = X$ and $\text{pr}_2(X \times Y) = Y$. We can then state: any function $g : X \to X \times Y$, for which it holds $\text{pr}_1 \circ g = \text{id}_X$, is the graph of a unique function $\text{pr}_2 \circ g$. The graph concept is thus well-defined.

As noted in [266], this view of functions has two advantages: conceptually, a function can be thought of as *field* (precisely in the sense of physics) where for each point x in the domain X there is a copy $\{x\} \times Y$ of the range Y and a single point in that copy gives the value of the field at x. Furthermore, the concept can be generalized to total spaces that are not diffeomorphic to a product of a base space and another space. For such a generalization to be useful, there must nevertheless be a local product structure, in the sense that each point in the total space must have a neighborhood that looks like a product of spaces. We are now ready to define these structures in some more detail.[94]

3.12.2 Fibered manifolds, bundles and fiber bundles

A *fibered manifold* is given by three objects, denoted a *triple*: (E, π, M) where E, the *total space*, and M, the *base space*, are manifolds and π is a map $\pi : E \to M$ that projects the full space onto the base manifold (see figure 3.3). For each point $p \in M$, the subset $\pi^{-1}(p) \equiv F_p$ of E is called the *fiber* over p.[95]

This definition offers enough structure to show that a fibered manifold has a local product structure. This means: around every point p of E there is a neighborhood U_p and another manifold F_p and a diffeomorphism $t_p : U_p \to \pi(U_p) \times F_p$ such that $\text{pr}_1(t_p(q)) = \pi(q)$ for all points q in U_p.

---

93 According to the set theoretic definition of functions, a function and its graph are precisely the same thing, simply because a function from X to Y is a subset of $X \times Y$.

94 The language becomes a bit stiff in the next section, but we are far from stringent in the mathematical sense. Some references treating the subject in a way useful for our topic are [258, 266, 251].

95 The meaning of $\pi^{-1}(p)$ is the set $\pi^{-1}(\{p\}) = \{e \in E : \pi(e) = p\}$, that is, the inverse image of the one-element subset $\{p\}$ of M.

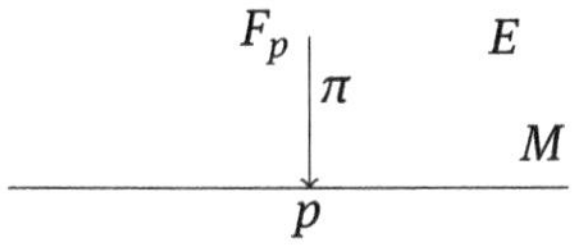

Figure 3.3: Intuitive picture of a fibered manifold.

However, the fibers of these local products may differ: the manifolds F_p may not be the same or even homeomorphic for different neighborhoods U_p. Having the fibers "look the same" is in general desirable. In order to achieve that, one defines local trivializations that make the product structure more uniform.

For (E, π, M), a fibered manifold and x a point in the base space M, a *local trivialization* around x is a triple (W_x, F_x, t_x) where W_x is a neighborhood of the point x, F_x a manifold (the fiber) and $t_x : \pi^{-1}(W_x) \to W_x \times F_x$ is a diffeomorphism with the property $\mathrm{pr}_1 \circ t_x = \pi|_{\pi^{-1}(W_x)}$.

When at least one such local trivialization exists around every point of the base space, the fibered manifold is said to *locally trivial*. Such a fibered manifold is a *bundle*.[96] It is then possible to show that the local fibers F_x are diffeomorphic to a *typical fiber* F for all $x \in M$. If one wants to record all the data for a bundle, one could write (E, M, π, F).

In applications to physics, the fibers may be, for instance: vector spaces, tangent and/or cotangent spaces or Lie groups. Thus one often has some additional structure in the fibers formalized as the action of a group. One then speaks of fiber bundles.

A *fiber bundle* (E, M, π, F, G) is a bundle (E, M, π, F) together with a group G of diffeomorphisms (or homeomorphisms) in the typical fiber F, and a set of local trivializations (U_i, t_i, F) where U_i is a cover of the base space M with i running over some index set. The group G is called the *structure group* of the fiber.

The bundle properties ensure that there are at least one local trivialization (W_p, t_p, F_p) around each point p of the base space. The extra requirement for being a fiber bundle is the existence of a cover of special local trivializations (U_i, t_i, F), all to the product space $M \times F$.

In conclusion, starting with the basic concept of a fibered manifold, one arrives at the concept of a fiber bundle by successively adding structure. First, the typical fiber, then the group acting in the typical fiber and the cover of local trivializations. Fiber bundles have enough structure to serve as models of the field theories of physics.

It is important to realize that – even though the fibers F_x are diffeomorphic to the typical fiber F – there is no "canonical" relation between fibers F_x at different points x. For that, one needs to add further structure, namely: connections, a concept that we will elaborate in Section 3.13 and in Chapter 4.

96 Note that "local" in the trivializations refer to the base space M.

Of particular importance are the *principal fiber bundles*. These are fiber bundles for which the typical fiber is identical to the structure group. Yang–Mills theories can be described in this way. The fiber is the gauge group, meaning that the Yang–Mills gauge fields are valued in the group.

Furthermore, we can now see how the concepts of tangent bundles and cotangent bundles of Section 3.10.1, naturally fit into this general scheme. These bundles play an important role in the theory of gravity, of which we will have more to say in Chapter 4.

3.12.3 Tangent and cotangent bundles

Let us elaborate a little on the definitions of Section 3.10.1. The *tangent bundle TM* is the union of all tangent spaces T_p at all points p in the manifold M. This concept allows for clear notion of a vector field varying from point to point in the manifold.

First, introduce a projection map π from the tangent bundle to the manifold. This map associates a point p with every vector $V(p)$ in the tangent bundle. In formulas, we have

$$TM = \bigcup_{p\in M} T_p \tag{3.397}$$

$$\pi : TM \to M, \;\; V(p) \mapsto p \tag{3.398}$$

The inverse π^{-1} map, maps points p in the manifold to tangent spaces T_p (i. e., the set of all tangent vectors at the point). In formulas,

$$\pi^{-1} : M \to TM, \;\; p \mapsto T_p. \tag{3.399}$$

It is important to realize that π^{-1} maps a specific point p to the vector space of all tangent vectors at that point, and not to a particular tangent vector. The tangent space at p is an example of a fiber at p. The dual concept of the *cotangent bundle* can then be introduced in the analogous way.

Interesting and important examples of these concepts arise in classical mechanics. Thinking of the configuration space of coordinates q_n of a mechanical system as a base space, one can consider the tangent spaces formed by the velocities $\dot{q}_n$ at each point q_n. Together they form a tangent bundle. Correspondingly, there is a cotangent bundle with fibers formed out of the conjugate momenta p^n. Then one may consider mechanical systems on phase spaces with nontrivial topology.

3.12.4 Fields as cross-sections of bundles

The physics concept of a field defined on a space M valued in some other space F, can now be modeled as cross-sections of bundles, generalizing the graph of a function. The definition – for a fiber bundle – is as follows.

A map $\phi : M \to E$ is called a *cross-section* (or a *section*) of the fiber bundle (E, π, M, F) if it satisfies the condition $\pi \circ \phi = \mathrm{id}_M$.

The concept of section can also defined for a bundle, or even for a fibered manifold, and it reads in exactly the same way in these cases. If all the sections F_x are the same, that is, if F_x is "constant", then sections reduce to ordinary functions. Vector fields and 1-forms can be viewed as sections of the tangent bundle and the cotangent bundle, respectively.

3.13 Infinitesimal coordinate transformations

The mathematical theory so briefly reviewed is very neat, but we will now return to a more physics oriented point of view. Let us here reconsider general coordinate transformations $x^\mu \to x'^\mu(x)$ as we do in general relativity oriented presentations. The purpose is to introduce the Lie derivative and the affine connection in a natural way. An infinitesimal coordinate transformation then reads[97]

$$\Delta_\varepsilon x^\mu = x'^\mu - x^\mu = \varepsilon^\mu(x) \tag{3.400}$$

We are interested in the behavior of fields under such transformations. According to standard general relativity we have the transformations laws for scalar and covariant and contravariant vector fields, respectively,

$$\varphi'(x') = \varphi(x) \tag{3.401a}$$

$$V'_\mu(x') = V_\alpha(x)\frac{\partial x^\alpha}{\partial x'^\mu} \tag{3.401b}$$

$$V'^\mu(x') = V^\alpha(x)\frac{\partial x'^\mu}{\partial x^\alpha} \tag{3.401c}$$

The transformation laws for covariant and contravariant vectors are the same as the transformation laws for partial derivatives ∂_μ and coordinate differentials dx^μ, respectively. All this is essentially applications of the chain rule (see formula (3.369)).[98]

The transformation rules (3.401) generalize to general (p, q)-tensors in a natural way [243]. Let us compute the infinitesimal transformation rule for a scalar field in detail

$$\begin{aligned}\delta_\varepsilon\varphi(x) &= \varphi'(x) - \varphi(x) = \varphi'(x' - \varepsilon) - \varphi(x) \\ &= \varphi'(x') - \varepsilon^\mu\partial_\mu\varphi'(x) - \varphi(x) = -\varepsilon^\mu\partial_\mu\varphi'(x) \\ &= -\varepsilon^\mu\partial_\mu(\varphi(x) + \delta_\varepsilon\varphi(x)) = -\varepsilon^\mu\partial_\mu\varphi(x)\end{aligned} \tag{3.402}$$

97 One can think of it as a passive point of view.

98 It works due to the assumed existence of the smooth transitions functions between overlapping coordinate charts; see Section 3.10.

where in the last approximate equality the term of second order in ϵ is discarded. Performing the corresponding detailed analysis for a covariant and contravariant vector fields, using (3.401), yields

$$\delta_\varepsilon V_\mu(x) = -\varepsilon^\alpha \partial_\alpha V_\mu(x) - (\partial_\mu \varepsilon^\alpha) V_\alpha \tag{3.403a}$$

$$\delta_\varepsilon V^\mu(x) = -\varepsilon^\alpha \partial_\alpha V^\mu(x) + (\partial_\alpha \varepsilon^\mu) V^\alpha \tag{3.403b}$$

where one should note the signs and index contractions in the second term in the equations. These transformation laws also generalize to tensor fields in a natural way. For each covariant or contravariant index, there is a term of the type $\partial\varepsilon$ corresponding to the second terms in (3.403a) and (3.403b), respectively. For instance, for $V^\rho_{\mu\nu}$ we get

$$\delta_\varepsilon V^\rho_{\mu\nu}(x) = -\varepsilon^\alpha \partial_\alpha V^\rho_{\mu\nu} - (\partial_\mu \varepsilon^\alpha) V^\rho_{\alpha\nu} - (\partial_\nu \varepsilon^\alpha) V^\rho_{\mu\alpha} + (\partial_\alpha \varepsilon^\rho) V^\alpha_{\mu\nu} \tag{3.404}$$

The generalization to higher order tensors should be obvious.

3.13.1 The Lie derivative

The transformation formulas for tensors suggest introducing a certain differential operator called the *Lie derivative* L_ε. Its action on a tensor $V^\rho_{\mu\nu}$ is defined by

$$L_\varepsilon V^\rho_{\mu\nu} = -\delta_\varepsilon V^\rho_{\mu\nu} \equiv \varepsilon^\alpha \partial_\alpha V^\rho_{\mu\nu} + (\partial_\mu \varepsilon^\alpha) V^\rho_{\alpha\nu} + (\partial_\nu \varepsilon^\alpha) V^\rho_{\mu\alpha} - (\partial_\alpha \varepsilon^\rho) V^\alpha_{\mu\nu} \tag{3.405}$$

Clearly, the Lie derivative is simply defined to give the infinitesimal form of a general coordinate transformation. The Lie derivative transforms a tensor of given type into a tensor of the same type. It is, as is obvious from the definition, linear in ϵ. By direct computation, it can be shown that it obeys the Leibniz rule and commutes with index contractions. Therefore, we can consistently compute (although we have not introduced the metric as yet)

$$L_\varepsilon V^\mu(x) = (L_\varepsilon g^{\mu\nu}) V_\nu + g^{\mu\nu} L_\varepsilon V_\nu \tag{3.406}$$

As would be expected, the Lie derivative satisfies the infinite dimensional Lie algebra of general coordinate transformations. Its action on a contravariant vector ξ^μ motivates the following definition of the *Lie bracket:*

$$[\epsilon, \xi]^\mu = L_\epsilon \xi^\mu \tag{3.407}$$

It now follows

$$[L_\xi, L_\eta] = L_{[\xi,\eta]} \tag{3.408}$$

Furthermore, the Lie derivative as well as the Lie bracket, also satisfy the Jacobi identity. Acting on scalars ϕ, the Lie derivative can be represented simply as $L_\epsilon \phi = \epsilon^\mu \partial_\mu \phi$ and (3.408) computes to

$$[\eta^\nu \partial_\nu, \xi^\mu \partial_\mu]\phi = (\eta^\nu \partial_\nu \xi^\mu - \xi^\nu \partial_\nu \eta^\mu)\partial_\mu \phi = [\eta, \xi]^\mu \partial_\mu \phi \tag{3.409}$$

and it is natural to think of the Lie bracket in (3.407) as "structure functions" for the infinite dimensional Lie algebra of general coordinate transformations.

The Lie derivative also subsumes familiar results on Poincaré transformations. Let ε^μ be an infinitesimal Poincaré transformation $\varepsilon^\mu = a^\mu + \lambda^\mu{}_\nu x^\nu$ where a^μ and $\lambda^\mu{}_\nu$ are constant (and infinitesimal). Then we get the Poincaré transformations of vector fields as

$$\delta_\varepsilon V_\mu(x) = -L_\varepsilon V_\mu(x) = -\varepsilon^\alpha \partial_\alpha V_\mu(x) - \lambda^\alpha{}_\mu V_\alpha \tag{3.410a}$$

$$\delta_\varepsilon V^\mu(x) = -L_\varepsilon V_\mu(x) = -\varepsilon^\alpha \partial_\alpha V^\mu(x) + \lambda^\mu{}_\alpha V^\alpha \tag{3.410b}$$

where we recognize the second terms as the spin part of the Lorentz transformations.

3.13.2 Covariant derivative and connection

It is a generic phenomenon pertaining to local symmetries that the derivative of a field does not transform as the field itself. Here, the derivative of a tensor do not transform as a tensor under coordinate transformations. Perform the following sample calculation:[99]

$$\begin{aligned}\delta(\partial_\mu V_\nu) = \partial_\mu(\delta V_\nu) &= -\epsilon^\alpha \partial_\alpha \partial_\mu V_\nu - \partial_\mu \epsilon^\alpha \partial_\alpha V_\nu - \partial_\nu \epsilon^\alpha \partial_\mu V_\nu - \partial_\mu \partial_\nu \epsilon^\alpha V_\alpha \\ &= -L_\epsilon(\partial_\mu V_\nu) - (\partial_\mu \partial_\nu \epsilon^\alpha) V_\alpha\end{aligned} \tag{3.411}$$

The offending term is elegantly compensated for by introducing the *covariant derivative* ∇_μ according to

$$\nabla_\mu V_\nu = \partial_\mu V_\nu - V_\alpha \Gamma_{\mu\nu}{}^\alpha \tag{3.412}$$

where the *affine connection* $\Gamma_{\mu\nu}{}^\alpha$ transforms inhomogeneously as

$$\delta \Gamma_{\mu\nu}{}^\alpha = -L_\epsilon \Gamma_{\mu\nu}{}^\alpha - \partial_\mu \partial_\nu \epsilon^\alpha \tag{3.413}$$

Due to the properties of the Lie derivative, the term $V_\alpha \Gamma_{\mu\nu}{}^\alpha$ transforms as a tensor apart from an inhomogeneous term that precisely cancels the corresponding term in (3.411).

99 The first equality is indeed correct since δ denotes a local transformation at the same space-time point.

All in all, $\nabla_\mu V_\nu$ transforms as a tensor in both indices. The corresponding formula for a contravariant vector is

$$\nabla_\mu V^\nu = \partial_\mu V^\nu + \Gamma_{\mu\alpha}{}^\nu V^\alpha \tag{3.414}$$

The formulas (3.412) and (3.414) generalize to general tensors in the obvious way.

The transformation law (3.413) suggests splitting the connection in a nontensor, symmetric part and a tensor, antisymmetric part according to

$$\Gamma_{\mu\nu}{}^\alpha = \frac{1}{2}\left(\Gamma_{(\mu\nu)}{}^\alpha + \Gamma_{[\mu\nu]}{}^\alpha\right) \tag{3.415}$$

where the antisymmetric part defines the *torsion tensor*

$$T_{\mu\nu}{}^\alpha = \Gamma_{[\mu\nu]}{}^\alpha \tag{3.416}$$

In standard general relativity, the torsion is set to zero, but it plays an interesting role in gauge theory approaches to gravity, as we will see. In this context, it is important to realize that the affine connection – or any other connection – is a structure that is added to the manifold. So far it can be any field with the transformation property (3.413). As we proceed, we will encounter other kinds of connections. We return to these questions in Section 4.5.

The covariant derivative acting on a tensor maps (p, q)-tensors to $(p+1, q)$-tensors. It satisfies the Leibniz rule and the Jacobi identity. It is closely related to the Lie derivative which can be written in terms of covariant derivatives and the torsion tensor. So, for instance, we have

$$L_\epsilon V^\mu = \epsilon^\alpha \nabla_\alpha V^\mu - \nabla_\alpha \epsilon^\mu V^\alpha + \epsilon^\alpha T_{\alpha\beta}{}^\mu V^\beta \tag{3.417}$$

This rewriting notwithstanding, the Lie derivative is independent on any connection.[100]

3.14 Lagrangian field theory

It is a curious fact that the classical action has dimension of angular momentum. In quantum mechanics, it is therefore naturally measured in terms of $\hbar$. For the purpose of setting up the basic theory, we will consider models where the dynamical variables are fields and their first derivatives, although in higher spin theory we will have to work with higher derivative theories, albeit presumably only in the interaction terms and not in the free theory kinetic term.

100 What happens is that the torsion term cancels the antisymmetric connection terms that arise from the covariant derivative terms. In that way, the rewriting is fairly trivial.

Consider therefore a set of classical fields $\varphi_i(x)$ labeled by a generic index i, possibly related to a symmetry transformation of some kind. We write for the action S, Lagrangian L and Lagrangian density $\mathcal{L}$

$$S = \int_{\tau_1}^{\tau_2} dt L(\varphi_i(x), \dot{\varphi}_i(x)) = \int_{\Omega} d^4x\, \mathcal{L}(\varphi_i(x), \partial_\mu \varphi_i(x)) \tag{3.418}$$

where Ω is the region of space-time integration (possibly bounded at spatial and/or temporal infinity). The first equality is the definition of the action as the time integral of the Lagrangian from which one can develop the Hamiltonian formulation if one wish. The second equality involves the assumption that the Lagrangian can be expressed as a space integral of a scalar *Lagrangian density* depending on the fields and their space and time derivatives. This is a way of building Poincaré invariance into the theory. By studying the behavior of S and $\mathcal{L}$ under *variations* and *transformations* of the fields, we can derive the field equations and Noether's theorem.

Terminology and notation for transformations

Consider a space-time point p that in one coordinate system is labeled by x and in another by x'. The smooth coordinate transformation $x \to x'$ is then called *passive*. For the *active* viewpoint, the transformation $x \to x'$ is thought of as mapping the point p to another point $x' = f(p)$ in the same coordinate system. A function ϕ defined on the space-time will be denoted by $\phi(x)$ and $\phi'(x')$, respectively, in the two systems. For an infinitesimal transformation, we define the *total variation* as $\Delta\phi = \phi'(x') - \phi(x)$ and the *local variation* as $\delta\phi = \phi'(x) - \phi(x)$. Neglecting terms second order in infinitesimals, we have

$$\Delta\phi = \delta\phi + \Delta x^\mu \partial_\mu \phi \quad \text{where } \Delta x = x' - x = \delta x \tag{3.419}$$

Note that for local variations: $\delta(\partial_\mu \phi) = \partial_\mu \delta\phi$.[101] Finally, for internal transformations (like matter field gauge transformations where no space-time transformation is involved) we have $\Delta\phi = \delta\phi$.

In order to clearly understand such notions as "on-shell" and "off-shell", Noether currents and the Noether method, it is useful to make a conceptual distinction between *variations* and *transformations* although the notation mixes them by using the same symbols δ and Δ. Transformations are always given by specific rules for the fields and coordinates. Variations are not governed by such rules, they are essentially arbitrary.

101 Note that if one adopts a passive viewpoint, then the variation $\phi'(x') - \phi(x)$ can be thought of as "local" in that the x and x' denote the same point in different coordinate systems. This seems not to be a common choice, but it is done in [23].

3.14.1 The action principle

Consider first an arbitrary local infinitesimal *variation* $\delta\varphi_i(x)$ in the field. The coordinates are not varied. The variation of the action becomes (writing just φ for the fields)

$$\delta S = \int_\Omega d^4x\delta\mathcal{L} = \int_\Omega d^4x\left[\frac{\partial\mathcal{L}}{\partial\varphi}\delta\varphi + \frac{\partial\mathcal{L}}{\partial(\partial_\mu\varphi)}\delta(\partial_\mu\varphi)\right]$$
$$= \int_\Omega d^4x\left[\frac{\partial\mathcal{L}}{\partial\varphi} - \partial_\mu\frac{\partial\mathcal{L}}{\partial(\partial_\mu\varphi)}\right]\delta\varphi + \int_\Omega d^4x\partial_\mu\left[\frac{\partial\mathcal{L}}{\partial(\partial_\mu\varphi)}\delta\varphi\right] \tag{3.420}$$

The last term, a total derivative, can be written as an integral over the surface σ bounding Ω

$$\oint_\sigma \frac{\partial\mathcal{L}}{\partial(\partial_\mu\varphi)}\delta\varphi \tag{3.421}$$

Requiring the action to be stationary under variations $\delta\varphi$ that vanish on the boundary σ we get the *Euler–Lagrange field equations*

$$\frac{\partial\mathcal{L}}{\partial\varphi} - \partial_\mu\frac{\partial\mathcal{L}}{\partial(\partial_\mu\varphi)} = 0 \tag{3.422}$$

One can identify the Euler–Lagrange equations with the *functional derivative* of the action, if we perform the computation in more detail (discarding the surface term)

$$\frac{\delta S}{\delta\varphi_i(x)} = \int_\Omega d^4y\frac{\delta\mathcal{L}(y)}{\delta\varphi_i(x)}$$
$$= \int_\Omega d^4y\left[\frac{\partial\mathcal{L}(y)}{\partial\varphi_j(y)}\frac{\delta\varphi_j(y)}{\delta\varphi_i(x)} + \frac{\mathcal{L}(y)}{\partial(\partial_\mu\varphi_j(y))}\frac{\delta(\partial_\mu\varphi_j(y))}{\delta\varphi_i(x)}\right]$$
$$= \frac{\partial\mathcal{L}(x)}{\partial\varphi_i(x)} - \partial_\mu\frac{\partial\mathcal{L}(x)}{\partial(\partial_\mu\varphi_i(x))} \tag{3.423}$$

where we have used

$$\frac{\delta\varphi_j(y)}{\delta\varphi_i(x)} = \delta^i_j\delta^4(y-x) \tag{3.424}$$

This computation is a generalization of the corresponding one in Section 3.1.2.

Derivatives revisited

---

i The action S is a *functional* (a function of functions) from the vector space of functions to the real numbers. The Lagrangian density $\mathcal{L}$ is an ordinary composite function, composed of fields and derivatives of fields. Therefore, it makes sense to write $\mathcal{L}(x)$. It also makes sense to write and compute partial *variational derivatives* of $\mathcal{L}$ with respect to the fields $\varphi_i(x)$. Writing $S = \int d^4x\, \mathcal{L}(x)$ we can think of S as depending on the functions $\varphi(x)$ building up $\mathcal{L}(x)$. The partial derivatives of the action with respect to the fields are *functional derivatives*. Via the chain rule, the variational derivatives of $\mathcal{L}(x)$ may be computed in terms of ordinary partial derivatives of $\mathcal{L}(x)$ with respect to the fields $\varphi_i(x)$, as indicated in the second line of the computation (3.423); compare to the analogous discussion in Section 3.1.2 on mechanics.

This is the *Hamilton action principle*, that is $\delta S = 0$ for variations $\delta\varphi_i(x)$ that vanish on the boundary of Ω. For variations that do not vanish on the boundary, we have the *generalized Hamilton action principle*

$$\delta S(\Omega) = \int_\Omega d^4x \partial_\mu K^\mu(x) \neq 0 \tag{3.425}$$

where the remaining integral represents the boundary contributions to the variation.

3.14.2 General transformations

Consider now *transformations* Δx^μ and $\Delta\varphi_i$. For Noether's first theorem, we are interested in global transformations, that is, where Δx^μ and $\Delta\varphi_i$ are constant. However, we may just as well perform the initial calculations with local space-time dependent transformations, as that will anyway be needed for the second theorem. That will also help clarify the exact nature of the second theorem as compared to the first.

We want to compute ΔS which is related to δS through the following calculation:

$$\begin{aligned}\Delta S &= \int \Delta(d^4x\, \mathcal{L}) = \int_\Omega ((\Delta d^4x)\, \mathcal{L} + d^4x \Delta\, \mathcal{L}) = \\ &= \int d^4x(\partial_\mu \Delta x^\mu\, \mathcal{L} + \delta\, \mathcal{L} + \Delta x^\mu \partial_\mu\, \mathcal{L}) = \delta S + \int d^4x \partial_\mu (\Delta x^\mu\, \mathcal{L})\end{aligned} \tag{3.426}$$

where we have used

$$\Delta d^4x = d^4x \partial_\mu \Delta x^\mu \tag{3.427}$$

as well as $\Delta\, \mathcal{L} = \delta\, \mathcal{L} + \Delta x^\mu \partial_\mu\, \mathcal{L}$ (see formula (3.419)). Then using δS from (3.420), retaining the surface term, and expressing the Euler–Lagrange term as the functional derivative of S according to (3.423), we get

$$\begin{aligned}\Delta S &= \int_\Omega d^4x \partial_\mu \left(\Delta x^\mu\, \mathcal{L} + \frac{\partial\, \mathcal{L}}{\partial(\partial_\mu \varphi_i)} \delta\varphi_i \right) + \int_\Omega d^4x \frac{\delta S}{\delta\varphi_i} \delta\varphi_i \\ &\equiv \int_\Omega d^4x \partial_\mu J^\mu + \int_\Omega d^4x \frac{\delta S}{\delta\varphi_i} \delta\varphi_i\end{aligned} \tag{3.428}$$

where the "not yet Noether current" J^μ is defined by this formula. It is convenient, in order to consider two important cases, to rewrite it in terms of total variations

$$J^\mu = \frac{\partial \mathcal{L}}{\partial(\partial_\mu \varphi_i)} \Delta\varphi_i + \left(\delta^\mu_\nu \mathcal{L} - \frac{\partial \mathcal{L}}{\partial(\partial_\mu \varphi_i)} \partial_\nu \varphi_i \right) \Delta x^\nu \tag{3.429}$$

We are interested in transformations that leave the action invariant, i. e., $\Delta S = 0$. It is then useful to write (3.428) as[102]

$$\int_\Omega d^4x \frac{\delta S}{\delta \varphi_i} \delta\varphi_i = - \int_\Omega d^4x \partial_\mu J^\mu \tag{3.430}$$

3.14.3 The first Noether theorem

To arrive at the *first Noether theorem*, we demand the field equations to hold, so that the left-hand side of (3.430) is zero. We are now interested in global transformations. Then the derivatives in $\partial_\mu J^\mu$ will be zero on the parameters of the transformation and it will be possible to define parameter independent Noether currents. Two important cases can be discerned.

The first case is global gauge transformations. Then $\Delta x^\mu = 0$, and in order to be able to write a "not to abstract" formula for the Noether current we need to fix the detailed form of the transformation in some more detail. In analogy to what we did in Section 3.1.3, we now write for a global gauge transformation

$$\delta\varphi_i = R_i^a \theta_a \tag{3.431}$$

with the θ_a infinitesimal constant parameters and R_i^a capturing the details of the transformation depending on the fields and their derivatives.[103] Then we immediately get the *on-shell conservation law* or *continuity equation* in terms of the *Noether current* $J^{a\mu}$

$$\partial_\mu J^{a\mu} = 0 \quad \text{where} \quad J^{a\mu} = \frac{\partial \mathcal{L}}{\partial(\partial_\mu \varphi_i)} R_i^a \tag{3.432}$$

The second case concerns space-time transformation $\Delta x^\mu \neq 0$, of which there are two important subcases:

$$\text{Translations} \qquad \Delta x^\mu = \epsilon^\mu \quad \text{and} \quad \Delta\varphi_i = 0 \tag{3.433}$$

$$\text{Lorentz transformations} \qquad \Delta x^\mu = \lambda^\mu{}_\nu x^\nu \quad \text{and} \quad \Delta\varphi_i = \frac{1}{2}\lambda_{\mu\nu} M_{ij}^{\mu\nu} \varphi^j \tag{3.434}$$

---

102 If the action is only invariant up to a boundary term, it can naturally be added to the right-hand side; compare to (3.425).

103 A concrete example are global gauge transformations of a Yang–Mills charged scalar field, where $\delta\varphi_i(x) = -\theta^a (T^a)_{ij} \varphi_j(x)$ (see formula (4.14) in Chapter 4).

The first symmetry of the field equations leads to the conserved energy-momentum tensor and the second to the conserved angular momentum tensor. The energy-momentum tensor can be copied from (3.429). Raising one index it is conventional to write it as

$$T^{\mu\nu} = \frac{\partial \mathcal{L}}{\partial(\partial_\mu \varphi_i)} \partial^\nu \varphi_i - \eta^{\mu\nu} \mathcal{L} \tag{3.435}$$

This defines the *canonical energy-momentum tensor*. It is not symmetric, but can be made so by taking advantage of the possibility to add total derivatives to ΔS.

The Noether theorems in words

Rather than stating the *first Noether theorem* more exactly, we capture its essence as:[104]

> There is a one-to-one correspondence between symmetry groups of an action and conservation laws of its Euler–Lagrange equations.

In a similar fashion, the essence of the *second Noether theorem* can be phrased as:

> An infinite-dimensional symmetry of an action, depending on arbitrary functions of space-time, corresponds to nontrivial differential relations among the Euler–Lagrange equations.

3.14.4 Internal gauge symmetries

For the *second Noether theorem*, we change the perspective and consider transformations that leave the action invariant without using the field equations. By an *infinite-dimensional symmetry transformation*, we mean a variation of the fields that leaves the action invariant even when the Euler–Lagrange equations are not satisfied.

Consider then internal symmetries, that is, symmetries with $\Delta x^\mu = 0$. In particular, we have in mind gauge symmetries. For the general theory, it is then customary to write[105] as we did in Section 3.1.3

$$\delta\varphi_i(x) = R_i^a(\varphi)\xi_a(x) \tag{3.436}$$

with $\xi_a(x)$ infinitesimal and space-time dependent, and $R_i^a(\varphi)$ capturing the details of the transformation depending on the fields and their derivatives. We are here employing a condensed notation. In field theory, we have

$$R_i^a(x)\xi_a(x) = \int d^4y R_i^a(x,y)\xi_a(y) \tag{3.437}$$

---

104 Adapted from a talk by P. Olver, "Noether's Two Theorems", Perimeter Institute, 2015.
105 See, for instance, [229].

where (derivatives with respect to x)

$$R_i^a(x,y) = \bar{R}_i^a \delta(x-y) + \bar{R}_i^{a\mu_1} \partial_{\mu_1} \delta(x-y) + \bar{R}_i^{a\mu_1\mu_2} \partial_{\mu_1} \partial_{\mu_2} \delta(x-y) + \cdots \tag{3.438}$$

The gauge functions themselves $\bar{R}_i^{a\mu_1\mu_2\cdots} \partial_{\mu_1} \partial_{\mu_2}$ may be polynomials in the field φ_i.

When there is no risk for confusion, we write $\delta\varphi_i = R_i^a \xi_a$.[106] Note also that the fields φ_i may not be just matter fields, but also the gauge fields themselves, which is our main interest.

The basic variational equation can now be read off immediately from the formula (3.428). We explicate it again in the form relevant for local symmetries

$$\delta S = \int\limits_\Omega d^4x \frac{\delta S}{\delta \varphi_i(x)} R_i^a(\varphi) \xi_a(x) + \int\limits_\Omega d^4x \partial_\mu J^\mu = 0 \tag{3.439}$$

where from (3.429) we have

$$J^\mu = \frac{\partial \mathcal{L}(x)}{\partial(\partial_\mu \varphi_i(x))} R_i^a(\varphi) \xi_a(x) \tag{3.440}$$

Let us now focus on the interpretation of this fundamental equation.

Interpreting the variational equation

? According to the general formula $\delta S = \int \delta \mathcal{L}\, d^4x$ we can write the variational equation for the Lagrangian density

$$\delta \mathcal{L}(x) = \frac{\delta S}{\delta \varphi_i(x)} \delta\varphi_i(x) + \partial_\mu \left[\frac{\partial \mathcal{L}(x)}{\partial(\partial_\mu \varphi_i(x))} \delta\varphi_i(x) \right] \tag{3.441}$$

The expression within the square bracket is the (parameter dependent) Noether current J^μ

$$J^\mu = J^{a\mu} \xi_a = \frac{\partial \mathcal{L}(x)}{\partial(\partial_\mu \varphi_i(x))} R_i^a(\varphi) \xi_a(x) \tag{3.442}$$

In terms of J^μ, we now have

$$\delta \mathcal{L}(x) = \partial_\mu J^\mu + \frac{\delta S}{\delta \varphi_i(x)} \delta\varphi_i(x) \tag{3.443}$$

This fundamental equation tells us several things.

- If the Lagrangian density is invariant under the transformations, that is, $\delta \mathcal{L}(x) = 0$, then (as we already know) the Noether current is conserved *on-shell*.
- We say that we have a *symmetry of the action* if, without using the equations of motion, we can write $\delta \mathcal{L} = \partial_\mu j^\mu$.
- For symmetries where j^μ vanish on the boundary, we get *gauge identities*.

The last point will be explicated in the next section.

---

106 In Section 4.2 we will study the case of non-Abelian gauge transformations.

3.14.5 Gauge identities – source constraints

From (3.443) we get for currents vanishing on the boundary

$$\frac{\delta S}{\delta \varphi_i} R_i^a = 0 \tag{3.444}$$

These identities, called *gauge identities*, or Noether identities (see Section 3.1.4) hold independent of the equations of motion. The simplest example is for electromagnetic field theory where the variation of the action is $\partial_\mu F^{\mu\nu} = \Box A_\nu - \partial_\nu \partial \cdot A$ and $R_\nu = \partial_\nu$, where we get $\partial_\nu(\Box A_\nu - \partial_\nu \partial \cdot A) = 0$ as an identity. Gauge identities are also called *source constraints* as we have seen in the historical chapter (see Sections 2.7, 2.8.1). The reason for this terminology is that if $S^{(0)}$ denotes the action of free fields φ_i, and if we attempt to couple these free fields to a current J^i through an invariant contribution $\varphi_i J^i$ to the action, then the gauge identity (3.444) reads

$$\left(\frac{\delta S^{(0)}}{\delta \varphi_i} - J^i \right) R_i^a = 0 \tag{3.445}$$

Since the free action itself satisfies the identity, we get $J^i R_i^a = 0$ which is a constraint on the source current.

3.14.6 Noether coupling method

The Noether coupling method is another name for the deformation theoretic approach to deriving interactions in initially free field theories. It was first analyzed in detail for higher spin by Berends, Burgers and van Dam in the mid 1980s [123, 212]. The deformation theoretic program for higher spin interactions was explicitly formulated by Fang and Fronsdal [8] based on earlier work on Yang–Mills and gravity. Parts of this history is told in Chapter 2 (see Sections 2.8.1 and 2.12.2).

The starting point is a free field theory with action $S^{(0)}$, quadratic in the fields φ_i, and transformations $\delta_\xi^{(0)} \varphi_i$ linear in derivatives on the parameters but independent of the fields if we think of the gauge transformations of conventional higher spin theory. The idea is next to think of the – yet to be constructed – full interacting theory as given by a (weak field) expansion

$$S = S^{(0)} + S^{(1)} + S^{(2)} + \cdots \tag{3.446}$$

$$\delta_\xi = \delta_\xi^{(0)} + \delta_\xi^{(1)} + \delta_\xi^{(2)} + \cdots \tag{3.447}$$

where the subscript $^{(n)}$ denotes the power of some expansion parameter g. The power of fields is $n + 2$ in the action terms and n in the transformation terms. Demanding invariance of the action $\delta_\xi S = 0$ then leads to the set of iterative equations

$$\begin{aligned} \delta_\xi^{(0)} S^{(0)} &= 0 \\ \delta_\xi^{(0)} S^{(1)} + \delta_\xi^{(1)} S^{(0)} &= 0 \\ \delta_\xi^{(0)} S^{(2)} + \delta_\xi^{(1)} S^{(1)} + \delta_\xi^{(2)} S^{(0)} &= 0 \\ &\vdots \end{aligned} \tag{3.448}$$

These equations are analogous to the equations one gets by expanding the gauge identities (3.444) in powers of the fields

$$\left(\frac{\delta S^{(0)}}{\delta \varphi_i} + \frac{\delta S^{(1)}}{\delta \varphi_i} + \frac{\delta S^{(2)}}{\delta \varphi_i} + \cdots\right)\left(R_i^{(0)a} + R_i^{(1)a} + R_i^{(2)a} + \cdots\right) = 0 \tag{3.449}$$

There is one difference though. While equations (3.448) are on the level of the action (so that partial integrations can and must be done) and involve the gauge parameters ξ, the equations (3.449) are on the level of field equations and are actual identities and do not involve the gauge parameters. Partial integrations can be done.[107]

A typical attempt to solve these equations would start with an initially given spectrum of higher spin fields and explicit expressions for $S^{(0)}$ and $\delta_\xi^{(0)} \varphi_i$ satisfying the first, zeroth-order equation. To solve the second, first-order equation (cubic in fields), one can look for an on-shell solution where $S^{(0)}$ is stationary. Then the cubic action term must be invariant under the free theory gauge transformations, i. e., $\delta_\xi^{(0)} S^{(1)} = 0$, as is seen from the second of the equations (3.448). In the next step, one determines $\delta_\xi^{(1)}$. We will see how this works out for Yang–Mills theory where the iteration stops at $n = 2$ corresponding to quartic order in the action (see Section 4.4). We will in the course of our development of the subject, encounter this system of equations in several guises.

As the system of equations (3.448) and (3.449) stands here, they do not constitute a complete set of consistency conditions on a higher spin theory. One still must investigate the gauge algebra, as it is captured be commutators of gauge transformations and Jacobi identities, and possibly higher order commutators.

It is also apparent from the discussion so far that the fields and the gauge parameters are treated in a very different way. A systematic way of capturing all the structure of a gauge theory; invariance of the action, closure of the gauge algebra, Jacobi identities and possible higher order structure, is to reformulate the theory in terms of the BRST-BV field-antifield language. This, however, belongs to Volume 2.

107 There are implicit integrations in (3.449) (hidden in the abstract index i), the field theoretic analogue of the type explained in Section 3.1.4 on mechanics. These can all be done since they involve delta functions and derivatives on delta functions.

3.15 Chapter 3 epilogue

This chapter has covered quite a lot of material, some in detail, some superficially. As stated at the outset, the objective has been to collect in one place concepts and methods needed in higher spin field theory. There are certainly more, and we will have to introduce some more in the second volume. What can be found above should be enough to study the free field theory of higher spin fields, as well as the known lower spin theory for Yang–Mills and gravity.

4 Lower spin theory

In this chapter, we will work through some aspects of the spin 1 and spin 2 gauge theories as these may serve as templates for higher spin. Yang–Mills is the cardinal example where a few, quite different approaches all produce the same end result. It is also the best understood example of a gauge theory. For gravity, which originally and conceptually is best understood as a geometrical theory, we will see that problems arise as soon as we try to view it as a gauge theory.

From a higher spin point of view, Yang–Mills theory is indeed the prototypical gauge theory of massless fields, in this case for spin 1 fields. The next step up, to spin 2, is considerably more complex both technically and conceptually. The problem of a gauge theory of gravity, was first approached by R. Utiyama [115] in a paper generalizing the Yang–Mills construction to arbitrary gauge groups. It was followed up by D. W. Sciama [129, 130] and T. W. B. Kibble [131], and from there on a large literature has grown.[1] The next step again, up to spin 3, requires an infinite tower of higher spin fields – as surmised by C. Fronsdal in [213] – and is still an open area of research.

There have been many attempts to model spin 2 theory on spin 1 theory – we will review some of them in due time – and it is only natural to seek clues for a higher spin theory from the Yang–Mills example. One can indeed say that Yang–Mills theory provides the role model for the *gauge principle*. It can be approached either with a minimum of mathematical apparatus, or with much more sophisticated concepts. In the first two sections to follow, we will treat Yang–Mills theory as an example of the gauging method. This approach has two aspects to it: a kinematical part which is fairly straightforward and can be formalized for a wide class of theories, and a nontrivial dynamical part. As far as known at the present, the dynamical part can only be approached one theory at a time. To dampen hope – for a simple route from spin 1 via spin 2 to higher spin – it is a fact that the only physical interesting theory that has been fully constructed[2] by the gauging method – without knowing the end result beforehand – is Yang–Mills theory itself. Nevertheless, it is one of our few solid stepping stones toward higher spin theory.

4.1 Gauge fixing and counting degrees of freedom

Let us start with the free field theories of massless lower spin particles in order to prepare the ground for higher spin fields in the next chapter. In particular, we need to clarify certain points about counting number of degrees of freedom and gauge choices. To begin, a massless spin zero particle is represented by a scalar field satisfying the

1 See the reprint volume [135] which also contains useful introductory essays on the subject.

2 By which is here meant: there is a gauge invariant Lagrangian.

https://doi.org/10.1515/9783110451771-004

massless Klein–Gordon wave equation

$$\Box\phi = 0 \tag{4.1}$$

There is of course no gauge invariance in this case. The field carries one dynamical degree of freedom. However, the equation is a second-order PDE and in order to solve it completely one would need to specify Cauchy initial value data. Essentially, this involves specifying the space variation of the field and its first time derivative at some initial time.[3] Having done this, in principle, we think of the field ϕ as propagating one dynamical degree of "field" freedom.

Next, consider the spin one Maxwell field A_μ with wave equation

$$\Box A_\mu - \partial_\mu \partial \cdot A = 0 \tag{4.2}$$

invariant under the gauge transformation

$$\delta A_\mu = \partial_\mu \xi \tag{4.3}$$

The vector field contains four components. The gauge can be partly fixed by the covariant *Lorenz condition*

$$\mathcal{G} = \partial \cdot A = 0 \tag{4.4}$$

To fix a gauge, the freedom in the arbitrary gauge function ξ must be used, at least partly. Under a gauge transformation $A_\mu \to A_\mu + \partial_\mu \xi$, the gauge condition transforms as $\partial \cdot A \to \partial \cdot A + \Box\xi$. Thus, to stay in the gauge, the gauge parameter must satisfy a wave equation $\Box\xi = 0$ by itself. This is a very weak condition on ξ, in fact still leaving it with one propagating degree of freedom. This d. o. f. can be used to "gauge away" one more component of the vector field. This is called *regauging*. The situation is now the following. The covariantly gauge fixed field is subject to the two equations

$$\Box A_\mu = 0 \qquad \partial \cdot A = 0 \tag{4.5}$$

both of which are invariant under gauge transformations with a parameter satisfying $\Box\xi = 0$. The gauge condition removes one d. o. f. while the regauging removes one more d. o. f., all in all leaving two propagating components corresponding to the two helicity components.

It is interesting to repeat the analysis for a free spin two field. Now the wave equation reads

$$\Box h_{\mu\nu} - \partial_\mu \partial \cdot h_\nu - \partial_\nu \partial \cdot h_\mu + \partial_\mu \partial_\nu h' = 0 \tag{4.6}$$

3 Subtleties that may occur is not important for this general discussion. In the case of a bounded region of space, boundary data has to be supplied also.

where h' is the trace of $h_{\mu\nu}$. The wave equation is invariant under the gauge transformations

$$\delta h_{\mu\nu} = \partial_\mu \xi_\nu + \partial_\nu \xi_\mu \tag{4.7}$$

A covariant gauge choice is now given by the *de Donder* condition

$$\mathcal{G}_\mu = \partial \cdot h_u - \frac{1}{2}\partial_\mu h' = 0 \tag{4.8}$$

The gauge variation of $\mathcal{G}_\mu$ is $\delta\mathcal{G}_\mu = \Box\,\xi_\mu$. Therefore, to stay in the gauge, it is enough for the parameters to satisfy wave equations $\Box\,\xi_\mu = 0$. We thus still have the possibility to regauge four components of the spin 2 field. The gauge condition removes four degrees of freedom and so does the regauging, leaving us with two propagating components. We will see (Section 5.1.1) that this pattern continues for higher spin fields with certain complications.

Spin 1 and 2 Fronsdal tensors and Bianchi identities

Let us streamline the equations in terms of tensors ϕ_μ and $\phi_{\mu\nu}$, where $\phi_{\mu\nu}$ is symmetric in its indices. For these fields, we introduce *Fronsdal tensors*

$$\mathcal{F}_\mu = \Box\phi_\mu - \partial_\mu \partial \cdot \phi \tag{4.9}$$

$$\mathcal{F}_{\mu\nu} = \Box\phi_{\mu\nu} - \partial_\mu \partial \cdot \phi_\nu - \partial_\nu \partial \cdot \phi_\mu + \partial_\mu \partial_\nu \phi' \tag{4.10}$$

where $\phi' = \eta^{\mu\nu}\phi_{\mu\nu}$. These tensors are invariant under the gauge transformations

$$\delta\phi_\mu = \partial_\mu \xi \tag{4.11}$$

$$\delta\phi_{\mu\nu} = \partial_\mu \xi_\nu + \partial_\nu \xi_\mu \tag{4.12}$$

respectively. For free field equations, we have $\mathcal{F}_\mu = 0$ and $\mathcal{F}_{\mu\nu} = 0$. For spin 1, we note that $\mathcal{F}_\mu$ is subject to a differential identity $\partial^\mu \mathcal{F}_\mu = 0$. For spin 2, we first introduce the "Einstein" tensor (the linearized Einstein tensor of GR)

$$G_{\mu\nu} = \mathcal{F}_{\mu\nu} - \frac{1}{2}\eta_{\mu\nu}\mathcal{F}' \tag{4.13}$$

In terms of this tensor we have the spin 2 "Bianchi" identity $\partial^\mu G_{\mu\nu} = 0$.

4.2 Yang–Mills theory (I) – algebraic version

The non-Abelian SU(2) gauge theory of isotopic spin was constructed in 1954 by C. N. Yang and R. L. Mills [114]. It subsequently developed over several decades into becoming the backbone of the Standard Model. We will now reconstruct Yang–Mills theory with a minimum of apparatus. We are aiming for a field theory of self-interacting massless spin 1 fields. Assume therefore that the gauge field is valued in a simple compact Lie group G.[4] We can represent it either explicitly with a Lie algebra index a as $A^a_\mu(x)$ or as $A_\mu = A^a_\mu(x)T^a$ where T^a are the anti-Hermitian traceless $N \times N$ matrices of the algebra. In analogy with electromagnetism (see formula (2.5)), we expect to introduce covariant derivatives $D_\mu = \partial_\mu + A_\mu$ encoding minimal coupling. The gauge coupling constant you would perhaps expect here is absorbed into the field.[5] The field strength $F_{\mu\nu}$ should be given by the commutator of covariant derivatives. Let us now derive this, in the standard fashion, by promoting a global symmetry to a local one by the method of *gauging*.

To start with, there are not any massless spin-1 gauge fields; instead, we think of a Lagrangian density $\mathcal{L}_M(\varphi_i, \partial_\mu\varphi_i)$ for a set of matter fields $\varphi_i(x)$, invariant under a global symmetry transformation

$$\delta\varphi_i(x) = -\xi^a(T^a)_{ij}\varphi_j(x) \tag{4.14}$$

which is the infinitesimal form of the group rotations

$$\varphi(x) \mapsto \varphi'(x) = U(\xi)\varphi(x) = \exp(-\xi^a T^a)\varphi(x) \tag{4.15}$$

The matrices T are particular to the representation that the matter fields transform under, but they always satisfy the commutation relations (3.382).

Typically we would have a matter Lagrangian

$$\mathcal{L}_M(\varphi_i, \partial_\mu\varphi_i) = -\frac{1}{2}(\partial_\mu\bar{\varphi}_i\partial^\mu\varphi_i + m^2\bar{\varphi}_i\varphi_i) \tag{4.16}$$

When the parameters ξ^a are taken as arbitrary functions of position, the kinetic term will not be invariant since the derivative of the fields will transform as

$$\delta(\partial_\mu\varphi_i(x)) = -\xi^a(x)(T^a)_{ij}\partial_\mu\varphi_j(x) - \partial_\mu\xi^a(x)(T^a)_{ij}\varphi_j(x) \tag{4.17}$$

The offending terms can be compensated for by introducing[6] a new field $A^a_\mu(x)$ and the corresponding *covariant derivative*

$$D_\mu = \partial_\mu + A^a_\mu T^a \equiv \partial_\mu + A_\mu \tag{4.18}$$

---

4 A thorough discussion on the restrictions on the groups and concomitant details of the formalism can be found in [139].

5 The relative sign between the derivative and the field, is a matter of convention. Furthermore, the derivative term should also be thought of as being multiplied by a unit matrix of the same dimension as the Lie algebra matrices.

6 For a critical discussion of the necessity of this step, see [267].

Then we require the covariant derivative to transform in the same way as the field in order to restore invariance of the kinetic term in the action. That is, we require

$$\delta(D_\mu\varphi_i(x)) = -\xi^a(T^a)_{ij}(D_\mu\varphi_j(x)) \tag{4.19}$$

From this follows the transformation law for the gauge field

$$\delta A^a_\mu = \partial_\mu\xi^a + f^{abc}A^b_\mu\xi^c \quad \text{or} \quad \delta A_\mu = \partial_\mu\xi + [A_\mu, \xi] \tag{4.20}$$

It is clear from the computations that in contrast to electromagnetism, the Yang–Mills field must be a self-interacting field which is reflected in the transformation law. Physically it must be so since it carries non-Abelian charge. This is also evident from computing the field strength as the commutator of two covariant derivatives

$$F_{\mu\nu} = [D_\mu, D_\nu] = \partial_\mu A_\nu - \partial_\nu A_\mu + [A_\mu, A_\nu] \tag{4.21}$$

This non-Abelian field strength $F_{\mu\nu} = F^a_{\mu\nu}T^a$ can also be expressed explicitly as

$$F^a_{\mu\nu} = \partial_\mu A^a_\nu - \partial_\nu A^a_\mu + f^{abc}A^b_\mu A^c_\nu \tag{4.22}$$

The field strength transform homogeneously as

$$\delta F^a_{\mu\nu} = f^{abc}F^b_{\mu\nu}\xi^c \quad \text{or} \quad \delta F_{\mu\nu} = [F_{\mu\nu}, \xi] \tag{4.23}$$

The gauge invariant action for the gauge field is

$$\mathcal{L}_{YM} = -\frac{1}{4g^2}F^{a\mu\nu}F^a_{\mu\nu} = \frac{1}{2g^2}\operatorname{Tr}(F^{\mu\nu}F_{\mu\nu}) \tag{4.24}$$

where g is the gauge coupling constant. It is interesting to expand the Yang–Mills Lagrangian density in powers of the fields. For this, we redefine the gauge fields as $A \to gA$. Then

$$\begin{aligned}\mathcal{L}_{YM} = &-\frac{1}{2}\left(\partial^\mu A^{a\nu}\partial_\mu A^a_\nu - \partial^\mu A^{a\nu}\partial_\nu A^a_\mu\right)\\ &- gf^{abc}A^{a\mu}A^{b\nu}\partial_\mu A^c_\nu - \frac{1}{4}g^2 f^{abe}f^{cde}A^{a\mu}A^{b\nu}A^c_\mu A^d_\nu\end{aligned} \tag{4.25}$$

Even expanded like this, it is a very simple and beautiful Lagrangian. We also record the field equations that follow from this form of the Lagrangian,

$$\begin{aligned}F^a_\nu = {}&\partial^\mu(\partial_\mu A^a_\nu - \partial_\nu A^a_\mu) - gf^{abc}\left(A^{b\mu}\partial_\nu A^c_\mu + A^b_\nu\partial\cdot A^c - 2A^{b\mu}\partial_\mu A^c_\nu\right)\\ &- g^2 f^{abe}f^{cde}A^{b\mu}A^c_\nu A^d_\mu = 0\end{aligned} \tag{4.26}$$

The equations can be written in a compact way as

$$F^a_\nu = (D^\mu F_{\mu\nu})^a \equiv \partial^\mu F^a_{\mu\nu} + gf^{abc}A^{b\mu}F^c_{\mu\nu} = 0 \tag{4.27}$$

where we have rescaled $f^{abc} \to gf^{abc}$ in the formulas for D_μ and $F^a_{\mu\nu}$.

4.3 Yang–Mills theory (II) – geometric version

We will do the geometric approach in two stages: as a direct "calculus" application of exterior algebra, and then indicate a more sophisticated "analysis" approach.

4.3.1 Yang–Mills on differential forms – calculus

As we will see, when Yang–Mills theory is formulated based on differential forms, there is no formal difference between the theory in curved space-time as compared to flat space-time.[7] Let us start with the "calculus" of forms.

The basic idea is to think of a covariant vector V_μ as 1-form V

$$V = V_\mu dx^\mu \tag{4.28}$$

where the differentials dx^μ are basis elements in cotangent space. In particular, the derivative operator ∂_μ, which transforms as a covariant vector, is represented as $d = dx^\mu \partial_\mu$. Likewise, any antisymmetric covariant tensor $V_{\mu\nu}$ is represented as

$$V = V_{\mu\nu} dx^\mu \wedge dx^\nu = V_{\mu\nu} dx^\mu dx^\nu \tag{4.29}$$

where in the second equality the wedge product $\wedge$ is suppressed, as is often done when confusion is not likely to occur.

With respect to Lie algebras, there is an extra bonus to this formalism. Consider a 1-form A valued in a Lie algebra $A = A_\mu^a T_a dx^\mu$. Then perform the following computation:

$$\begin{aligned} A^2 = A \wedge A = A_\mu^a T_a dx^\mu \wedge A_\nu^b T_b dx^\nu &= \frac{1}{2}(A_\mu^a T_a A_\nu^b T_b - A_\nu^b T_b A_\mu^a T_a) dx^\mu \wedge dx^\nu \\ &= \frac{1}{2} A_\mu^a A_\nu^b [T_a, T_b] dx^\mu \wedge dx^\nu = \frac{1}{2} f_{ab}^{\ \ c} A_\mu^a A_\nu^b T_c dx^\mu \wedge dx^\nu \\ &= \frac{1}{2}[A_\mu, A_\nu] dx^\mu \wedge dx^\nu \end{aligned} \tag{4.30}$$

The formula can be derived directly without going through the explicit T-matrix algebra. Now applying this formula to Yang–Mills we immediately get

$$dA + A^2 = d \wedge A + A \wedge A = \frac{1}{2} F_{\mu\nu}^a T_a dx^\mu \wedge dx^\nu \equiv \frac{1}{2} F \tag{4.31}$$

If the gauge group is Abelian, then $A^2 = 0$. Although, potentially confusing, one can write the quadratic term in the field strength as $\frac{1}{2}[A, A]$ or perhaps better $\frac{1}{2}[A, A]_\wedge$ in-

7 When it comes to solutions of the field equations, the geometry and topology of the underlying manifold is of course of crucial importance. For reviews, see [268, 269].

dicating that the commutator product is the wedge product (so that the bracket is formally symmetric). The following formula is useful:

$$A \wedge B = \frac{1}{2} A^a \wedge B^b [T_a, T_b] = \frac{1}{2} A^a_\mu B^b_\nu [T_a, T_b] dx^\mu \wedge dx^\nu \equiv \frac{1}{2} [A, B]_\wedge \tag{4.32}$$

The gauge transformations can be written as

$$\delta A = d\xi + [A, \xi]_\wedge \tag{4.33}$$

where the wedge product is understood in the first term, although it is trivial there since the gauge parameter is a 0-form given by $\xi = \xi^a T_a$. The field strength transforms as follows in this formalism:

$$\delta F = [F, \xi]_\wedge \tag{4.34}$$

Referring back to example 6 and equation (3.378) on page 199, we can write the Yang–Mills action in form language as

$$S_{YM} = -\frac{1}{4g^2} \operatorname{Tr} \int_M {}^*F \wedge F \tag{4.35}$$

This way of writing the action is valid also in Riemannian spaces.

4.3.2 Fiber bundles for Yang–Mills – analysis

We now turn to the reformulation of Yang–Mills theory in terms of fiber bundles. In the calculus version of geometric Yang–Mills theory, we do not worry about where the gauge fields are "valued". This can be remedied by introducing the concept of a fiber bundle, with space-time as the base manifold, and the gauge group as the typical fiber. What we have then is a *principal fiber bundle*, meaning that the structure group of the typical fiber is the fiber itself (see Section 3.12.2). The connection on this fiber bundle is the gauge field A^a_μ itself. This should be evident from its index structure and the role it plays in the gauge covariant derivative.

The basic algebraic formulation of the theory is precisely the same as in the calculus version. The reason we now call it "analysis" rests with the greater mathematical strength provided by the fiber bundle formulation. For instance, global topological properties may be addressed since the principal bundle actually is a topological space. Although locally it has a direct product structure, that need not be so globally.[8] Furthermore, the theory can be formulated on curved Riemannian space-times, not just Minkowski space-time.

8 An old but readable introductory review of Yang–Mills geometry is M. F. Atiyah's [268].

The gauging method – bottom line on kinematics

From $[T^a, T^b] = f^{abc}T^c$ with $\xi = \xi^a T^a$, the gauge algebra can be expressed as $[\xi_1, \xi_2]^a = f^{abc}\xi_1^b\xi_2^c$. The Jacobi identity holds for the bracket $[\xi_1, \xi_2]^a$. The *kinematic part of gauging* amounts to associating a gauge field A_μ^a to each generator T^a, transforming as $\delta A_\mu^a = \partial_\mu \xi^a + [A_\mu, \xi]^a$. Field strengths $F_{\mu\nu}^a$, or curvatures, are defined as $F_{\mu\nu}^a = \partial_\mu A_\nu^a - \partial_\nu A_\mu^a + [A_\mu, A_\nu]^a$ transforming homogeneously as $\delta F_{\mu\nu}^a = [F_{\mu\nu}, \xi]^a$. Note particularly that all brackets in these expressions goes back to the gauge algebra bracket $[\xi_1, \xi_2]^a$. Thus the structure of the algebra governs the transformations of the gauge fields, the definition of the curvatures and their transformations, as well as the covariant derivative $D_\mu = \partial_\mu + A_\mu$.

This kinematic gauge theory structure must be tailored to the kind of theory at hand, but it remains essentially a basic ingredient in any gauge theory. Dynamics – for the gauge fields themselves – is however more difficult to introduce.

The reader may have noted that we did not provide much in the way of motivations for the Lagrangian (4.24) for the gauge field itself. What one can offer is that it is the simplest object that is Poincaré invariant, gauge invariant and with a kinetic – free field theory – term of second order in derivatives. This, and some basic intuition and experience, was enough to guess the Yang–Mills Lagrangian. This may give the impression that the dynamics step is easy. That is not so.

4.4 Yang–Mills theory (III) – Noether coupling version

As a prototypical example of the Noether coupling (deformation theoretic) approach to gauge theory we will now work through Yang–Mills theory in some detail. We will explicate the calculation in the formalism (3.449), leaving the corresponding computations in the formalism (3.448) to the reader.

Since we already have the Yang–Mills equations of motion we will start by checking the known solution. To zeroth order in the coupling we have the equation

$$\frac{\delta S^{(0)}}{\delta A_\mu^a} R_\mu^{(0)ab} = 0 \tag{4.36}$$

with

$$\begin{cases} \delta S^{(0)}/\delta A_\mu^a = \Box A_\mu^a - \partial_\mu \partial \cdot A^a \\ R_\mu^{(0)ab} = \delta^{ab}\partial_\mu \delta(x-y) \end{cases} \tag{4.37}$$

Then using the implicit integration over the space-time variable y in (4.36) we find that the equation is identically fulfilled. Then to first order in the coupling we have

$$\frac{\delta S^{(1)}}{\delta A_\mu^a} R_\mu^{(0)ab} + \frac{\delta S^{(0)}}{\delta A_\mu^a} R_\mu^{(1)ab} = 0 \tag{4.38}$$

with the known term from (4.26)

$$\frac{\delta S^{(1)}}{\delta A^{\mu a}} = -gf^{abc}(A^{\nu b}\partial_\mu A^c_\nu - A^{\nu b}\partial_\nu A^c_\mu + \partial^\nu(A^b_\mu A^c_\nu)) \tag{4.39}$$

we now get (remember partial integration of ∂_μ)

$$\frac{\delta S^{(1)}}{\delta A^a_\mu} R^{(0)ab}_\mu = gf^{bcd}A^{\nu c}(\Box A^d_\nu - \partial_\nu \partial \cdot A^d) \tag{4.40}$$

Then inserting this expression in the equation (4.38) we can identify the first-order gauge transformation $R^{(1)}$ according to the scheme

$$\delta A^a_\mu = R^{(0)ab}_\mu \xi^b + R^{(1)ab}_\mu \xi^b \tag{4.41}$$

That is

$$\frac{\delta S^{(0)}}{\delta A^a_\mu} R^{(1)ab}_\mu = -gf^{bcd}A^{\nu c}(\Box A^d_\nu - \partial_\nu \partial \cdot A^d) \tag{4.42}$$

which gives

$$R^{(1)ab}_\mu = -gf^{abc}A^c_\mu \tag{4.43}$$

This gives the correct first-order gauge transformations.

In the literature, this procedure is often described as solving the equations $\delta^{(0)}S^{(1)} = 0$ since $\delta^{(1)}S^{(0)} = 0$ when the free field equations are satisfied. Then "reading off" the first-order gauge transformations. That is confusing as a description of the procedure. A better description is to say that one takes an ansatz for the first-order interaction and computes the zero-order gauge variation on the ansatz and then tries to rewrite the result as a transformation of the free field equations. Then the first-order gauge transformations can indeed be "read off".

To the next order, we have

$$\frac{\delta S^{(2)}}{\delta A^a_\mu} R^{(0)ab}_\mu + \frac{\delta S^{(1)}}{\delta A^a_\mu} R^{(1)ab}_\mu + \frac{\delta S^{(0)}}{\delta A^a_\mu} R^{(2)ab}_\mu = 0 \tag{4.44}$$

The first two terms can now be computed. They turn out to add up to zero due to the Jacobi identity for the structure constants (and total antisymmetry). Therefore, there are no second order contribution to the gauge transformations and we have $R^{(2)ab}_\mu = 0$.

One may now perform these computations on an ansatz for the cubic couplings. Referring back to the Ogievetskij–Polubarinov analysis (see formula (2.187) in Section 2.9.3), there is actually just one type of term to write down; however, without any assumption as to the nature of the coupling constants. For the second-order equation (4.44), one also needs an ansatz for the $\delta^{(2)}S/\delta A$ term. The computations get a bit involved, but one may anticipate what will happen from dimensional analysis. The two first terms in (4.44) have the overall structure of ∂A^3. Since $\delta^{(0)}S/\delta A$ already contain two powers of derivatives, there is no possible local $R^{(2)}$ term to write down. It should also be clear that the Jacobi identity for the coupling constants will be a result of the computations. We leave it to the reader to sort out the details.

4.5 General relativity and its generalizations

There are many good texts on general relativity[9] and a standard review will not be attempted here. Instead, we will pursue the subject as it sits, so to speak, between Yang–Mills theory and higher spin theory while emphasizing topics that seem to be particularly relevant for higher spin.

To avoid confusion in the sequel when we discuss various types of space-times, we will reserve *general relativity* with the abbreviation GR for standard Einstein general relativity in the terminology to follow: Riemannian space-time, that is, space-time with a covariantly constant metric and no torsion, and where the metric is dynamically determined by the Einstein field equations.[10] This is in accordance with the terminology of [135]. Occasionally, the wordings "standard general relativity" or "Einstein general relativity" will be used for GR.

The purpose of this section is to discuss various different space-times and structures defined on them in relation to the gauge theory approach to gravity. The backdrop to this enterprise is the concept of a manifold in the differential geometric sense. It comes with its infinite set of coordinate systems and its tangent and cotangent spaces, but nothing more. To get going, let us first define the metric and the vierbeins and investigate some properties of these objects.

4.5.1 The metric and the vierbeins

The metric is a symmetric covariant tensor $g_{\mu\nu}$. Under a coordinate transformation $x^\mu \to x'^\mu(x)$, it transforms as

$$g'_{\mu\nu}(x') = \frac{\partial x^\alpha}{\partial x'^\mu}\frac{\partial x^\beta}{\partial x'^\nu} g_{\alpha\beta}(x) \tag{4.45}$$

The metric measures the distance $d\tau$ between two infinitesimally nearby points x^μ and $x^\mu + dx^\mu$

$$d\tau^2 = -g_{\mu\nu}(x)dx^\mu dx^\nu \tag{4.46}$$

9 Some of which have been useful in my own writing are [270, 243, 127, 271].

10 Some would say that *Lorentzian* (or pseudo-Riemannian) is a better designation than Riemannian, but the qualifier *space-time* should be enough to remove any confusion as to the signature of the metric.

At each point P in space-time, there is – according to the Equivalence Principle (EP) – a freely falling inertial coordinate system with coordinates $\xi_P^\alpha(x)$. Special relativity is valid in this system, and the proper time is given by the corresponding formula (3.84). Of the infinitely many possible coordinate systems available at the point P, let x^μ denote a generic one. Consider the coordinate transformation $\xi_P^\alpha \to x^\mu(\xi)$ from the inertial system ξ to the "curved" system x. Then applying the transformation formula (4.45), we get

$$g_{\mu\nu}(x) = \frac{\partial \xi_P^\alpha}{\partial x^\mu}\frac{\partial \xi_P^\beta}{\partial x^\nu} g_{\alpha\beta}(\xi) \tag{4.47}$$

On the right-hand side, the metric $g_{\alpha\beta}(\xi)$ is then the constant Minkowski metric $\eta_{\alpha\beta}$. This formula can be used to introduce vierbein fields. In order to do that, we first trade the label P for "flat" Minkowski coordinate labels $a, b, c, \ldots$ thus writing the formula

$$g_{\mu\nu} = \frac{\partial \xi^a}{\partial x^\mu}\frac{\partial \xi^b}{\partial x^\nu}\eta_{ab} \tag{4.48}$$

The derivatives that occur in this formula define the *vierbein fields* or *tetrads*

$$e_\mu{}^a(x) = \frac{\partial \xi^a(x)}{\partial x^\mu} \tag{4.49}$$

and we can write the well-known formula for the metric in factors of the vierbeins

$$g_{\mu\nu} = e_\mu{}^a e_\nu{}^b \eta_{ab} \tag{4.50}$$

The vierbeins can be inverted through the formula

$$e^\mu{}_a = \eta_{ab} g^{\mu\nu} e_\nu{}^b \tag{4.51}$$

and we get

$$e^\mu{}_a e^\nu{}_b g_{\mu\nu} = \eta_{ab} \tag{4.52}$$

an equation that expresses the *orthonormality* of the tetrad basis $e^\mu{}_a$.

The inverse metric

The metric $g_{\mu\sigma}$ (a covariant tensor) is invertible, and its inverse $g^{\sigma\nu}$ (a contravariant tensor) is defined through the formula

$$g_{\mu\sigma} g^{\sigma\nu} = \delta_\mu^\nu \tag{4.53}$$

where on the right-hand side we have the Kronecker tensor δ_μ^ν. It is the only tensor, apart from constants and scalars, that are the same in all coordinate systems. It has nothing in particular to do with the Minkowski metric.

The vierbeins can be used to – as it is expressed – convert between “world” (“curved”) indices μ and “inertial” (“tangent”) indices a. What it actually does is to transform the coordinate basis dx^μ of the cotangent space to an orthonormal noncoordinate basis e^a according to

$$e^a = e_\mu{}^a dx^\mu \tag{4.54}$$

The established mathematical terminology for the 1-form field e^a is *coframe* (“co” as in cotangent).[11] Likewise, the inverse vierbein $e^\mu{}_a$ transforms the coordinate basis ∂_μ in tangent space to a noncoordinate orthonormal basis, or *frame* e_a

$$e_a = e^\mu{}_a \partial_\mu \tag{4.55}$$

The frames and coframes are vector space duals in the ordinary sense of linear algebra, as the following short calculation shows

$$\langle e^a, e_b \rangle = \langle e_\mu{}^a dx^\mu, e^\nu{}_b \partial_\nu \rangle = e_\mu{}^a e^\nu{}_b \langle dx^\mu, \partial_\nu \rangle = e_\mu{}^a e^\nu{}_b \delta^\mu_\nu = \delta^a_b \tag{4.56}$$

where we have used (3.372).

The vierbeins $e_\mu{}^a$ themselves, being 4×4 asymmetric matrices, have 16 components, whereas the symmetric metric $g_{\mu\nu}$ has only 10 components. The 6 extra components of the vierbeins correspond to the possibility of making Lorentz transformations in the tangent space. There are therefore many noncoordinate bases e^a that yield the same metric, related by the transformation

$$e^a \to e'^a = \Lambda^a{}_b e^b \tag{4.57}$$

where the transformation matrix $\Lambda^a{}_b$ depends on x. Since the dual basis transforms as

$$e_a \to e'_a = (\Lambda^{-1})^b{}_a e_b \tag{4.58}$$

it is clear that the metric is invariant under such local Lorentz transformations.

Pause for thinking

?

It is important to realize that the local inertial systems are in fact local. Their role is to transform away the effects of the gravitational field locally. There is just one inertial system, modulo Lorentz transformations, at each point in space-time. The terminology alluded to in text: “curved” or “world” indices and “inertial” or “tangent” indices also needs clarifying. It is a very common usage in theoretical

11 In the higher spin literature, the 1-form description of higher spin fields goes under the name “frame-like” formulation. ‘Coframe-like” formulation would perhaps be more appropriate.

physics, and while the meaning is fairly clear, it is potentially confusing if one pauses for thinking. According to the differential geometric view, all coordinate systems are in fact mappings into "flat" space-times (see Section 3.10). For any particular such coordinate system, one can contemplate the tangent and cotangent space-times. These are vector spaces with natural coordinate bases ∂_μ and dx^μ, respectively. Gravity may still be felt in these spaces. In fact, the existence of tangent and cotangent spaces have nothing to do with the existence of a metric. All tensors live in tangent- and cotangent space-time.

Then a coordinate transformation can be done into the inertial system (at the point) and the basis in this system is precisely given by $e^a = e_\mu{}^a dx^\mu$. The special role played by the inertial system motivates the use of special indices $a, b, c, \ldots$. As we move around in space-time, the inertial coordinates change. Within an inertial system, one can make Lorentz transformations. Thus understood, the wordings "world" indices and "tangent" indices, and the like, are convenient shorthand.

Note that the basis vectors e_a do not commute in general. A short computation yields

$$[e_a, e_b] = -\Omega_{ab}{}^c e_c \quad \text{where } \Omega_{ab}{}^c = e^\mu{}_a e^\nu{}_b(\partial_\mu e^c{}_\nu - \partial_\nu e^c{}_\mu) \tag{4.59}$$

where $\Omega_{ab}{}^c$ are the *anholonomy coefficients*, also called *Ricci rotation coefficients*. The coordinate basis is holonomic whereas the orthonormal basis is not. It is possible to set up other tetrad bases with $e^\mu{}_a e^\nu{}_b g_{\mu\nu} = g_{ab}$ for some desired metric g_{ab}. Such a basis is not orthonormal unless $g_{ab} = \eta_{ab}$.

A notational issue

In this context, we note a potentially confusing notational issue that may arise if one thinks of the coordinate basis ∂_μ as a particular frame, namely with $e^\mu{}_a = \delta^\mu{}_a$. Then one might be tempted to write $e_a = \partial_a$, thinking of $\delta_a{}^\mu$ as Kronecker delta. Then it seems that one loses the distinction between "world" and "tangent" indices/spaces. This, however, may be understood from the discussion in the box above. The derivatives ∂_μ are actually computed locally in tangent space-time, as are the ∂_a derivatives. On the other hand, $\delta^\mu{}_a$ cannot be a Kronecker delta. The Kronecker deltas δ^μ_ν and δ^b_a are the only tensors, apart from scalars, that are the same in all coordinate systems. Then the formula $\delta^\mu{}_a = e^\nu{}_a \delta^\mu_\nu$ show that $\delta^\mu{}_a$ is a vierbein, and not a fixed Kronecker delta.

On the other hand, one may be tempted to use $\delta_\mu{}^a$ for a Minkowski vierbein, and it seems that we get a notational clash. That is indeed the case. It may be resolved by the following reasoning. In the first case, when one thinks of the coordinate basis as a particular frame, we have no coordinate transformation, just the coordinate map to flat space-time, and therefore the vierbeins are $\delta_\mu{}^a$. For this, we may not need any notation and we may choose not to use δ in this way, reserving $\delta_\mu{}^a$ for the Minkowski vierbein But then one must be aware of the fact that the whole space is Minkowski, not just locally. In this case, $\delta^\mu{}_a$ is a Kronecker delta and there is no distinction between "curved" and "flat" indices. Local inertial frames can obviously not be denoted by $\delta^\mu{}_a$.

In a perturbative approach to gravity, where one thinks in terms of a weak spin 2 field propagation in a space-time background, we will use the symbol h for the background metric or vierbeins. From here on, $e_\mu{}^a$ and $g_{\mu\nu}$ will be used for a general tetrad and metric, respectively. Anticipating working in AdS space-time, we will use the symbol h for a background vierbein or metric (with appropriate indices). In case of Minkowski background, we use $h = \eta$.

This discussion is not entirely without bearing on the higher spin problem since in setting up a basic machinery for maintaining an infinite spectrum of higher spin fields, one has to make choice as to how to represent these fields. The Vasiliev theory makes one such choice, to which we will come to in Volume 2.[12]

Mind bending questions in the elevator

The frames can be thought of as setting up freely falling, local inertial systems. Such a freely falling inertial system at a point P is sometimes pictorially referred to as the *Einstein elevator*. The formulas (4.57) and (4.58) tell us that we can perform Lorentz transformations in the elevator. Physically, it is clear that rotating the coordinate system makes no difference. Also, boosting it must be allowed. It will fall a little faster, or slower, but adding or subtracting a constant velocity make no difference.

But what about translations? Would not a translation of the elevator bring it to another point where the gravitational field is different? And come to think of it, will not the falling elevator anyway move into new points with different gravitational fields? Is everything going haywire? But must not one think infinitesimally, and not too literally? The elevator is indeed falling into different points, but is not that supposed to be taken care of by the frame field that defines new elevators as it falls? The acceleration is not constant unless the gravitational field is homogeneous. As for translations, an active translation would bring the elevator to another point, but would not a passive translation – amounting to a change of origin in the elevator – be just fine? These questions do arise again when we consider local Poincaré transformations in gauge theory approaches to gravity.

4.5.2 Connections

Connections on a manifold is a concept initially independent of the existence of a metric or not. It has to do with comparing vectors at different points in the manifold, or parallel transporting them. Consider a covariant vector V_ν. Using infinitesimal linearity, displacing the vector from the point x^μ to the point $x'^\mu = x^\mu + dx^\mu$, the change dV_ν (at the point x) is expected to be

$$dV_\nu = \Gamma_{\mu\nu}{}^{\lambda} V_\lambda dx^\mu \tag{4.60}$$

for some object $\Gamma_{\mu\nu}{}^{\lambda}$ – the *affine connection* – that parametrizes the way the vector change. Then computing the difference between the vector at the point x', that is, $V_\nu(x')$, and the parallel transported value $V_\nu(x) + dV_\nu(x)$, we get

$$V_\nu(x') - (V_\nu(x) + dV_\nu(x)) = (\partial_\mu V_\nu - \Gamma_{\mu\nu}{}^{\lambda} V_\lambda) dx^\mu = (\nabla_\mu V_\nu) dx^\mu \tag{4.61}$$

Sign and index conventions in (4.60), are chosen so that the result is consistent with the action of the covariant derivative on the vector.

12 One cannot escape the feeling that the technical problems that face all approaches to higher spin theory, has deep conceptual roots, possibly in relation to the structure of space-time.

As we will have occasion to discuss in more detail in the following, the affine connection is a structure that is added to the manifold. This may appear a little odd at first thought. After all, vectors live in the tangent spaces of the manifold – in the tangent bundle to be precise – and should not that structure be determined by the manifold in question? That is true, but the point is that each fiber contains an infinite set of vectors, and we still have to decide how a certain vector in a fiber gets parallel transported to another fiber. That is what the affine connection tells.

4.5.3 Curvature and the Riemann tensor

Once there is a connection on the space-time, we can define covariant derivatives as in equations (3.412) and (3.414). The *Riemann curvature tensor* can be computed from the commutator of covariant derivatives. Explicitly, one gets

$$[\nabla_\mu, \nabla_\nu] V_\sigma = -R_{\mu\nu\sigma}{}^{\alpha} V_\alpha - T_{\mu\nu}{}^{\alpha} \nabla_\alpha V_\sigma \tag{4.62}$$

$$[\nabla_\mu, \nabla_\nu] V^\sigma = R_{\mu\nu\alpha}{}^{\sigma} V^\alpha - T_{\mu\nu}{}^{\alpha} \nabla_\alpha V^\sigma \tag{4.63}$$

The torsion terms drop out for a symmetric connection.[13] In any way, the curvature tensor works out to

$$R_{\mu\nu\rho}{}^{\sigma} = \partial_\mu \Gamma_{\nu\rho}{}^{\sigma} - \partial_\nu \Gamma_{\mu\rho}{}^{\sigma} + \Gamma_{\mu\lambda}{}^{\sigma} \Gamma_{\nu\rho}{}^{\lambda} - \Gamma_{\nu\lambda}{}^{\sigma} \Gamma_{\mu\rho}{}^{\lambda} \tag{4.64}$$

where Γ still denotes the affine connection, possibly containing a torsion term.

The covariance of the covariant derivative and the curvature tensor follows from the nonhomogeneous transformation law (3.413) for the affine connection. This is independent of any metric on the manifold. Given the formal similarity to the corresponding Yang–Mills concepts, this fact is clearly interesting from a higher spin perspective. Can one think of the connection as a gauge field and the curvature as the field strength? Let us pause this question and take it up in Section 4.6.

It is clear from the formulas (4.62) and (4.63) for the commutators of covariant derivatives on vectors, that the curvature is antisymmetric in the first two indices μ and ν

$$R_{(\mu\nu)\rho}{}^{\sigma} = 0 \tag{4.65}$$

This is also obvious from the explicit formula (4.64). Since the commutator of covariant derivatives must satisfy the Jacobi identity, one can derive further identities[14]

$$R_{[\mu\nu\rho]}{}^{\sigma} - \nabla_{[\mu} T_{\nu\rho]}{}^{\sigma} + T_{[\mu\nu}{}^{\alpha} T_{\rho]\alpha}{}^{\sigma} = 0 \tag{4.66}$$

---

13 Torsion is defined in Section 3.13.2. What actually comes out of the computation of, for instance, (4.62), is just $-(\Gamma_{\mu\nu}{}^{\alpha} - \Gamma_{\nu\mu}{}^{\alpha})\nabla_\alpha V_\sigma$. No more explicit formula for the torsion can result from this type of calculation.

14 To derive the second identity, the first is needed.

$$\nabla_{[\mu} R_{\nu\rho]\sigma}{}^{\lambda} - T_{[\mu\nu}{}^{\alpha} R_{\rho]\alpha\sigma}{}^{\lambda} = 0 \tag{4.67}$$

In the case of zero torsion, the two equations (4.66) and (4.67) are referred to as the *first (algebraic) Bianchi identity* and the *second (differential) Bianchi identity*, respectively.

Antisymmetrization and cyclic sums

The notation [...] enclosing a set of indices means total antisymmetrization with weight 1 in analogy to how (...) denotes total symmetrization. For instance, $A_{[\mu}B_{\nu\rho]} = A_\mu B_{\nu\rho} - A_\mu B_{\rho\nu} + A_\rho B_{\mu\nu} - A_\rho B_{\nu\mu} + A_\nu B_{\rho\mu} - A_\nu B_{\mu\rho}$. If the tensor $B_{\mu\nu}$ happens to be antisymmetric, then the antisymmetrization is equal to twice the cyclic sum, that is, $A_{[\mu}B_{\nu\rho]} = 2(A_\mu B_{\nu\rho} + A_\nu B_{\rho\mu} + A_\rho B_{\mu\nu})$.

By lowering the contravariant index on the curvature tensor, it can be written in covariant form as $R_{\mu\nu\rho\sigma}$. The question then arises as to further index symmetry properties of the curvature tensor. These are the facts: the curvature tensor can be defined without any reference to a metric, it only needs a connection, then no further general index symmetry properties follow. When a metric is present to raise and lower indices, one can compare the two equations (4.62) and (4.63).[15] It then follows, under the assumption that the metric is covariantly constant $\nabla_\rho g_{\mu\nu} = 0$, that the curvature tensor $R_{\mu\nu\rho\sigma}$ is antisymmetric also in the second set of indices $\rho\sigma$.

From the Riemann tensor, one can compute the Ricci tensor and the curvature scalar by contracting indices. The *Ricci tensor* $R_{\mu\rho}$ is defined as

$$R_{\mu\rho} = R_{\mu\nu\rho}{}^{\nu} \tag{4.68}$$

In the case of antisymmetry in both sets of indices on the curvature tensor, the Ricci tensor is essentially the unique trace of the curvature tensor. The Ricci tensor is in general not symmetric. Contracting once more with the metric yields the curvature scalar R

$$R = R_{\mu\rho} g^{\mu\rho} \tag{4.69}$$

If the torsion is zero, then the curvature tensor $R_{\mu\nu,\rho\sigma}$ is symmetric in interchanging the index groups $\mu\nu$ and $\rho\sigma$. The tensor therefore has the symmetry of the Young tableaux (see Section 5.7.4)

$$\begin{array}{|c|c|}\hline \mu & \rho \\ \hline \nu & \sigma \\ \hline \end{array} \tag{4.70}$$

In this case, it follows that the Ricci tensor is symmetric.

15 Note the conspicuous different signs in front of the curvature terms.

Index groups

When later on generalizing to higher spin, it will be convenient to explicitly distinguish different sets of indices, and we will have occasion to write the curvature tensor as $R_{\mu\nu,\rho\sigma} = R_{\mu\nu\rho\sigma}$ with a comma to separate the index groups. These index groups may be subject to relations. For entirely unrelated index groups, we will write for instance $T_{\mu\nu|\rho\sigma\lambda}$. Never, in this book, is a comma to be interpreted as denoting a derivative.

4.5.4 Space-times

In Sections 3.10 and 3.13, we recapitulated the basic facts about manifolds and coordinate transformations. In order to "do physics" in such a manifold, we need more structure. For this structure, there are choices to be made. We will go trough them in some detail here. These are choices that may have bearing on the higher spin problem. The basic four-dimensional manifold will be denoted by X_4 or just X. It comes with its tangent and cotangent spaces, but to begin with, no further structure: no metric, no connection, no curvature.[16] There is a sequence of space-times: L_4, $L_{4,g}$, U_4, V_4, A_4 and M_4, depending on the structure added, that are often discussed in this context.

Affinely connected space-time L_4

Covariant vector fields, varying from point to point, take their values in the tangent spaces. But since the tangent spaces vary from point to point, there is no notion of parallelism, or of a vector field being constant in X_4, or even of comparing vectors at different points. The extra structure needed for that purpose is the *affine connection* Γ. As we have already seen in Section 3.13.2, the affine connection also enters the covariant derivative. This is only natural since computing derivatives entails forming infinitesimal differences involving neighboring points. Connections allow us to construct well-behaved notions of differentiation. Endowing the manifold X_4 with an affine connection Γ yields an *affinely connected space-time* L_4.

The affine connection should be thought of as structure added to the manifold. At this stage then, the only information about the connection is that it provides for a covariant derivative and parallel transport of vectors. It can be any field with these properties and the transformation law of equation (3.413). The affine connection $\Gamma_{\mu\nu}{}^{\lambda}$ is asymmetric in its covariant indices $\mu\nu$ so we split it in a symmetric $\Gamma_{(\mu\nu)}{}^{\lambda}$ and an antisymmetric part $\Gamma_{[\mu\nu]}{}^{\lambda}$ according to $\Gamma_{\mu\nu}{}^{\lambda} = \frac{1}{2}(\Gamma_{(\mu\nu)}{}^{\lambda} + \Gamma_{[\mu\nu]}{}^{\lambda})$. The anti-symmetric

---

16 For readable texts, with historical comments, on the material presented here, see Chapter 2 of [128] and the commented reprint volume [135]. More details are provided by [133], which is partly written from the gauge theory of gravity perspective. See also Chapter 1 of [127].

part defines the *torsion tensor*

$$T_{\mu\nu}{}^{\lambda} = \Gamma_{[\mu\nu]}{}^{\lambda} = \Gamma_{\mu\nu}{}^{\lambda} - \Gamma_{\nu\mu}{}^{\lambda} \tag{4.71}$$

The torsion is indeed a tensor as the nonhomogeneous part of the transformation law (3.413) drops out of the transformation for the antisymmetric part.

Affinely connected metric space-time L_{4g}

The next structure to add to the manifold is the *metric* $g_{\mu\nu}$. It is introduced, as we have seen, by considering the invariant *proper time* $d\tau$ expressed in a freely falling coordinate system ξ^{α} and in an arbitrary coordinate system x^{μ} through

$$d\tau^2 = -\eta_{ab} d\xi^a d\xi^b = -g_{\mu\nu} dx^{\mu} dx^{\nu} \tag{4.72}$$

which leads to the formula (4.48) for the metric.

So far, the metric is not dynamic. It is a function of the arbitrary coordinates. On the other hand, in general relativity where the metric is determined by the Einstein equations, the coordinates are determined by only up to general covariance. We are not yet there. What we have, for now, is an *affinely connected metric space-time* denoted by $L_{4,g}$.

In standard Einstein general relativity, where the torsion is zero, the relation between the affine connection and the metric is derived by a well-known procedure that is interesting to dwell on a little bit. The metric is then of course dynamical, but the derivation does not depend on that.

Example 7 (The relation between the Christoffel symbols and the metric). There are a number of steps involved in the computation. First, in an inertial system we have the straight line equation for free fall (actually the shortest length line)

$$\frac{d^2\xi^a}{d\tau^2} = 0 \tag{4.73}$$

in terms of the inertial coordinates ξ^a. As these are functions of any other system of coordinates x^{μ}, one can reexpress the free fall equation in x^{μ} according to the geodesic equation

$$\frac{d^2x^{\sigma}}{d\tau^2} + \Gamma_{\mu\nu}{}^{\sigma} \frac{dx^{\mu}}{d\tau} \frac{dx^{\nu}}{d\tau} \tag{4.74}$$

where $\Gamma_{\mu\nu}{}^{\sigma}$ is given by the formula

$$\Gamma_{\mu\nu}{}^{\sigma} = \frac{\partial x^{\sigma}}{\partial \xi^a} \frac{\partial^2 \xi^a}{\partial x^{\mu} \partial x^{\nu}} \tag{4.75}$$

Note that this is an object that comes out of the computation of the derivatives in (4.73) when the freely falling coordinates ξ^{α} are expressed in terms of the coordinates x^{μ}

(see, for instance, [243], Chapter 3 for details). Second, contracting this formula with $\partial\xi^b/\partial x^\sigma$ and using the chain rule, one gets

$$\frac{\partial^2 \xi^a}{\partial x^\mu \partial x^\nu} = \Gamma_{\mu\nu}{}^\sigma \frac{\partial \xi^a}{\partial x^\sigma} \tag{4.76}$$

Third, differentiating the expression in equation (4.48) for the metric, yields after using precisely the formula (4.76),

$$\frac{\partial g_{\mu\nu}}{\partial x^\lambda} = \Gamma_{\lambda\mu}{}^\sigma g_{\sigma\nu} + \Gamma_{\lambda\nu}{}^\sigma g_{\sigma\mu} \tag{4.77}$$

Fourth, and finally a purely algebraic step, adding the formula (4.77) to itself with $\mu \leftrightarrow \lambda$ and subtracting it with $\nu \leftrightarrow \lambda$, one arrives at, observing that $\Gamma^\sigma{}_{\mu\nu}$ is symmetric in the covariant indices μ and ν

$$\frac{\partial g_{\mu\nu}}{\partial x^\lambda} + \frac{\partial g_{\lambda\nu}}{\partial x^\mu} - \frac{\partial g_{\mu\lambda}}{\partial x^\nu} = 2 g_{\sigma\nu} \Gamma_{\lambda\mu}{}^\sigma \tag{4.78}$$

from which the familiar formula

$$\Gamma_{\mu\nu}{}^\sigma = \frac{1}{2} g^{\sigma\rho} \left(\frac{\partial g_{\nu\rho}}{\partial x^\mu} + \frac{\partial g_{\mu\rho}}{\partial x^\nu} - \frac{\partial g_{\mu\nu}}{\partial x^\rho} \right) \equiv \left\{ {\sigma \atop \mu\nu} \right\} \tag{4.79}$$

results, defining the *Christoffel symbols* $\left\{ {\sigma \atop \mu\nu} \right\}$. ◂

Let us now combine this with what we already know. Perhaps equation (4.77) rings a bell. It looks conspicuously like requiring the covariant derivative of the metric to be zero. That is indeed the case, the equation expresses $\nabla_\lambda g_{\mu\nu} = 0$.

This may raise the question of whether the formula for the connection in terms of the metric can be derived from the *covariant constancy of the metric* only. We see from formula (4.75) that we have also worked with a symmetric connection, so this might be a necessary condition. As we will see, covariant constancy of the metric and symmetry of the connection, are not sufficient conditions. It turns out that the connection must also be torsion-free.

On the other hand, the assumptions made in example 7: the existence of a local inertial system where geodesics (shortest length lines) are "straight lines" and the metric is Minkowskian, leads directly to the covariant constancy of the metric. This is therefore a necessary condition for space-time to be locally Minkowskian.

The effects of gravity in GR

Reflecting on what we have done so far, one could say that we are halfway toward standard general relativity. Three steps have been taken. Based on the EP, the metric has been introduced to measure the proper time between nearby points. Also based on the EP, the equation for a freely falling particle has been deduced. These equations capture the effects of the gravitational field. Then we have derived the formula expressing the affine connection in terms of derivatives of the metric. The metric can be viewed as a gravitational potential.

In the more general setting of an affinely connected metric space-time, the procedure for relating the connection to the metric, can be generalized. Introduce the *tensor of nonmetricity* $Q_{\alpha\beta\gamma}$ as

$$Q_{\alpha\beta\gamma} = -\nabla_\alpha g_{\beta\gamma} \tag{4.80}$$

Normally, $Q_{\alpha\beta\gamma}$ is set to zero so that the metric can be said to be covariantly constant, thus generalizing the constancy of the metric of Minkowski space-time. This is very convenient as it makes the operations of raising and lowering indices commute with covariant differentiation, which is a nontrivial operation in curved space-times. However, taking it nonzero, one can still compute the following combination, in analogy with the example 7:

$$-\Delta^{\alpha\beta\gamma}_{\nu\mu\sigma} Q_{\alpha\beta\gamma} = \nabla_\nu g_{\mu\sigma} + \nabla_\mu g_{\sigma\mu} - \nabla_\sigma g_{\nu\mu} \tag{4.81}$$

where the convenient permutation symbol is defined by

$$\Delta^{\alpha\beta\gamma}_{\nu\mu\sigma} = \delta^\alpha_\nu \delta^\beta_\mu \delta^\gamma_\sigma + \delta^\alpha_\mu \delta^\beta_\sigma \delta^\gamma_\nu - \delta^\alpha_\sigma \delta^\beta_\nu \delta^\gamma_\mu \tag{4.82}$$

From the expressions appearing in the computation of the covariant derivatives in (4.81), one can extract the asymmetric affine connection $\Gamma_{\mu\nu}{}^{\lambda}$ and the antisymmetric torsion part $T_{\mu\nu}{}^{\lambda}$ of $\Gamma_{\mu\nu}{}^{\lambda}$. What results is the general formula for the connection

$$\Gamma_{\mu\nu}{}^{\lambda} = \left\{ {\sigma \atop \mu\nu} \right\} + K_{\mu\nu}{}^{\lambda} + L_{\mu\nu}{}^{\lambda} \tag{4.83}$$

where the first term is the Christoffel symbol. The second term is the *contorsion* tensor $K_{\mu\nu}{}^{\lambda}$ that computes to

$$K_{\mu\nu}{}^{\lambda} = \frac{1}{2}(T_{\mu\nu}{}^{\lambda} - T_{\mu}{}^{\lambda}{}_{\nu} - T_{\nu}{}^{\lambda}{}_{\mu}) \tag{4.84}$$

The contorsion tensor is asymmetric (since it is the sum of an antisymmetric tensor and a symmetric) and its antisymmetric part is proportional to the torsion tensor

$$K_{[\mu\nu]}{}^{\lambda} = \frac{1}{2} T_{\mu\nu}{}^{\lambda} \tag{4.85}$$

The third term $L_{\mu\nu}{}^{\lambda}$ in the formula (4.83), is a combination of the nonmetricity tensor that can be computed to

$$L_{\mu\nu}{}^{\lambda} = \frac{1}{2} g^{\lambda\sigma} \Delta^{\alpha\beta\gamma}_{\nu\mu\sigma} Q_{\alpha\beta\gamma} \tag{4.86}$$

Extracting the asymmetric affine connection

It is quite interesting to see how the formula (4.83) appears out of the computation of (4.81). It may, at first, seem strange that one can extract a formula for the general affine connection, yet having torsion related terms on the right-hand side. On second thoughts, however, this is precisely what the formula achieves: it clearly separates different contributions to the general affine connection. Given that the covariant derivative of the metric is $\nabla_\nu g_{\mu\sigma} = \partial_\nu g_{\mu\sigma} - \Gamma_{\nu\mu}{}^{\alpha} g_{\alpha\sigma} - \Gamma_{\nu\sigma}{}^{\alpha} g_{\mu\alpha}$, one computes $-\Delta^{\alpha\beta\gamma}_{\nu\mu\sigma} Q_{\alpha\beta\gamma}$ with the result

$$-\Delta^{\alpha\beta\gamma}_{\nu\mu\sigma} Q_{\alpha\beta\gamma} = \partial_\nu g_{\mu\sigma} + \partial_\mu g_{\nu\sigma} - \partial_\sigma g_{\nu\mu} - g_{\mu\alpha}\Gamma_{[\nu\sigma]}{}^{\alpha} - g_{\sigma\alpha}\Gamma_{(\nu\mu)}{}^{\alpha} - g_{\nu\alpha}\Gamma_{[\mu\sigma]}{}^{\alpha}. \tag{4.87}$$

Then one takes advantage of the fact that $\Gamma_{(\nu\mu)}{}^{\alpha} = 2\Gamma_{\mu\nu}{}^{\alpha} - \Gamma_{[\nu\mu]}{}^{\alpha}$. Furthermore, by definition $T_{\mu\nu}{}^{\alpha} = \Gamma_{[\mu\nu]}{}^{\alpha}$. This is enough to see how the formula (4.83) emerges.

This makes it explicit that *metricity*, that is, covariant constancy of the metric, and zero torsion leads to the standard formula for the *Levi–Civita connection* $\Gamma(g)$ given by the Christoffel symbols in (4.79). Only requiring covariant constancy of the metric, yields a constraint on the allowed connections that then take the general form

$$\Gamma_{\mu\nu}{}^{\lambda} = \left\{{\lambda \atop \mu\nu}\right\} + K_{\mu\nu}{}^{\lambda} \tag{4.88}$$

It is clear that in order for the affine connection to be entirely expressed in terms of the metric as the Christoffel symbols, we need both covariant constancy of the metric and zero torsion. Symmetry is not enough, since the contorsion has a symmetric part.

Auto-parallel curves versus geodesics

We will not have much need for the concepts of auto-parallel curves and geodesics, except as underlining the properties of different connections. The auto-parallel curve equation reads

$$\ddot{x}^{\mu} + \dot{x}^{\rho}\dot{x}^{\sigma}\Gamma_{\rho\sigma}{}^{\mu} = 0 \tag{4.89}$$

where $\Gamma_{\rho\sigma}{}^{\mu}$ is any affine connection. It generalizes the concept of "straightest line" from Minkowski space-time: it is the equation of parallel transport. It does not require a metric, only a connection. Even though only the symmetric part of the connection enters the equation, there may be – even in space-times with a metric – a contribution from the symmetric part of the contorsion tensor according to equation (4.88).

The geodesic equation (4.74) has the same form, but then $\Gamma_{\rho\sigma}{}^{\mu}$ is the Christoffel connection $\left\{{\mu \atop \rho\sigma}\right\}$ defined in terms of the metric. It is the line of "shortest length", and as such, requires a metric. The two concepts: auto-parallel and geodesic, coincide if the torsion – if it is present at all – is totally antisymmetric because then the contorsion tensor vanishes.

Riemann–Cartan space-time U_4: $Q = 0$

The first step toward the space-time of GR is to demand *metricity* $Q_{\alpha\beta\gamma} = 0$ or the *metric postulate*. This results in a general relativistic space-time with torsion, a so-called *Riemann–Cartan space-time* denoted by U_4. As we will see, this is the type of space-time that results from gauging the Poincaré group.

Riemann space-time V_4: $Q = 0, T = 0$

Setting the torsion to zero (as well as demanding metricity) yields *Riemann space-time* denoted by V_4. This is the space-time of GR. With zero torsion, the familiar Bianchi identities of the GR curvature tensor, can then be deduced from the formulas of Section 4.5.3.

Minkowski space-time M_4: $Q = 0, T = 0, R = 0$

Minkowski space-time results from demanding zero curvature.

Weitzenböck space-time A_4: $Q = 0, R = 0$

To reach Riemann space-time from an affinely connected space-time, we first set $Q = 0$ and then $T = 0$. One can also contemplate first setting $Q = 0$ and then $R = 0$, keeping torsion nonzero. The space-time that then results is called *Weitzenböck space-time*.

4.5.5 Frames in tangent space

The tangent, and cotangent spaces, are of course always there, but they are made concrete using vierbeins. As is well known, vierbeins are needed in order to study spinor (half-integer spin) fields in general curved spaces. The vierbeins can be thought of as setting up local coordinate systems in tangent space, or more correctly, in the tangent bundle. These local coordinate systems explicitly "behave" as Minkowski space-time. In particular, Lorentz transformations can be performed.

We then have general coordinate transformations (GCT) in the base manifold, and linear transformations in the fibers (which are vector spaces). The covariant derivative must correspondingly compensate for local GCTs and local linear transformations, that is, Lorentz transformations. We thus define the *total covariant derivative*

$$\mathcal{D}_\mu V^\nu_a = \partial_\mu V^\nu_a + \Gamma_{\mu\sigma}{}^\nu V^\sigma_a - \omega_{\mu a}{}^b V^\nu_b \tag{4.90}$$

$$\mathcal{D}_\mu V^a_\nu = \partial_\mu V^a_\nu - \Gamma_{\mu\nu}{}^\sigma V^a_\sigma + \omega_{\mu b}{}^a V^b_\nu \tag{4.91}$$

and following this pattern for more general mixed tensors. What we now have corresponds to an affinely connected metric space-time, only we have two unrelated connections, the affine connection Γ and the *spin connection* ω. To compare to GR, we must relate the connections, and the connections to the metric and vierbeins. This is done through the two vierbein postulates.

The first vierbein postulate concerns commutativity of taking covariant derivatives and converting between tangent and world indices. We would like the following formula to hold:

$$e^a{}_\nu \mathcal{D}_\mu V^\nu = \mathcal{D}_\mu(e^a{}_\nu V^\nu) \tag{4.92}$$

Computing the indicated covariant derivative on the right-hand side using the Leibniz rule yields

$$\mathcal{D}_\mu(e^a{}_\nu V^\nu) = (\mathcal{D}_\mu e^a{}_\nu)V^\nu + e^a{}_\nu \mathcal{D}_\mu V^\nu \tag{4.93}$$

If we demand $\mathcal{D}_\mu e^a{}_\nu = 0$, then we can indeed "convert indices" inside a covariant derivative. This is therefore the *first vierbein postulate*:

$$\mathcal{D}_\mu e^a{}_\nu = 0 \tag{4.94}$$

Assuming the first vierbein postulate, the formula (4.90) applied to the vierbein $e^\mu{}_a$, yields a relation between the two connections

$$\omega_{\mu a}{}^b = \Gamma_{\mu a}{}^b - e_a{}^\sigma \partial_\mu e^b{}_\sigma \quad \text{or} \quad \Gamma_{\mu a}{}^b = \omega_{\mu a}{}^b + e_a{}^\sigma \partial_\mu e^b{}_\sigma \tag{4.95}$$

where $\Gamma_{\mu a}{}^b$ is the affine connection[17] with indicated indices converted to tangent space using the appropriate vierbeins. An equation is an equation, but the second way of writing the relation between the connections may remind us of the fact that the Lorentz connection ω is what it is, it is determined by the Lorentz transformation, while the affine connection Γ maintains a certain freedom of choice that is here taken advantage of. Formulas like (4.90) and (4.91) for the total covariant derivative are of course still correct, but the affine connection Γ and the Lorentz connection ω are related under the first vierbein postulate by the equation (4.95).

The first vierbein postulate leads to an interesting equation for the torsion that is related to the gauge theory of gravity. Antisymmetrizing the first vierbein postulate (4.94) and using formula (4.91) applied to the vierbein $e_\mu{}^a$ yields

$$T^a{}_{\mu\nu} = \partial_\mu e_\nu{}^a - \partial_\nu e_\mu{}^a + \omega_{\mu b}{}^a e_\nu{}^b - \omega_{\nu b}{}^a e_\mu{}^b = D_{[\mu} e_{\nu]}^a = 0 \tag{4.96}$$

We will encounter this formula again – in form language – in equation (4.127) when we discuss gauge theory of gravity. Then the torsion appears as the field strengths of local translations. Here, we see that upon extracting the antisymmetric part of the first vierbein postulate, we get a formula for the torsion where the torsion is expressed as an antisymmetrized Lorentz covariant derivative of the coframe field.

Catalogue of covariant derivatives

i To forestall confusion over notation for covariant derivatives, these are our conventions:

- ∇_μ denotes the standard "differential geometric" covariant derivative – with or without the inclusion of torsion as circumstances may dictate.

17 No restrictions on Γ are imposed as yet.

- $\mathcal{D}_\mu$ denotes the total covariant derivative including affine connection and Lorentz connection.
- $\mathcal{D}_a = e^\mu{}_a \mathcal{D}_\mu$ denotes the total covariant derivative transferred to cotangent space.
- D_μ or D_a denote Yang–Mills covariant derivatives (indices are dictated by circumstances). Here, we also include the case that the gauge group is the Lorentz group. This case will appear below in Section 4.6.1.

Note in particular that in the sequel, the notation $\mathcal{D}_\mu$ or $\mathcal{D}_a$ will be used for the total covariant derivative under the assumption of the first vierbein postulate, if not otherwise explicitly stated.

The first vierbein postulate does not in general imply the metric postulate unless one considers orthonormal frames. For general frames, where the formula (4.50) is replaced by $g_{\mu\nu} = e_\mu{}^a e_\nu{}^b g_{ab}$, it is clear that covariant constancy of the metric does not follow from covariant constancy of the vierbeins, since g_{ab} may very well be itself covariant nonconstant. However, the Minkowski frame metric η_{ab} is covariant constant due to the antisymmetry of the Lorentz connection $\omega_\mu{}^{ab}$. Let us therefore continue to investigate how the various kinds of space-times considered above, comes about in a tetrad approach.

Consistent covariant derivatives and curvatures

Even having made up ones mind about notation for the various covariant derivatives that occur in the subject, it might be a good idea to make a few consistency checks. Certainly, one would like formulas such as

$$\mathcal{D}_a V_b = \partial_a - \omega_{ab}{}^c V_c \quad \text{and} \quad \mathcal{D}_a V^b = \partial_a + \omega_{ac}{}^b V^c \tag{4.97}$$

to hold and be consistent with (4.90) and (4.91). Furthermore, it is quite interesting to compute

$$[\mathcal{D}_\mu, \mathcal{D}_\nu] V_a = -(\partial_\mu \omega_{\nu a}{}^b - \partial_\nu \omega_{\mu a}{}^b - \omega_{\mu a}{}^c \omega_{\nu c}{}^b + \omega_{\nu a}{}^c \omega_{\mu c}{}^b) V_b \equiv -R_{\mu\nu a}{}^b V_b \tag{4.98}$$

with no torsion terms appearing. Also, when the equation (4.95) relating the affine connection and the Lorentz connection holds, the two corresponding curvatures should be related. It can be checked that

$$R_{\mu\nu\rho}{}^\sigma(\Gamma) = e_\rho{}^a e^\sigma{}_b R_{\mu\nu a}{}^b(\omega) \tag{4.99}$$

4.5.6 Space-times reconsidered

Consider an arbitrary frame, that is one in which we have $g_{\mu\nu} = e_\mu{}^a e_\nu{}^b g_{ab}$. The first vierbein postulate is supposed to hold, but not the metric postulate and torsion is nonzero. Then we have the general formula (4.83) for the affine connection Γ. The computation can be redone with all indices in the frame. The result is

$$\omega_{ab}{}^c = \omega_{ab}{}^c(e) + K_{ab}{}^c + L_{ab}{}^c \tag{4.100}$$

where

$$\omega_{ab}{}^{c}(e) = \left\{{}^{c}_{ab}\right\} - \frac{1}{2}(\Omega_{ab}{}^{c} - \Omega_{a}{}^{c}{}_{b} - \Omega_{b}{}^{c}{}_{a}) \tag{4.101}$$

Here, the Christoffel symbols are computed as in formula (4.79) with all indices taken as frame indices. Thus, $\omega(e)$ only depends on the metric and the vierbeins through the anholonomy coefficients (4.59). Furthermore, $K_{ab}{}^{c} = e^{\mu}{}_{a}e^{\nu}{}_{b}e^{c}{}_{\rho}K_{\mu\nu}{}^{\rho}$ and similarly for L. In the formula (4.100), the general (affine) connection ω corresponds to the general affine connection Γ, while the connection $\omega(e)$ corresponds to the Levi–Civita connection $\Gamma(g)$.

Now one can put $Q = 0 \Rightarrow L = 0$ (metricity) and $K = 0$ (no torsion) and then one gets back to the Riemann space-time V_4 but now with an arbitrary frame. The connections then reduce to the Cartan and Levi-Civita connections, but they are still not equal. Equality results in the coordinate basis or in a holonomic basis.

The metric postulate $Q = 0$ is in this context sometimes called the *second vierbein postulate*. Only imposing this, while keeping torsion nonzero, results in the space-time U_4. Covariant derivatives commute with all kinds of index raising and lowering, thus computations can be done as we are used to. U_4 is the space-time of the so-called *Cartan–Sciama–Kibble theory* which is related to the gauge theory of the Poincaré group. We will return to it below in Section 4.6.3.

4.5.7 Standard general relativity and the Weyl tensor

In Einstein general relativity, the affine connection is solved for in terms of the metric as in example 7 in Section 4.5.4. The calculations amount to demanding the covariant constancy of the metric. This is very convenient in that it makes covariant derivation commute with index contraction using the metric. The connection used in the covariant derivative becomes the standard Levi–Civita connection.

The *Einstein field equations* for GR (without matter or cosmological constant) are

$$R_{\mu\nu} - \frac{1}{2}g_{\mu\nu}R = 0 \tag{4.102}$$

Contracting once more with the metric implies $R = 0$, so that an equivalent form of the field equations are

$$R_{\mu\nu} = 0 \tag{4.103}$$

The Einstein field equations follow from an action principle that we will review below in Section 4.7.1.

The field equations set some components of the Riemann curvature tensor to zero, and it is interesting to disentangle those components that are nonzero. This is what the Weyl tensor achieves.

The *Weyl tensor* $W_{\mu\nu,\rho\sigma}$ can be defined by subtracting components the Ricci tensor and the scalar curvature from the Riemann tensor. An ansatz subject to the symmetry properties of (4.70) is

$$W_{\mu\nu,\rho\sigma} = R_{\mu\nu,\rho\sigma} - a(g_{\mu\rho}R_{\nu\sigma} - g_{\nu\rho}R_{\mu\sigma} - g_{\mu\sigma}R_{\nu\rho} + g_{\nu\sigma}R_{\mu\rho}) \\ + b(g_{\mu\rho}g_{\nu\sigma} - g_{\nu\rho}g_{\mu\sigma})R \tag{4.104}$$

Contracting with $g^{\nu\rho}$ and using that the Weyl tensor is itself traceless over the $\nu\rho$ indices, yields $a = 1/2$ and $b = -1/6$. The Weyl tensor satisfies the same index symmetries and Bianchi identities as the Riemann tensor, as well as being traceless in all index pairs.

Shift of perspective

The introduction of the Weyl tensors allows for a shift of perspective that will become important in the Vasiliev theory of higher spin. Instead of writing Einstein's equations as $R\mu\nu = 0$, we can equivalently write them as $R_{\mu\nu,\rho\sigma} = W_{\mu\nu,\rho\sigma}$. Instead of specifying which components of the curvature that vanish, we specify those that do not vanish in terms of new fields.

Connection terminology

Quite a few connections occur in the theory, and although which one is meant is often clear from the context, one may get momentarily confused. Here is an attempt to set the record straight.[18] Controlling two circumstances may help to navigate. (i) Are we talking about connections as independent geometric objects or derivative of a metric? (ii) Are we in a purely metric context, or are we in a frame/coframe context?

An affine connection is always a general connection as defined in Section 4.5.2. In general, it can be expressed as in formula (4.83) if there is a metric and torsion. The Levi–Civita connection is always given by the Christoffel symbols.

A general (affine) spin connection is given by the formula (4.100). It is on the same level of generality as an affine connection. It is related to the affine connection in equation (4.95) through the first vierbein postulate. The *Cartan connection* results when the metric postulate holds but there may be torsion.

Finally, the *Lorentz connection*, is always the connection corresponding to local Lorentz transformations in the frames. There is then no torsion terms. "Lorentz" connection and "spin" connection are often used interchangeably when there is no torsion. They may differ by the anholonomicity coefficients.

18 Maybe not every author would agree on the following distinctions.

4.6 Gauge theory of gravity

Let us begin with some general comments. It is useful to be aware of a distinction that eventually comes out of a gauge treatment of gravity – and in retrospect is present in any gauge theory – even spin 1 theory, but which is not so often explicitly noted. This is the distinction between the *the effects of the presence* of the gauge field and *the self-consistent theory* of the gauge field *itself*. Or phrased more concretely: the *matter action* as contrasted to the *gauge field action*. One can regard the matter action as a crutch to motivate the introduction of the gauge field and the covariant derivative.

We saw this already for spin 1 Yang–Mills theory in Section 4.2. The reason why we had to introduce a gauge field into the theory when we had local symmetries is the need to have derivatives in the matter theory. This is in the last analysis an experimental fact. From Newton's mechanics and onwards, all fundamental theories of physical systems has turned out to be described in terms of differential equations. In field theory, these are partial differential equations. In free field theory, we have linear wave equations, in interacting field theory we have nonlinear wave equations, preferably deriving from least action principles. Now, it is precisely the presence of derivatives that forces the need to introduce gauge fields and the corresponding covariant derivatives, in order to promote rigid, that is, global, symmetries to local symmetries.

This comment stresses a point that we have already made: before introducing an action for the gauge field itself, we have only done the kinematical part of gauging.[19] We have taken into account the *effects of the gauge field*, but not the *dynamics of the gauge field* itself. Metaphorically, this can perhaps be phrased as follows. Stepping into Einstein's elevator and letting it drop, we are free of the gravitational field we detect before the drop starts. But if we did not understand how the gravitational field came about before the drop, we may have even less understanding of it during the drop.

There is a trap one should try to avoid falling into when considering gauge approaches to gravity. Since we know what we want to arrive at: Einstein gravity or some variant of it, it is all to easy to put too much of that structure into the discussion too early. If one does that, one risks missing the subtleties. It is much more interesting to force oneself into a mindset where flat Minkowski space-time and its Poincaré invariance group is all there is.

Avoiding the trap of knowing to much

? Gauge approaches to gravity were all initialized after the Einstein theory had been studied for a long time. Imagine a context where the gauge approach to electromagnetism had been applied to weak- and strong interactions, but where one had been content with the special theory of relativity, Minkowski

---

19 This is also discussed in [132], page 34.

space-time and Newtonian gravity. Furthermore, imagine that differential geometry had not been invented, so there were no clear concepts of curved spaces and their tangent and cotangent spaces. Under such circumstances, what theory of gravity would theoretical physicists arrive at if they undertook the gauging of the Poincaré group?

With these caveats, we will now study two ways of approaching the problem of gauging the Poincaré group. We focus on the kinematical part. For dynamics, we will rely on "well-known results" as well as what has been reviewed above. Actions for gravity will be considered in Section 4.7. We will be quite naive as regards mathematical apparatus, not wanting to risk submerging the problems in formalism.[20]

4.6.1 Poincaré gauge theory (I)

As already mentioned in Section 2.9.2, R. Utiyama [115] proposed a gauge theory of gravity where the Lorentz group played the role of the gauge group. This was just a couple of years after the construction of the SU(2) gauge theory of isotopic spin by C. N. Yang and R. L. Mills in 1954 [114]. Only gauging the Lorentz group provides no motivation for the vierbein fields, and these are indeed needed for the Lorentz group to make any sense in a gravitational theory.

The problem was reconsidered by T. W. B. Kibble [131], who by gauging the inhomogeneous Lorentz group, found a way of providing a motivation for the vierbeins. As we noted in the historical chapter, there is an extensive literature on the subject of "gauge theory of gravity". The problem is both technically and conceptually complicated. The bottom line is: what group to gauge? The translation group, the Lorentz group, the full Poincaré group or some larger group containing the Poincaré group? As the Kibble and K. S. Stelle write in the review [134]:

> The basic problem is that the analogy between gravity and other gauge theory is necessarily less than perfect.

This description of the situation is not an exaggeration. In the example below, we will follow [134] in an attempt to generalize the standard Yang–Mills gauging procedure, from an internal gauge group to the Poincaré group, to see where it leads.

Before starting, it may be useful to ask oneself the question: why is it so relatively simple to generalize the gauge approach of electromagnetism to an arbitrary internal semisimple group? No conceptual problems seem to arise, if one does not count the effort to imagine a copy of the internal group residing at every space-time point. Indeed, being in Minkowski space-time and having mastered quantum mechanics, we

---

20 As the reader will surely suspect, there is an extensive literature on the subject of gauge approaches to gravity and supergravity. What follows below is at best an introduction to the subject. Again, the focus is on intuition, not rigor or formalism.

simply think of the matter fields as carrying an internal quantum number and being subject to a concomitant symmetry transformation.

The point seems to be that the mathematics of this symmetry appears to be entirely separate from space-time itself. Of course, it can be elegantly formulated in terms of fiber bundles – and then more subtle topological phenomena can be investigated – but we are not forced to do that. A simple-minded approach using matrices $(T^a)_{ij}$ is sufficient to set up the theory and carry through the gauging recipe. The matrices T and the indices a have nothing to do with space-time. It is all separate and clean. This is not the case when gauging the space-time symmetry group, in particular not the translation part. At some stage one must introduce a distinction between "world" and "tangent" space indices. The questions are: when and with what motivation? What from a geometric perspective is a strength of gravitational theory, is from a gauging perspective a weakness: the fibers of gravity are just tangent spaces, not truly independent internal spaces.

The effect of this is the impression that it seems almost impossible to rederive gravity from the gauge principle without already knowing the answer. This is in contrast to Yang–Mills: none of the inventors of that theory knew the equations beforehand.

Example 8 (Naive gauging the Poincaré group). Start with an active infinitesimal Poincaré transformation

$$\delta x^\mu = \lambda^\mu{}_\nu x^\nu + \epsilon^\mu \equiv \xi^\mu \tag{4.105}$$

where we imagine being in Minkowski space-time. Correspondingly, a set of matter fields ψ will transform as (compare Section 4.2)

$$\delta\psi = -\frac{1}{2}\lambda^{ab} S_{ab}\psi - \xi^\mu \partial_\mu \psi \tag{4.106}$$

In the first term, we have changed indices from space-time indices μ, ν to – what will eventually be interpreted as – tangent space indices a, b. However, as that would be running in advance of one-self, for now these indices indicate a particular representation of the Lorentz group on the matter fields. This first term in the transformation is analogous to the internal transformation of formula (4.14). Alternatively, the different indexing can be thought of as distinguishing between terms involving the parameters $\lambda^{\mu\nu}$ and ξ^μ.[21]

Already at this stage, we see a difference as compared to spin 1. True, the role of the Yang–Mills T^a matrices are now played by the Lorentz $\mathfrak{so}(3,1)$ matrices S_{ab}. But the internal field rotation in group space is now induced by the space-time transformation (4.105). In the second term of (4.106), we see the effect of these transformations. We see that the generators involve space-time derivatives.

21 Note that, following [131, 133] we parametrize the Poincaré transformations with these parameters, rather than with $\lambda^{\mu\nu}$ and ϵ^μ.

The next step is to let the parameters $\lambda^\mu{}_\nu$ and ξ^μ become local functions of the coordinates x^μ. Precisely here, one runs into the first problem. The distinction between local translations and local Lorentz transformations in (4.105) becomes blurred. Translations with a local $\xi^\mu(x)$ already contains all local coordinate transformations generated by the vector field $\xi^\mu(x)\partial_\mu$. On the other hand, if the corresponding local Lorentz transformations $\lambda^{ab}(x)S_{ab}$ are discarded, the transformations on the fields (4.106) become ambiguous [134].

However, this problem may very well be due to an overreliance on mathematical formalism. As argued in [133], we know that there are local Lorentz transformations on matter fields. We need a formalism that can maintain precisely that structure. So let us pause this line of inquiry and instead compute the transformation of the derivative of the matter field

$$\delta\partial_\mu\psi = -\frac{1}{2}\lambda^{ab}S_{ab}\partial_\mu\psi - \xi^\nu\partial_\nu\partial_\mu\psi - \frac{1}{2}(\partial_\mu\lambda^{ab})S_{ab}\psi - (\partial_\mu\xi^\nu)\partial_\nu\psi \tag{4.107}$$

As in Yang–Mills, we find inhomogeneous terms. The third term can be taken care of by introducing a gauge field $\omega_\mu{}^{ab}$ with an inhomogeneous transformation term $\partial_\mu\lambda^{ab}$. Correspondingly, we have a covariant derivative[22]

$$D_\mu = \partial_\mu + \frac{1}{2}\omega_\mu{}^{ab}S_{ab} \equiv \partial_\mu + \omega_\mu \tag{4.108}$$

in close analogy to the formalism of Yang–Mills theory, defining $\omega_\mu \equiv \frac{1}{2}\omega_\mu{}^{ab}S_{ab}$. Since the Lorentz group is non-Abelian, the transformation for the law for the gauge field is

$$\delta\omega_\mu{}^{ab} = \partial_\mu\lambda^{ab} + [\omega_\mu, \lambda]^{ab} \tag{4.109}$$

which is exactly as in Yang–Mills theory.

The last term in (4.107) must be treated in a different way since it involves derivatives ∂_μ instead of matrices. Focusing on this term by itself, one could introduce a gauge field $h_\mu{}^\nu$ transforming as $\delta h_\mu{}^\nu = \partial_\mu\xi^\nu + \mathcal{O}(\xi, h)$ and a covariant derivative $D_\mu = \partial_\mu + h_\mu{}^\nu\partial_\nu = e_\mu{}^\nu\partial_\nu$ with $e_\mu{}^\nu = \delta_\mu{}^\nu + h_\mu{}^\nu$. Although it is not quite clear what we are doing here physically, the mathematics indicates a "multiplicative" covariant derivative rather than an "additive". This is the second problem, or at least a new phenomenon, that has to be accommodated in the gauge theory. ◂

We see that a gauge theory for spin 2 is not likely to be a simple rewriting of the gauge theory for spin 1. In a loose language, we could say that Yang–Mills theory is basically very algebraic whereas gravity is very geometric. One way to equalize the differences would be to make Yang–Mills theory more geometric. This is precisely what one does in the modern fiber bundle approach. Another way would be to make gravity more algebraic. Let us analyze the second problem from example 8.

---

22 We are still "thinking" Yang–Mills, so the notation is consistent with the note above about covariant derivatives.

Example 9 (Naive gauging of the translations). Consider a space-time dependent translation $\delta x^\mu = \xi^\mu(x)$. The matter field transforms as $\delta\psi = -\xi^\mu\partial_\mu\psi$. We want to construct a covariant derivative D_μ such that the covariant derivative on the field transforms as the field itself, that is, we want $\delta D_\mu\psi = -\xi^\nu\partial_\nu D_\mu\psi$. Following the recipe, we introduce a gauge field $h_\mu{}^\nu$ and write the covariant derivative according to $D_\mu = \partial_\mu + h_\mu{}^\nu\partial_\nu = e_\mu{}^\nu\partial_\nu$ with $e_\mu{}^\nu = \delta_\mu^\nu + h_\mu{}^\nu$. Then we compute $\delta D_\mu\psi = (\delta e_\mu{}^\nu)\partial_\nu\psi + e_\mu{}^\nu\delta(\partial_\nu\psi)$ and demand that the result be equal to $-\xi^\nu\partial_\nu D_\mu\psi$. This yields the transformation rule

$$\delta e_\mu{}^\nu = -\xi^\sigma\partial_\sigma e_\mu{}^\nu + \partial_\sigma\xi^\nu e_\mu{}^\sigma \tag{4.110}$$

for the field $e_\mu{}^\nu$.

This looks familiar, but it is not quite right. Assuming that we know about differential geometry, this is how a contravariant vector would transform (see Section 3.13). But the covariant index on $e_\mu{}^\nu$ does not transform, we lack the expected term $-\partial_\mu\xi^\sigma e_\sigma{}^\nu$. ◂

Taking this result at face value, the conclusion is that the covariant index on the field $e_\mu{}^\nu$ plays a different role than the contravariant. Let us therefore replace it and write the field as $e^\nu{}_a$, thinking about it as a set of contravariant vectors parametrized by the index a. Furthermore, it makes intuitive sense that local translations corresponds to general coordinate transformations. But we have to remember that we are making local translations in a, to begin with, flat space-time. The coordinate transformations are therefore between a flat space-time and a curved. Indeed, transforming the initially flat coordinates x^μ locally as $\delta x^\mu = \xi^\mu(x)$ does make the coordinates curvilinear. Again, assuming we know differential geometry, we can consider the tangent space-time with basis ∂_μ. Then repeating the discussion in Section 4.5.1 we can set up a local inertial system of coordinates $e^\mu{}_a\partial_\mu$. In this way, one may motivate introducing the two sets of coordinates, indexed by different indices. Phrased pictorially, making a global translation ξ^μ local, that is, $\xi^\mu \to \xi^\mu(x)$, will "deform" space-time. It is then restored locally by the gauge fields $e^\mu{}_a$.

After these preliminaries, we can now approach the problem of constructing a derivative covariant under local Poincaré transformations. The matter fields $\psi(x)$ still transform as in equation (4.106). We want to construct a covariant derivative D_a acting on a matter field ψ such that the $D_a\psi$ transforms in the same way as the field itself, that is, without derivatives on the parameters. We will do it in two steps: first writing a derivative $D_\mu\psi$ covariant under local Lorentz transformation. Second, writing the derivative $D_a\psi$ that is also covariant under local translations.

Note that we continue, for a while, to use Yang–Mills-like notation D for the covariant derivative. At the end of our deliberations, we will see that what we have got is actually the total covariant derivative $\mathcal{D}$.

For the first step, we take the covariant derivative D_μ from equation (4.108) but the transformation law for the gauge field $\omega_\mu{}^{ab}$ (a covariant vector) must be amended by general coordinate transformation terms

$$\delta\omega_\mu{}^{ab} = \partial_\mu\lambda^{ab} + [\omega_\mu, \lambda]^{ab} - \xi^\nu\partial_\nu\omega_\mu{}^{ab} - \partial_\mu\xi^\nu\omega_\nu{}^{ab} \tag{4.111}$$

Then we get the transformation

$$\delta(D_\mu\psi) = -\frac{1}{2}\lambda^{ab}S_{ab}D_\mu\psi - \xi^\nu(\partial_\nu D_\mu)\psi - (\partial_\mu\xi^\nu)D_\nu\psi \tag{4.112}$$

For the second step, we introduce the multiplicative covariant derivative

$$D_a = e^\mu{}_a D_\mu \tag{4.113}$$

Computing $\delta(D_a\psi)$, we find a transformation free of derivatives on parameters

$$\begin{aligned}\delta(D_a\psi) &= (\delta e^\mu{}_a)D_\mu\psi + e^\mu{}_a\delta(D_\mu\psi)\\ &= -\frac{1}{2}\lambda^{bc}S_{bc}D_a\psi - \xi^\nu(\partial_\nu D_a)\psi + \lambda_a{}^b D_b\psi\end{aligned} \tag{4.114}$$

where we have used the transformation law (4.110) augmented by a Lorentz transformation on the covariant tangent space index.

$$\delta e^\mu{}_a = -\xi^\sigma\partial_\sigma e^\mu{}_a + \partial_\sigma\xi^\mu e_a{}^\sigma + \lambda_a{}^b e_b{}^\mu \tag{4.115}$$

The calculation shows that it is possible to carry out the kinematical part of the gauging procedure for the Poincaré group: introducing gauge fields and a suitable covariant derivative. However, as compared to Yang–Mills theory, some jury-rigging is needed. We have to amend the gauge transformation formula for the gauge field ω with a general coordinate transformation as seen in equation (4.111). True, it can be motivated with what we learned in example 8, but it is still rather ad hoc. Or more to the point, we are using what we already know is true.

We can now compute commutators of covariant derivatives, in that way arriving at expressions for the "field strengths" of the theory, or what we will interpret as curvature and torsion. First, we get by commuting the Lorentz covariant derivatives

$$[D_\mu, D_\nu]\psi = \frac{1}{2}R_{\mu\nu}{}^{ab}S_{ab}\psi \tag{4.116}$$

where we read off

$$R_{\mu\nu}{}^{ab} = \partial_\mu\omega_\nu{}^{ab} - \partial_\nu\omega_\mu{}^{ab} + (\omega_\mu{}^{ac}\omega_\nu{}^{db} - \omega_\nu{}^{ac}\omega_\mu{}^{db})\eta_{cd} \tag{4.117}$$

Next, commuting the Poincaré covariant derivatives we get

$$[D_a, D_b]\psi = [e^\mu{}_a D_\mu, e^\nu{}_b D_\nu]\psi = \frac{1}{2}R_{ab}{}^{cd}S_{cd}\psi + T_{ab}{}^c D_c\psi \tag{4.118}$$

where the Lorentz curvature is reproduced as

$$R_{ab}{}^{cd} = e^{\mu}{}_{a} e^{\nu}{}_{b} R_{\mu\nu}{}^{cd} \tag{4.119}$$

We also get, from the action of the D covariant derivatives on the vierbeins,

$$T_{ab}{}^{c} = (e^{\mu}{}_{a} D_{\mu} e^{\nu}{}_{b} - e_{b}{}^{\mu} D_{\mu} e_{a}{}^{\nu}) e_{\nu}{}^{c} \tag{4.120}$$

This is the *torsion* where the Lorentz covariant derivative on the vierbeins is computed as

$$D_{\mu} e_{a}{}^{\nu} = \partial_{\mu} e_{a}{}^{\nu} - \omega_{\mu a}{}^{b} e_{b}{}^{\nu} \tag{4.121}$$

Frames of coframes?

? Readers, as well as the author, may suffer from some index confusion at this point. We started out with a flat space-time with coordinates x^{μ}. We deformed it by making local Poincaré transformations. To restore order as far as the local Lorentz transformations went, it was natural to introduce gauge fields $\omega_{\mu}{}^{ab}$. So far the story is analogous to Yang–Mills. But for the local translations, it was natural to introduce the frame fields $e^{\mu}{}_{a}$. Why not the coframe fields $e_{\mu}{}^{a}$?

Technically, one could say that the vierbein field $e^{\mu}{}_{a}$ is an invertible matrix, so it really does not matter that much which of the fields $e^{\mu}{}_{a}$ or $e_{\mu}{}^{a}$ are introduced first. From a physical point of view, it is perhaps better to say that the coordinates x^{μ} start out as flat and that gauging the Lorentz transformations can be partly accommodated by introducing the gauge fields $\omega_{\mu}{}^{ab}$. Then the gauging of the translations make the coordinates x^{μ} curved, or the tangent space basis vectors ∂_{μ} curved. However, it is still possible to set up local tangent spaces with basis vectors $e^{\mu}{}_{a}\partial_{\mu}$. As we will see, a shift of perspective brings in the coframe fields as basic translation gauge fields.

4.6.2 Thinking through the Cartan–Sciama–Kibble theory

The theory outlined in the previous section, when supplied with an action for the gravitational field as well as for the matter field, is often referred to as the *Cartan–Sciama–Kibble theory* (CSK), perhaps with a couple of other names attached such as Einstein and Weyl (or some names detached). We have not done the details, but we know that vierbeins have to be introduced in general relativity in order to accommodate half-integer spin matter fields.[23] In fact, we have seen this in our treatment of the gauge theory approach above. The "matter crutch" could very well be, and most naturally is, a half-integer spin field.

In the standard approach to introducing half-integer matter into general relativity, one starts by introducing the vierbein fields. This provides for local inertial frames

23 As always, the history is convoluted. Original literature as well as comments can found in either of the reprint volumes [128] or [135].

while maintaining general covariance. Then the principle of equivalence requires that special relativity should apply in the local frames. This means that we should be able to perform Lorentz transformations in every local frame, that is, the Lorentz transformations are local. The Lorentz transformations come with the possibility to transform not just tensors, but also spinors. Then since the matter action always involves derivatives of the fields, we must introduce covariant derivatives in order to make derivation compatible with local Lorentz invariance. This entails introducing spin connections. Note that this construction can be argued without actually invoking any idea of "gauging" the Poincaré group. We are simply introducing a Lorentz covariant derivative in the theory. The vierbeins transfer indices between world and tangent spaces according to (4.113). The theory so obtained allows for torsion.

Theories flowing naturally from principles

In much of the discussion on gauge approaches to gravity, there seems to be an implicit notion of "theories flowing naturally from principles" such as in "the gauge principle leading to Yang–Mills theory" or the "equivalence principle leading to general relativity". However, upon examining what is actually done in such endeavors, it seems clear that quite a few number of choices have to be made along the way from the principle to the theory. The principle prompts problems that have to be solved by judicious choices or inventions. Is it perhaps a matter of basic outlook whether one views such choices as dictated by the principle, or just inspired by it? Einstein's own struggles (the story told in many places) is testimony to the fact that the road from principle to theory is not at all easy to travel the first time, or even the second or third time.

Be that as it may, let us instead count degrees of freedom, and do this from the point of view of gauge theory. The vierbein field has 16 components, of which 8 may be fixed by general coordinate transformations, interpreted as gauge transformations. The remaining 6 unphysical components are removed by local Lorentz-transformations. The Lorentz connection contains 24 components. Clearly, none of these can be physical if we want to recover Einstein general relativity. So the Lorentz connection must be fixed in terms of the vierbeins. That can be done at an early stage of theory development by imposing the vierbein postulates which allow for expressing the Lorentz connection in terms of the Levi–Civita connection and the contorsion (see Section 4.5.6). Upon setting torsion to zero, the Lorentz connection is completely determined in terms of the vierbeins. The "gauge" transformations of the spin connection, needed for local Lorentz invariance of the matter Lagrangian, are still in effect, but may now be seen as following from the local Lorentz transformations of the vierbeins. They do not remove any d. o. f. as there are none to remove, and the freedom in the local Lorentz transformations are anyway already used up (as we argued above). In this approach, when seeking an action for the gravitational fields, it is inherent in the approach that the metric is the only dynamical field.

On the other hand, if one remains with the connections as independent fields, then upon seeking an action, one is lead to consider first-order actions involving both vierbeins and spin connections or the metric and the affine connection. The equations relating the two kinds of fields must then follow from the action.

We will now run a little ahead of ourselves, since we are not yet finished with our investigations of the gauge approach to gravity, and we have not discussed action principles for gravity. It turns out that when gauging the Poincaré group, the vierbeins may be interpreted as the gauge fields of the local translations and the torsion is the corresponding field strength. The spin connection is the gauge field of the local Lorentz transformations, and the curvature tensor is the corresponding field strength. The dynamics (in matter free space-time) then force the torsion to be zero, that is, zero field strength for local translations. The dynamics also determine the spin connection to be auxiliary and expresses it in terms of the vierbeins. The spin connection suffer no independent gauge transformations. In the presence of matter with spin, then the torsion is algebraically related to the spin-current density of the matter fields. It is in any way a nonpropagating field.

4.6.3 Poincaré gauge theory (II). Formalizing and understanding

Let us now turn to another approach to gauging the Poincaré group that is more formal in its nature. We want to see how far we can stretch the analogy with Yang–Mills theory. As was announced at the start of the computations, we want to ascertain that the Yang–Mills resembling covariant derivative for the local Poincaré transformations, does turn out to be the total covariant derivative. Examining the above computations, we see that they were done relying on the matter crutch ψ.

Is the "Kibble–Stelle" covariant derivative equal to the total covariant derivative?

? The matter crutch ψ is assumed to live entirely in the tangent space, so the covariant derivative only contain the spin connection per definition. But suppose the matter is a vector V^a. Then we can transfer its tangent space index to the world by $V^\nu = e^\nu{}_a V^a$, and what happens then?

Again, we must be conscious about how much of standard vierbein GR we want to assume. What are our objectives? Do we want to derive GR by gauging the Poincaré group (Kibble) or are we content with accommodating local Lorentz invariance within vierbein GR (Sciama)?

We now start out afresh and introduce the coframe field $e^a = e_\mu{}^a dx^\mu$ (as in formula (4.54)) and the connection $\omega^{ab} = \omega_\mu{}^{ab} dx^\mu$. Then, without worrying about interpretative issues, we attempt to gauge the Poincaré algebra by combining these fields into one 1-form gauge field B valued in the Lie algebra according to[24]

$$B = e^a P_a + \frac{1}{2}\omega^{ab} M_{ab} \tag{4.122}$$

and the corresponding 0-form gauge parameter

$$\Xi = -\xi^a P_a - \frac{1}{2}\lambda^{ab} M_{ab} \tag{4.123}$$

The gauge transformation then reads

$$\delta B = d\Xi + [B, \Xi]_\wedge \tag{4.124}$$

In these formulas, P_a and M_{ab} are assumed to satisfy the Poincaré Lie algebra (3.114)–(3.115). All this is in analogy with Yang–Mills theory (see Section 4.3.1). The curvature (field strength) is defined as

$$G = d \wedge B + B \wedge B = dB + \frac{1}{2}[B, B]_\wedge \tag{4.125}$$

It can be computed with the result

$$G = T^a P_a + \frac{1}{2}R^{ab} M_{ab} \tag{4.126}$$

where

$$T^a = d \wedge e^a + \omega^a{}_b \wedge e^b = D \wedge e^a \tag{4.127}$$

$$R^{ab} = d \wedge \omega^{ab} + \omega^{ac} \wedge \omega_c{}^b = D \wedge \omega^{ab} \tag{4.128}$$

These two formulas – called the *Cartan structure equations* – clearly have the form of covariant derivatives acting on the fields, as indicated, and indeed it is the Lorentz covariant derivative $D = d + \omega$ that appears here. Compared to the formulas of the previous section, the curvature 2-form is exactly the same, and the torsion is the same up to a sign.

We can also compute the infinitesimal gauge transformations of the fields that follow from (4.124):

$$\delta e^a = -D\xi^a + \lambda^{ac} e_c = -d\xi^a - \omega^{ac}\xi_c + \lambda^{ac} e_c \tag{4.129}$$

24 In the previous section, the fields e, ω, T and R have their conventional meaning as vierbein, spin connection, torsion and curvature respectively. In this section, we will consider gauge fields and field strengths for the Poincaré group. I have resorted to denote these new fields with B for the connections (gauge fields) and G for the curvatures (field strengths). In that way, the traditional fields can keep their names.

$$\delta\omega^{ab} = -D\lambda^{ab} = -d\lambda^{ab} - \omega^{ac}\lambda_c{}^b + \omega^{bc}\lambda_c{}^a \tag{4.130}$$

We have now approached the kinematical gauging of the Poincaré group in two different ways. In the first "matter crutch approach", space-time was initially flat and we attempted to make the Poincaré transformations of a matter field local. This forced the introduction of the spin connection $\omega_\mu{}^{ab}$ and the vierbein $e_\mu{}^a$ as well as need to distinguish indices as "frame" and "world".

In the second "formal approach", we started in an arbitrary space-time as indicated by working in the form language. Therefore, general coordinate invariance – but not necessarily gravity – is built into the theory from the outset. One may therefore suspect that the transformations generated by P_a – what we may think of as translations in a local frame – may not be the same as local coordinate transformations. This is related to what we discussed above regarding the Einstein elevator in Section 4.5.1. There is no problem, in accordance with the equivalence principle, with rotating and boosting the elevator. But perhaps translating within the elevator to another point inside it may be problematic?

That something along these lines may be going on here can be seen from the transformation formulas in the two different approaches. Consider an infinitesimal GCT with $\delta x^\mu = \zeta^\mu$. Then we know how the vierbein $e_\mu{}^a$ transforms. This transformation is then rewritten in a way as to resemble the transformation (4.129). The result of such a rewriting is [133]

$$\delta e_\mu{}^a = -\zeta^\nu\partial_\nu e_\mu{}^a - \partial_\mu\zeta^\nu e_\nu{}^a = -D_\mu(\zeta^\nu e_\nu{}^a) + \zeta^\nu D_{[\mu}e_{\nu]}^a + (\zeta^\nu\omega_\nu{}^a{}_c)e_\mu{}^c \tag{4.131}$$

This shows that an infinitesimal GCT $\delta x^\mu = \zeta^\mu$ can be reproduced by a local translation with parameter $\xi^a = \zeta^\nu e_\nu{}^a$ and local Lorentz transformation with parameter $\lambda^{ab} = \zeta^\nu\omega_\nu^{ab}$ provided the torsion is set to zero.

Looking back at what we have done in this section, we see that we have treated the Poincaré group as an internal symmetry group just like any Yang–Mills group. This point of view offers another way to understand the discrepancy between infinitesimal GCTs and local Poincaré translations. Making them agree requires an additional step: the field strength of the "translation connection" $e_\mu{}^a$ must be set to zero.

There are two further observations that should be noted. First, the Poincaré group is not semisimple, having the translations as an Abelian invariant subgroup. Yang–Mills theory is normally done with simple or semisimple internal groups. Second, the arguments presented so far are all in the infinitesimal. The infinite-dimensional diffeomorphism group is structurally different from the local group of Poincaré translations.

4.7 Actions for gravity

We will review a few versions of the action for general relativity, starting with the standard Einstein–Hilbert action.

4.7.1 The Einstein–Hilbert action

The Einstein–Hilbert action reads

$$S_{EH} \sim \int_M d^4x \sqrt{g} R \tag{4.132}$$

where g denotes $-\det(g_{\mu\nu})$ and R is the curvature scalar. The constant in front of the integral depends on conventions chosen. With the metric dimensionless, the constant should have mass dimension 2 in units where $c = \hbar = 1$.

Varying the action with respect to the metric to get the Einstein field equations goes through a number of standard steps that we will just briefly indicate. We start with

$$\delta(\sqrt{g}R) = \sqrt{g}R_{\mu\nu}\delta g^{\mu\nu} + R\delta\sqrt{g} + \sqrt{g}g^{\mu\nu}\delta R_{\mu\nu} \tag{4.133}$$

In this formula, $\delta g^{\mu\nu}$ is computed from $\delta(g^{\mu\nu}g_{\nu\sigma}) = 0$ with the result

$$\delta g^{\mu\nu} = -g^{\mu\rho}g^{\nu\sigma}\delta g_{\rho\sigma} \tag{4.134}$$

Next, the variation of $\sqrt{g}$ is computed from the formula for the derivative of the determinant of a matrix, with the result

$$\delta\sqrt{g} = \frac{1}{2}\sqrt{g}g^{\mu\nu}\delta g_{\mu\nu} \tag{4.135}$$

And then $\delta R_{\mu\nu}$ is the so called *Palatini identity* which is computed by straightforward variation in the formula for the Ricci tensor

$$\delta R_{\mu\nu} = \nabla_\mu \delta\Gamma_{\rho\nu}{}^{\rho} - \nabla_\rho \delta\Gamma_{\mu\nu}{}^{\rho} \tag{4.136}$$

Now the first two terms in (4.133) combine into

$$-\sqrt{g}\,(R^{\mu\nu} - \frac{1}{2}g^{\mu\nu}R) \tag{4.137}$$

while the last term can be written as

$$\sqrt{g}\,\nabla_\mu v^\nu \quad \text{with} \quad v^\mu = g^{\mu\sigma}\delta\Gamma_{\rho\sigma}{}^{\rho} - g^{\rho\sigma}\delta\Gamma_{\rho\sigma}{}^{\mu} \tag{4.138}$$

Here, the covariant constancy of the metric is used. In general, a covariant divergence can be written

$$\nabla_\mu v^\mu = \frac{1}{\sqrt{g}}\partial_\mu(\sqrt{g}\,v^\mu) \tag{4.139}$$

Thus (4.138), the last term of (4.133), turns out to be a total derivative. Putting everything together, we get the variation of the Einstein–Hilbert action

$$\delta S_{EH} \sim -\int_M d^4x\sqrt{g}\,(R^{\mu\nu} - \frac{1}{2}g^{\mu\nu}R)\delta g_{\mu\nu} + \int_M d^4x\partial_\mu v^\mu \tag{4.140}$$

The total derivative term cannot be discarded off lightly. It can be converted to a surface integral – assuming the integration region M to be bounded as is appropriate – but it turns out not to be sufficient to assume a constant metric on the boundary ∂M of M (so that $\delta g_{\mu\nu}$ vanishes on ∂M). Also the first derivatives of the metric must be fixed on ∂M. This problem is often cured by adding a surface term to the Einstein–Hilbert action, such that it cancels the offending surface integral upon variation.[25] Here, we will simply assume that to be done, and conclude that demanding the action to be stationary (under arbitrary variations of the metric that vanish on the boundary) yields the Einstein field equations

$$G^{\mu\nu} \equiv R^{\mu\nu} - \frac{1}{2}g^{\mu\nu}R = 0 \tag{4.141}$$

4.7.2 First-order action

Our discussions of the gauge theory treatments of gravity have suggested that it should be possible, and interesting, to treat the metric and the affine connection as independent fields.[26] Since the connection, under certain assumptions (no torsion, metricity) eventually turn out to be expressible in terms of derivatives of the metric, this points toward a first-order formulation of the theory where the expression for the connection should be one of the field equations.

The action, as it turns out, can be taken to be the Einstein–Hilbert action, but now with the curvature scalar explicitly written as the metric contraction of the Ricci tensor

$$S_P \sim \int_M d^4x\sqrt{g}g^{\mu\nu}R_{\mu\nu}(\Gamma) \tag{4.142}$$

From the formula for the Riemann tensor (4.64), we get for the Ricci tensor

$$R_{\mu\nu} = R_{\mu\alpha\nu}{}^{\alpha} = \partial_\mu\Gamma_{\nu\alpha}{}^{\alpha} - \partial_\alpha\Gamma_{\mu\nu}{}^{\alpha} + \Gamma_{\mu\beta}{}^{\alpha}\Gamma_{\alpha\nu}{}^{\beta} - \Gamma_{\alpha\beta}{}^{\alpha}\Gamma_{\mu\nu}{}^{\beta} \tag{4.143}$$

25 Details of this can be found in [271, 127].

26 Although the Palatini identity is used in the computation, it seems that the first-order variational approach is due to Einstein himself, rather than to A. Palatini; see [272].

entirely expressed in terms of the affine connection $\Gamma_{\mu\nu}{}^{\rho}$ which is assumed to be free of torsion.[27] Its variation, with respect to a variation in the connection, is given by the Palatini identity (4.136).

Since there is such clear separation of the variables $\sqrt{g}g^{\mu\nu}$ and $\Gamma_{\mu\nu}{}^{\rho}$, it is more convenient to vary the action with respect to the tensor density $\mathfrak{g}^{\mu\nu} = \sqrt{g}g^{\mu\nu}$ rather than the metric itself. Without further ado, we then get

$$\delta S_P \sim \int_M d^4x (R_{\mu\sigma}\delta\mathfrak{g}^{\mu\sigma} + \mathfrak{g}^{\mu\sigma}(\nabla_\mu \delta\Gamma_{\rho\sigma}{}^{\rho} - \nabla_\rho \delta\Gamma_{\mu\sigma}{}^{\rho})) \tag{4.144}$$

The first term immediately gives the field equation for Γ

$$\frac{\delta S_P}{\delta \mathfrak{g}^{\mu\sigma}} \sim R_{\mu\sigma}(\Gamma) = 0 \tag{4.145}$$

The second term of (4.144) involving $\delta\Gamma$ needs some reworking to bring forth a field equation for the metric. Essentially, we want to get the derivatives off $\delta\Gamma$. First, it is rewritten using the Leibniz rule

$$\int_M d^4x[\nabla_\rho((\mathfrak{g}^{\rho\sigma}\delta^\alpha_\beta - \mathfrak{g}^{\alpha\sigma}\delta^\rho_\beta)\delta\Gamma_{\alpha\sigma}{}^{\beta}) + (\nabla_\beta \mathfrak{g}^{\alpha\sigma} - \nabla_\rho \mathfrak{g}^{\rho\sigma}\delta^\alpha_\beta)\delta\Gamma_{\alpha\sigma}{}^{\beta}] \tag{4.146}$$

Then, using a formula for the covariant divergence of a vector density that reads $\nabla_\mu \mathfrak{v}^\mu = \partial_\mu \mathfrak{v}^\mu$, the first term yields a total derivative that is discarded (with the same caveat as before), and a torsion term that goes with the second term of (4.146). Finally, one gets the following contribution to the variation of the action:

$$\int_M d^4x(\nabla_\beta \mathfrak{g}^{\alpha\sigma} - \nabla_\rho \mathfrak{g}^{\rho\sigma}\delta^\alpha_\beta)\delta\Gamma_{\alpha\sigma}{}^{\beta} \tag{4.147}$$

This may look like a complicated field equation for the metric density. But taking advantage of the symmetry of $\delta\Gamma_{\alpha\sigma}{}^{\beta}$ in the lower indices, the equation can be symmetrized and written

$$\nabla_\beta \mathfrak{g}^{\alpha\sigma} - \frac{1}{2}(\nabla_\rho \mathfrak{g}^{\rho\sigma}\delta^\alpha_\beta + \nabla_\rho \mathfrak{g}^{\rho\alpha}\delta^\sigma_\beta) = 0 \tag{4.148}$$

We have a homogeneous system of 40 equations for the 40 variables $\nabla_\beta \mathfrak{g}^{\alpha\sigma}$ and the only solution is $\nabla_\beta \mathfrak{g}^{\alpha\sigma} = 0$. From here follows the covariant constancy of the metric itself, and the formula for the Levi–Civita connection can be calculated as in example 7 above.

27 So far, $R_{\mu\nu}$ is asymmetric in its indices, but the metric, assumed to be symmetric, projects out any antisymmetric part in the Lagrangian density. A version with torsion and non-symmetric metric can be found in [127]. Compare also to S. Deser's work reviewed in Section 2.9.1.

4.7.3 First-order vierbein action

The first-order vierbein action can be written as

$$S \sim \int R_{\mu\nu}{}^{ab}(\omega) e_\rho{}^r e_\sigma{}^s \epsilon_{abrs} \epsilon^{\mu\nu\rho\sigma} \tag{4.149}$$

where $R_{\mu\nu}{}^{ab}$ is the curvature expressed as a function of the spin connection according to formula (4.98). The form of this action is dictated by what we already know about actions for gravity. The superficial resemblance to the Einstein–Hilbert action (4.132) and in particular to the first-order action (4.142) should be clear. Independent variation of this action with respect to e and ω yield the field equations

$$G^a{}_\mu = 0 \quad \text{and} \quad D_{[\mu} e_{\nu]}^a = 0 \tag{4.150}$$

Here, $G^a{}_\mu$ corresponds to the Einstein tensor calculated from $R_{\mu\nu}{}^{ab}$ using the vierbeins to contract indices appropriately. In the second equation, D_μ is the Lorentz covariant derivative, hence this equation expresses zero torsion (see equation (4.96)). The equivalence to standard GR follows upon requiring the first vierbein postulate. Then the spin connection and the Levi–Civita connection become equal.

The action principles reviewed here are essentially the available choices for the dynamical part of the gauging method, at least if one wants to reproduce Einstein gravity closely. Nonpropagating torsion can be accommodated, as in the Cartan–Sciama–Kibble U_4 theory. Torsion also occurs in supergravity, for instance, in the first-order formulation of the Deser–Zumino $N = 1$ supergravity [142].

4.7.4 Provisional summary of gauge Yang–Mills versus gauge gravity

It is quite clear that the resemblance to Yang–Mills theory is "less than perfect". I would dare say that it is not perfect at all. At the kinematical level of gauging, there are conceptual problems as regards the gauging of the translation part of the Poincaré group and its relation to general coordinate transformations. This shows up in the gauging procedure as the question of the nature of the tetrad fields: when and with what motivation should they be introduced and what is the nature of their indices? Formally, the tetrads become the gauge fields of local translations. The corresponding field strength is the torsion, which is forced to be zero in the absence of matter at least. The tetrads furthermore must be invertible viewed as matrices in order to retrieve standard GR. This is a feature not seen in Yang–Mills theory. The tetrad fields may formally be viewed as connections on the local tangent bundle. The invertibility, however, seems to introduce a novel concept of a kind of duality where the base space-time manifold may be viewed as a fiber bundle over the local Minkowski space-time.

The kinematical gauging of the Lorentz group works out a little better. No deep conceptual problems seem to appear. The gauge field is the spin connection and the field strength is the curvature. However, when we come to the dynamical part of the gauging procedure, the Lagrangian density is not quadratic in field strengths, but linear. The equation of motion for the gauge field will be a constraint: the covariant constancy of the metric. As we have seen, the end result is the Einstein field equations where the dynamical field indeed turn out to be the metric itself, or the vierbein fields. The connections, either the Levi–Civita or the spin connection are auxiliary and completely determined by the metric or the vierbeins.

One way of viewing all this is to realize that there is an extra tier to spin 2 theory as compared to spin 1 theory. For spin 1, there is just the gauge potential and the field strength which is also the gauge covariant curvature. For spin 2, there are three tiers: the gauge potential (the metric), a second tier (the connections) and a third which is the curvature. The three tiers are related in a special way as we have seen.

Higher spin gauge field theory will exhibit a generalization of this in that there are $s+1$ tiers for spin s. We have already seen this in the de Wit–Freedman elaboration of the Fronsdal theory (see Section 2.10.3). This structure – in the free field theory – has been investigated by D. Francia and A. Sagnotti, a topic that will be reviewed in Sections 5.3.2 and 5.5. It will also appear in the "frame-like" approach to higher spin field theory that is on the route to the Vasliev theory (see Section 5.7).

In this context, it should be mentioned that there is an approach to gauge theory of gravity that aims to circumvent the problems surrounding the special way that the translations has to be managed. It basically amounts to gauging the semisimple group $SO(3,2)$ instead of the Poincaré group, and then breaking the symmetry down to Poincaré. This method was much in use in the research into various supergravity theories. We will not go further here, neither with reviewing the material, which is outside our scope, nor with references, but rather refer the reader to the Kibble–Stelle review [134] from that time.

4.8 Chapter 4 epilogue

I am afraid I may have left the reader in a stage of confusion. It was not my intention. I must say I am somewhat confused myself. To see it from the bright side, let us agree that it is probably a healthy confusion. The step from spin 1 gauge theory to spin 2 gauge theory is fraught with conceptual and technical problems. In the light of this, one could just press ahead nonetheless, or try to understand the conceptual problems and the low spin cases better. Most likely one must do both. We will try to do so in the second volume. Pressing ahead using the gauge principle heuristically will lead to the Vasiliev theory. If it may lead to other kinds of higher spin theories, I, at least, do not know. It is my belief that trying to understand the conceptual problems will eventually be necessary. However, for now we turn to the free field theory of arbitrary spin.

5 Exploring the free field theory

In this chapter, we will analyze parts of the classical work on higher spin gauge fields. Contrary to Chapter 2 where we followed the historical route, we will here start with the Fronsdal theory in order to lay down the foundations of the theory and explain the notation that is needed in order to work efficiently with higher spin fields. It will also allow us to contrast what came before and after Fronsdal with the Fronsdal theory itself.

5.1 The Fronsdal theory

It is natural to use a symmetric tensor field $\varphi_{\mu_1\ldots\mu_s}$ with s indices for the field theoretic realization of a spin s representation of the Poincaré group. This formulation of the free theory has become known as the *metric-like formulation*. The *frame-like formulation* will be reviewed in Section 5.7. However, as we have already noted, the spin of a massless, or massive, representation of the Poincaré group has no simple relation to the number of indices on the space-time field realization. There are many choices of sets of fields available, and one must take field equations, subsidiary conditions and gauge invariances into account. This story will be unraveled as we proceed.

We studied the free field equations and Lagrangians for spin 1 and 2 in Section 4.1. Higher spin field equations turn out to be quite close to their lower spin counterparts. Let us first define the *Fronsdal tensor*[1]

$$\mathcal{F}_{\mu_1\ldots\mu_s} = \Box\varphi_{\mu_1\ldots\mu_s} - \partial_{(\mu_1}\partial\cdot\varphi_{\mu_2\ldots\mu_s)} + \partial_{(\mu_1}\partial_{\mu_2}\varphi'_{\mu_3\ldots\mu_s)} \tag{5.1}$$

In terms of this tensor, the wave equation reads

$$\mathcal{F}_{\mu_1\ldots\mu_s} = 0 \tag{5.2}$$

This equation naturally generalizes the wave equations for massless fields of spin 0, 1 and 2. Under a gauge transformation,

$$\delta\varphi_{\mu_1\ldots\mu_s} = \partial_{(\mu_1}\xi_{\mu_2\ldots\mu_s)} \tag{5.3}$$

we record, for the readers convenience, the transformations of the trace and divergence

$$\delta\varphi'_{\mu_3\ldots\mu_s} = 2\partial\cdot\xi_{\mu_3\ldots\mu_s} + \partial_{(\mu_3}\xi'_{\mu_4\ldots\mu_s)} \tag{5.4}$$

$$\delta\,\partial\cdot\varphi_{\mu_2\ldots\mu_s} = \Box\xi_{\mu_2\ldots\mu_s} + \partial_{(\mu_2}\partial\cdot\xi_{\mu_3\ldots\mu_s)} \tag{5.5}$$

1 B. de Wit and D. Z. Freedman [150] wrote $W_{\mu_1\ldots\mu_s}$ for this tensor, certainly a better choice, but F has become standard in the modern higher spin literature. To disambiguate its use from other prominent F's, I resort to write $\mathcal{F}$.

https://doi.org/10.1515/9783110451771-005

The three terms in the Fronsdal tensor transform as

$$\begin{aligned}\delta(\Box\varphi_{\mu_1\ldots\mu_s}) &= \partial_{(\mu_1}\Box\xi_{\mu_2\ldots\mu_s)}\\ \delta(-\partial_{(\mu_1}\partial\cdot\varphi_{\mu_2\ldots\mu_s)}) &= -\partial_{(\mu_1}\Box\xi_{\mu_2\ldots\mu_s)} - \partial_{(\mu_1}\partial_{(\mu_2}\partial\cdot\xi_{\mu_3\ldots\mu_s))}\\ \delta(\partial_{(\mu_1}\partial_{\mu_2}\varphi'_{\mu_3\ldots\mu_s)}) &= 2\partial_{(\mu_1}\partial_{\mu_2}\partial\cdot\xi_{\mu_3\ldots\mu_s)} + \partial_{(\mu_1}\partial_{\mu_2}\partial_{(\mu_3}\xi'_{\mu_4\ldots\mu_s))}\end{aligned} \tag{5.6}$$

The Fronsdal tensor itself transforms as

$$\delta\mathcal{F}_{\mu_1\ldots\mu_s} = 3\,\partial_{(\mu_1}\partial_{\mu_2}\partial_{\mu_3}\xi'_{\mu_4\ldots\mu_s)} \tag{5.7}$$

In order to have gauge invariant wave equations, we have to require the gauge parameters for spin 3 and higher to be traceless, that is, $\xi'_{\mu_4\ldots\mu_s} = 0$. This is however not the full story. We have to ensure that the number of dynamical components of the field come out right. In $D = 4$, the number of physical degrees of freedom[2] is 2 independent of spin, while a symmetric tensor field has $\binom{s+3}{3}$ components in $D = 4$.

For the moment, we assume that the higher spin gauge transformations remove twice the number of components of the gauge parameter just as for spin 1 and 2. Taking the tracelessness into account, this number then works out to $2\left(\binom{s+2}{3} - \binom{s}{3}\right) = 2s^2$ which is clearly not enough to reduce $\binom{s+3}{3}$ down to 2. A clue of what to do can be obtained by explicitly computing the number of components of the field and it first and second traces. We record the result in Table 5.1.

Table 5.1: Number of components of φ and ξ and their traces.

Field or Parameter	Number of components
$\varphi_{\mu_1\ldots\mu_s}$	$\binom{s+3}{3} = \frac{1}{6}(s^3 + 6s^2 + 11s + 6)$
$\xi_{\mu_2\ldots\mu_s}$	$\binom{s+2}{3} = \frac{1}{6}(s^3 + 3s^2 + 2s)$
$\varphi'_{\mu_3\ldots\mu_s}$	$\binom{s+1}{3} = \frac{1}{6}(s^3 - s)$
$\xi'_{\mu_4\ldots\mu_s}$	$\binom{s}{3} = \frac{1}{6}(s^3 - 3s^2 + 2s)$
$\varphi''_{\mu_5\ldots\mu_s}$	$\binom{s-1}{3} = \frac{1}{6}(s^3 - 6s^2 + 11s - 6)$

It is clear that a field subject to a vanishing double trace condition – *double tracelessness* – effective from spin 4 onwards

$$\varphi''_{\mu_5\ldots\mu_s} = \varphi^{\alpha\beta}{}_{\alpha\beta\mu_5\ldots\mu_s} = 0 \tag{5.8}$$

will carry $2s^2 + 2$ degrees of freedom. Subtracting the gauge freedom leaves us with the required 2 physical components. A symmetric, double traceless field may be referred to as a *Fronsdal field*. The number of field components, $2s^2 + 2$, is the same number

2 We are now counting field degrees of freedom; see Section 1.1.

as in two symmetric and traceless fields with s and $s-2$ indices, respectively. As we saw in Section 2.10.1, this combination of fields appeared in Fronsdal's analysis of the massless limit of massive higher spin theory. We will work through the counting of degrees of freedom in more detail in Section 5.1.1, but first some comments on notation for higher spin fields.

Index symmetrization

The $(\ldots)$ notation means symmetrization of the enclosed indices with weight 1. Thus, in the simplest case

$$\partial_{(\mu_1}\varphi_{\mu_2\ldots\mu_s)} = \partial_{\mu_1}\varphi_{\mu_2\ldots\mu_s} + \partial_{\mu_2}\varphi_{\mu_3\ldots\mu_s\mu_1} + \cdots + \partial_{\mu_s}\varphi_{\mu_1\ldots\mu_{s-1}} \tag{5.9}$$

including as many terms as needed (but not any more) to make the expression fully symmetric, in this case s terms. With this logic, the last term in the Fronsdal tensor (5.1) has $s(s-1)/2$ terms. In some computations, for instance in the computation of the gauge variation of the second term in the Fronsdal tensor there will appear the expression $\partial_{(\mu_1}\partial_{(\mu_2}\xi_{\mu_3\ldots\mu_s))}$, that is, a double symmetrization. This expression contains $s(s-1)$ terms and, therefore, overcounts the number of terms as compared to $\partial_{(\mu_1}\partial_{\mu_2}\xi_{\mu_3\ldots\mu_s)}$ with a factor of 2. Note that $\partial_{(\mu_1}\partial_{\mu_2)} = 2\partial_{\mu_1}\partial_{\mu_2}$. The following formula captures the general case:

$$\partial_{(\mu_1}\ldots\partial_{\mu_p}\partial_{(\mu_{p+1}}\ldots\partial_{\mu_{p+q}}\xi_{\mu_{p+q+1}\ldots\mu_n))} = \binom{p+q}{q}\partial_{(\mu_1}\ldots\partial_{\mu_{p+q}}\xi_{\mu_{p+q+1}\ldots\mu_n)} \tag{5.10}$$

To denote symmetrizations over two independent index groups, one writes $A_{(\mu_1(\nu_1\ldots\nu_n)\mu_2\ldots\mu_m)}$, each index group being enclosed by round brackets in a hopefully not too confusing manner. As with all shortened notation, a bit of care is needed in parsing the formulas.

As is soon discovered, unit weight symmetrization is very convenient in higher spin theory. Using full symmetrization by summing over all s permutations and dividing by $s!$ would produce lots of unnecessary terms as the expressions to be symmetrized often already possess a large index symmetry, as for instance in formula (5.9). Such factors of $1/s!$ would be a nuisance, and in writing a formula one would have to compute the number of terms in order to get the factor right.

There are various versions of condensed notation employed in the higher spin literature.[3] Common to most condensed notation are to use round brackets ( ) to enclose indices that are to be symmetrized, most often with unit weight as noted above.

Condensed notation

The following notation will be used here. A symmetric tensor with n indices $\varphi^{\mu_1\ldots\mu_n}$ is written $\varphi^{(n)}$ and correspondingly for tensors with lower indices. If symmetrization is needed, it is always done with unit weight. Traces of a tensor are decorated with a prime $'$ or a double prime $''$ and the number of

3 At least going back to some of the 1960s work cited in Chapter 2.

remaining symmetrized indices. Thus the trace of an n index tensor $\varphi^{(n)}$ is written as $\varphi'^{(n-2)}$. Multiple traces are denoted by a square bracket superscript as in $\varphi^{[p]}$.

A dot · and sometimes, a double dot : are used for divergences and double divergences as in $\partial\cdot\varphi^{(n-1)}$ and $\partial\cdot\partial\cdot\varphi^{(n-2)} = \partial\partial : \varphi^{(n-2)}$. The superscript (or subscript) (p) always denotes the number of symmetrized indices left after the indicated contraction operations (traces, divergences etc.). Contractions between tensors are often left implicit in writing for instance $A_{(n)}B_{(n)}$ where all indices are contracted. If care is exercised, it is possible to calculate reliably using this notation. As an example, computing a divergence as in $\partial\cdot\left(\partial^{(1}\varphi^{s-1)}\right) = \Box\varphi^{s} + (s-1)\partial\cdot\varphi^{(s-1)}$, we see that the total number of terms are preserved on both sides of the equation. The Fronsdal field equations can be written in this notation as

$$\Box\varphi^{(s)} - \partial^{(1}\partial\cdot\varphi^{s-1)} + \partial^{(1}\partial^{2}\varphi'^{s-2)} = 0 \tag{5.11}$$

We will also write integrals simply as $\int$ when the integration variables are obvious.

An even more condensed notation was introduced by D. Francia and A. Sagnotti in [273] where the decoration telling the number of indices are dropped. Unit weight symmetrization is always assumed and not explicitly written. Multiple partial derivatives are the written ∂^{m} denoting a product of m derivatives with different indices, for instance, $\partial\partial = \partial^{2}$. Divergences are still denoted by $\partial\,\cdot$. The multiple symmetrization rule (5.10) becomes $\partial^{p}\partial^{q} = \binom{p+q}{q}\partial^{p+q}$. The Fronsdal equations become

$$\Box\varphi - \partial\partial\cdot\varphi + \partial\partial\varphi' = 0 \tag{5.12}$$

We will occasionally use this simplified system of notation. It is very convenient for fast communication of the essentials of a situation. For detailed manipulations, one may prefer to carry more baggage.

We also record a few more equations involving traces and divergences of the Fronsdal tensor that will be useful later on. We will write them in condensed notation and keep terms with double traces on the field:

$$\mathcal{F}'_{(n-2)} = 2\Box\varphi'_{(n-2)} - 2\partial\cdot\partial\cdot\varphi_{(n-2)} + \partial_{(1}\partial\cdot\varphi'_{n-3)} + \partial_{(1}\partial_{2}\varphi''_{n-4)} \tag{5.13}$$

$$\mathcal{F}''_{(n-4)} = 3\partial_{(1}\partial\cdot\varphi''_{n-3)} + 3\Box\varphi''_{(n-4)} + \partial_{(1}\partial_{2}\varphi^{[3]}_{n-6)} \tag{5.14}$$

$$\partial\cdot\mathcal{F}_{(n-1)} = \partial_{(1}\Box\varphi'_{n-2)} - \partial_{(1}\partial\cdot\partial\cdot\varphi_{n-2)} + \partial_{(1}\partial_{2}\partial\cdot\varphi'_{n-3)} \tag{5.15}$$

$$\begin{aligned}\partial\cdot\mathcal{F}'_{(n-3)} &= 3\Box\partial\cdot\varphi'_{(n-3)} - 2\partial\cdot\partial\cdot\partial\cdot\varphi_{(n-3)} + \partial_{(1}\partial\cdot\partial\cdot\varphi'_{n-4)}\\ &\quad + \Box\partial_{(1}\varphi''_{n-4)} + \partial_{(1}\partial_{2}\partial\cdot\varphi''_{n-5)}\end{aligned} \tag{5.16}$$

5.1.1 Counting physical components

In order to correctly count the number of independent physical components of the higher spin field, we impose the covariant gauge condition [150]

$$\mathcal{G}_{\mu_2\ldots\mu_s} \equiv \partial\cdot\varphi_{\mu_2\ldots\mu_s} - \frac{1}{2}\partial_{(\mu_2}\varphi'_{\mu_3\ldots\mu_s)} = 0 \tag{5.17}$$

generalizing the Lorenz and de Donder gauge conditions for spin 1 and 2. We will call $\mathcal{G}_{\mu_2\ldots\mu_s}$ the *de Donder tensor*. A short computation shows

$$\partial_{(\mu_1}\mathcal{G}_{\mu_2\ldots\mu_s)} = \partial_{(\mu_1}\partial\cdot\varphi_{\mu_2\ldots\mu_s)} - \partial_{(\mu_1}\partial_{\mu_2}\varphi'_{\mu_3\ldots\mu_s)} \tag{5.18}$$

The Fronsdal tensor may therefore be written in terms of the de Donder tensor as

$$\mathcal{F}_{\mu_1\ldots\mu_s} = \Box\varphi_{\mu_1\ldots\mu_s} - \partial_{(\mu_1}\mathcal{G}_{\mu_2\ldots\mu_s)} \tag{5.19}$$

Imposing the covariant gauge condition (5.17) then reduces the wave equation to

$$\Box\varphi_{\mu_1\ldots\mu_s} = 0 \tag{5.20}$$

Thus the particles are massless. Next, we compute the trace of $\mathcal{G}_{\mu_2\ldots\mu_s}$ to find

$$\mathcal{G}'_{\mu_4\ldots\mu_s} = -\frac{1}{2}\partial_{(\mu_4}\varphi''_{\mu_5\ldots\mu_s)} = 0 \quad \text{if} \quad \varphi''_{\mu_5\ldots\mu_s} = 0 \tag{5.21}$$

We learn that if the field is double traceless then the gauge condition is traceless and, therefore, has as many components as the gauge parameter (namely s^2). Next, we compute the gauge variation of the gauge condition to find

$$\delta\mathcal{G}_{\mu_2\ldots\mu_s} = \Box\xi_{\mu_2\ldots\mu_s} - \frac{1}{2}\partial_{(\mu_2}\partial_{(\mu_3}\xi'_{\mu_4\ldots\mu_s)} = \Box\xi_{\mu_2\ldots\mu_s} \tag{5.22}$$

To stay in the gauge, that is have $\delta\mathcal{G}_{\mu_2\ldots\mu_s} = 0$, it is therefore enough to use a gauge parameter all of whose components satisfy the Klein–Gordon equation. The field equation $\Box\varphi = 0$ is gauge invariant under such a gauge transformation. This allows for *regauging* as many field components as components in the gauge parameter. All in all, covariant gauge fixing and regauging removes $2s^2$ components leaving just 2 components out of the $2s^2 + 2$ components of a double traceless symmetric tensor field.

Fixing gauges and regauging: The TT-gauge or Fierz–Pauli–Umezawa example

---

? One may perceive the argument above as a little bit too clever for its own good. What are we actually doing? What is a gauge choice and what does it mean to regauge? To clarify this, one may proceed as follows. We start with field equations $\mathcal{F}(\varphi) = 0$ invariant under gauge transformations $\delta\varphi = \partial\xi$, that is, $\delta\mathcal{F}(\varphi) = 0$. The gauge condition is $\mathcal{G}(\varphi) = 0$.

We then think of an initial field configuration φ_0 not satisfying the gauge condition. To fix the gauge, we perform a gauge transformation $\varphi_0 \to \varphi = \varphi_0 + \partial\xi_0$ so that $\mathcal{G}(\varphi) = 0$. This can be done with a gauge parameter ξ_0 satisfying $\mathcal{G}(\partial\xi_0) = -\mathcal{G}(\varphi_0)$. As we see from equation (5.22), this means that $\Box\xi_0 = -\mathcal{G}(\varphi_0)$ for the de Donder gauge condition and a traceless gauge parameter.

For the new field configuration φ, we now have the field equation $\Box\varphi = 0$ and the gauge condition $\mathcal{G}(\varphi) = 0$ removes half the number of unphysical field components. Now we may perform the regauge transformation $\delta\varphi = \partial\xi$ with $\Box\xi = 0$ and $\xi' = 0$. The wave equation for φ is clearly invariant as is the gauge condition! We remove the second-half of the unphysical field components. That the count works out correctly is assured by equation (5.21) and the argument following it.

Let us now contrast this gauge choice with the *TT-gauge* choice (Transverse -Traceless), that results in the system of equations

$$\Box\varphi = 0 \qquad \partial\cdot\varphi = 0 \qquad \varphi' = 0 \tag{5.23}$$

invariant under re-gauge transformations with a parameter Λ satisfying

$$\Box\Lambda = 0 \qquad \partial\cdot\Lambda = 0 \qquad \Lambda' = 0 \tag{5.24}$$

This system is sometimes called the *Fierz–Pauli system*, since it was first studied for spin 2 by Fierz and Pauli (see Chapter 2).[4] The counting of field components is exactly as for massive higher spin fields since the conditions on the fields are the same. There is no double tracelessness now. Thus the field contains $2s+1$ independent components. The regauge parameter Λ contains $2s-1$ independent components. Subtracting, we get the correct number of physical components equal to 2.

It remains to study the gauge transformations that enforce the TT-gauge conditions. Consider an initial field configuration φ_{00} that is neither traceless nor transverse. A first gauge transformation $\varphi_{00} \to \varphi_0 = \varphi_{00} + \partial\xi_{00}$, chosen so that $2\partial\cdot\xi_{00} + \partial\xi'_{00} = -\varphi'_{00}$ enforces $\varphi'_0 = 0$. Now in order to have an invariant wave equation $\Box\varphi_0 - \partial\partial\cdot\varphi_0$ and gauge condition $\varphi'_0 = 0$, any further gauge transformation $\delta\varphi = \partial\xi$ must satisfy $\partial\cdot\xi = 0$ and $\xi' = 0$. Then perform a second such transformation $\varphi_0 \to \varphi = \varphi_0 + \partial\xi_0$ to enforce $\partial\cdot\varphi = 0$. This requires the parameter to satisfy $\Box\xi_0 = -\partial\cdot\varphi_0$. The gauge conditions $\varphi' = 0$ and $\partial\cdot\varphi = 0$ are then invariant under any further transformation satisfying $\partial\cdot\xi = \xi' = \Box\xi = 0$. So is the wave equation $\Box\varphi = 0$. We have arrived at the TT system of equations.

5.1.2 The Fronsdal Lagrangian

Fronsdal rederived the Singh–Hagen Lagrangian for massive higher spin fields. As we reviewed in the historical Section 2.5, it was found that a decreasing spectrum of fields of spin $s, s-2, s-3, \ldots$ was needed in order to write a Lagrangian for a massive spin s field that yields the correct Euler–Lagrange equations (2.140)–(2.142). Fronsdal found that all the lower spin fields except the one with spin $s-2$ decouple when the mass was set to zero. He then combined the two traceless spin s and $s-2$ fields into a single spin s field with nonzero trace but with zero double trace. This field then becomes the higher spin massless gauge field. The Fronsdal Lagrangian is

$$\begin{aligned}\mathcal{L} &= \frac{1}{2}\Big(\varphi_{\mu_1\ldots\mu_s}\Box\varphi^{\mu_1\ldots\mu_s} - \frac{s(s-1)}{2}\varphi'_{\mu_3\ldots\mu_s}\Box\varphi'^{\mu_3\ldots\mu_s} + s\partial\cdot\varphi_{\mu_2\ldots\mu_s}\partial\cdot\varphi^{\mu_2\ldots\mu_s} \\ &\quad + s(s-1)\varphi'_{\mu_3\ldots\mu_s}\partial\cdot\partial\cdot\varphi^{\mu_3\ldots\mu_s} + \frac{s(s-1)(s-2)}{4}\partial\cdot\varphi'_{\mu_3\ldots\mu_s}\partial\cdot\varphi'^{\mu_3\ldots\mu_s}\Big) \\ &= \frac{1}{2}\Big(\varphi\cdot\Box\varphi - \frac{s(s-1)}{2}\varphi'\cdot\Box\varphi' + s(\partial\cdot\varphi)\cdot(\partial\cdot\varphi) \\ &\quad + s(s-1)\varphi'\cdot(\partial\partial\cdot\varphi) + \frac{s(s-1)(s-2)}{4}(\partial\cdot\varphi')\cdot(\partial\cdot\varphi')\Big)\end{aligned} \tag{5.25}$$

In the second expression, we have taken the opportunity to write the action in condensed notation. The potentially worrisome negative sign in front of the kinetic term $\varphi'\cdot\Box\varphi'$ is correct and does not render the theory unphysical. It appears even for spin 2.

4 The only pre-Fronsdal mentioning of this particular system for arbitrary spin – that I am aware of – is in H. Umezawa's textbook [83] from 1956 (see our Section 2.4.5). Umezawa performs the counting of degrees of freedom as done here.

As mentioned in Section 2.10.2, the Fronsdal theory was rederived by T. Curtright by an ansatz-verification method. Take the higher spin gauge transformation law (5.3) for granted and assume the most general form for the free Lagrangian with two derivatives and no double traces (or higher) on the field. Then the coefficients in the ansatz can be fixed so that the variation of the Lagrangian is a total derivative only if the gauge parameter is traceless. The result is the Fronsdal Lagrangian (5.25).

5.2 The de Wit–Freedman elaboration

In the paper [150] – appearing about a year after the Fronsdal paper [3] – de Wit and Freedman clarified the structure of the theory by introducing a hierarchy of "Christoffel symbols" generalizing the spin 1 electromagnetic field strength and spin 2 free theory Christoffel symbols and curvature tensor. The special case of spin 3 was further investigated by T. Damour and S. Deser in [152].

The first-order spin s *Christoffel symbol* is defined as

$$\Gamma^{(1)}_{\rho;\mu_1\dots\mu_s} = \partial_\rho \varphi_{\mu_1\dots\mu_s} - \partial_{(\mu_1}\varphi_{\rho\mu_2\dots\mu_s)} \tag{5.26}$$

Higher order symbols are then defined recursively

$$\Gamma^{(m)}_{\rho_1\dots\rho_m;\mu_1\dots\mu_s} = \partial_{\rho_1}\Gamma^{(m-1)}_{\rho_2\dots\rho_m;\mu_1\dots\mu_s} - \frac{1}{m}\partial_{(\mu_1}\Gamma^{(m-1)}_{\rho_2\dots\rho_m;\rho_1\mu_2\dots\mu_s)} \tag{5.27}$$

so that, for instance, $\Gamma^{(2)}$ comes out explicitly as

$$\Gamma^{(2)}_{\rho_1\rho_2;\mu_1\dots\mu_s} = \partial_{\rho_1}\partial_{\rho_2}\varphi_{\mu_1\dots\mu_s} - \frac{1}{2}\partial_{(\rho_1}\partial_{(\mu_1}\varphi_{\rho_2)\mu_2\dots\mu_s)} + \partial_{(\mu_1}\partial_{\mu_2}\varphi_{\rho_1\rho_2\mu_3\dots\mu_s)} \tag{5.28}$$

The symmetry in the index group ρ is clear, a result that is true also for the higher order symbols.

The coefficients in the definition of the Christoffel symbols are chosen to produce simple gauge transformation properties for the symbols. We get

$$\delta\Gamma^{(m)}_{\rho_1\dots\rho_m;\mu_1\dots\mu_s} = (-1)^m(m+1)\partial_{(\mu_1}\dots\partial_{\mu_{m+1}}\xi_{\rho_1\dots\rho_m\mu_{m+2}\dots\mu_s)} \tag{5.29}$$

with all the indices from the group ρ appearing only on the gauge parameter.

The *generalized Christoffel symbol* $\Gamma^{(m)}_{\rho_1\dots\rho_m;\mu_1\dots\mu_s}$ is a linear combination of m partial derivatives on the field φ, independently symmetric in the two index groups μ and ρ. This together with the gauge transformation property (5.29) make them unique.

The $m = s$ Christoffel symbol for spin s is gauge invariant, that is,

$$\delta\Gamma^{(s)}_{\rho_1\dots\rho_s;\mu_1\dots\mu_s} = 0 \tag{5.30}$$

for the very simple reason that the gauge parameters have one index less that the gauge fields.[5] The symbols $\Gamma^{(s)}_{\rho_1\ldots\rho_s;\mu_1\ldots\mu_s}$ are called *generalized curvature tensors* in [150] for this reason, and deserve the special notation

$$R_{\rho_1\ldots\rho_s;\mu_1\ldots\mu_s} = \Gamma^{(s)}_{\rho_1\ldots\rho_s;\mu_1\ldots\mu_s} \tag{5.31}$$

An explicit formula can be derived from recursive equation (5.27). It reads[6]

$$R_{\rho_1\ldots\rho_s;\mu_1\ldots\mu_s} = \sum_{k=0}^{s} \frac{(-1)^k}{\binom{s}{k}} \partial_{(\rho_1} \ldots \partial_{\rho_{s-k}} \partial_{(\mu_{s-k+1}} \ldots \partial_{\mu_s} \varphi_{\mu_1\ldots\mu_{s-k})\rho_{s-k+1}\ldots\rho_s)} \tag{5.32}$$

where again the ρ and μ indices should be symmetrized separately. The curvature tensors obey Bianchi-type relations (see Section 5.2.1).

The lower order Christoffel symbols, contracted with the Minkowski metric over two of the ρ indices, are gauge invariant under transformations with traceless gauge parameters, as is clear from (5.29). Since the second-order symbol is then a gauge invariant, second-order derivative object, it is a candidate for a wave equation. Indeed we get

$$\Gamma^{(2)\sigma}_{\sigma;\mu_1\ldots\mu_s} = \Box\varphi_{\mu_1\ldots\mu_s} - \partial_{(\mu_1} \partial\cdot\varphi_{\mu_2\ldots\mu_s)} + \partial_{(\mu_1}\partial_{\mu_2}\varphi'_{\mu_3\ldots\mu_s)} = \mathcal{F}_{\mu_1\ldots\mu_s} \tag{5.33}$$

which is precisely the Fronsdal tensor.

Relation to the spin 2 Riemann tensor

Already from the symmetry properties of the spin 2 tensor $\Gamma^{(2)}_{\rho_1\rho_2;\mu_1\mu_2}$ it is clear that it is not equal to the usual Riemann curvature R of equation (4.64). Linearizing and writing R with all indices lowered, we get

$$\begin{aligned} R_{\mu_1\mu_2;\rho_1\rho_2} &= \partial_{\mu_1}\Gamma_{\mu_2\rho_1;\rho_2} - \partial_{\mu_2}\Gamma_{\mu_1\rho_1;\rho_2} \\ &= \frac{1}{2}\left(\partial_{\mu_1}\partial_{\rho_1}\varphi_{\mu_2\rho_2} + \partial_{\mu_2}\partial_{\rho_2}\varphi_{\mu_1\rho_1} - \partial_{\mu_1}\partial_{\rho_2}\varphi_{\mu_2\rho_1} - \partial_{\mu_2}\partial_{\rho_1}\varphi_{\mu_1\rho_2}\right) \end{aligned} \tag{5.34}$$

clearly separately antisymmetric in both index pairs μ and ρ. A linear recombination yields $\Gamma^{(2)}$ according to the formula

$$\Gamma^{(2)}_{\rho_1\rho_2;\mu_1\mu_2} = R_{\rho_2\mu_2;\rho_1\mu_1} - R_{\rho_1\mu_2;\mu_1\rho_2} \tag{5.35}$$

The linearized gravitational field equations can be written as

$$\eta^{\rho_1\rho_2}\Gamma^{(2)}_{\rho_1\rho_2;\mu_1\mu_2} = 0 \tag{5.36}$$

---

5 In a nonlinear theory, one would only expect gauge covariance.

6 A formula for lower order symbols can be found in the deWit and Freedman paper.

At this stage, we can note a curiosity of spin 1. For spin 1, it is the first-order Christoffel symbol $\Gamma^{(1)}_{\rho;\mu}$ that is gauge invariant. It is equal to the field strength tensor $F_{\rho\mu}$, and the field equations are, as usual, $\partial^{\rho}\Gamma^{(1)}_{\rho;\mu} = 0$. In this sense, the higher spin Fronsdal equations could be said to be "Einstein-like" rather than "Maxwell-like". However, if one were to compute the second order Christoffel symbol for spin 1 according to the general formula (5.27) one would find

$$\Gamma^{(2)}_{\rho_1\rho_2;\mu} = \partial_{\rho_1}\partial_{\rho_2}\varphi_{\mu} - \frac{1}{2}\partial_{\mu}\partial_{(\rho_1}\varphi_{\rho_2)} \tag{5.37}$$

This object is gauge invariant, and one would find that the Maxwell equations can also be written as the trace

$$\eta^{\rho_1\rho_2}\Gamma^{(2)}_{\rho_1\rho_2;\mu} = 0 \tag{5.38}$$

There is actually quite a lot more that can be said about this, and we will review such work beginning in Section 5.3.2 and in more detail in Section 5.5.

A note on notation

When, in the sequel, we have occasion to refer back to these generalized Christoffel symbols, we will drop the superscript $^{(m)}$ instead letting the number of ρ indices indicate the order of the symbol. It will be convenient to define a notation for tensors involving several traces over the ρ indices. Thus we take a superscript $^{[n]}$ to denote n traces, so that for instance $\Gamma^{[2]} = \Gamma''$.

5.2.1 Bianchi identities

In analogy to lower spin, and due to their definition in terms of derivatives, it is to be expected that the generalized Christoffel symbols $\Gamma^{(m)}$ obey Bianchi-type identities. This is indeed the case, although as pointed out in [150], it is only for the $m = s$ gauge invariant "curvatures" that an unambiguous meaning can be given to the identities.[7]

The traditional meaning of Bianchi identities refers to general relativity. The free Einstein equations $G_{\mu\nu} = 0$ comprise ten algebraically independent differential equations. They are related by four differential identities $\nabla^{\mu}G_{\mu\nu} = 0$ that reduce the number of effective equations to six. This leaves four undetermined metric components in $g_{\mu\nu}$ which correspond precisely to the gauge arbitrariness of the coordinate functions x^{μ}. The term Bianchi identity also refers to the identity $\nabla_{\lambda}R_{\mu\nu\rho\sigma} + \nabla_{\rho}R_{\mu\nu\sigma\lambda} + \nabla_{\sigma}R_{\mu\nu\lambda\rho} = 0$ for the curvature tensor from which $\nabla^{\mu}G_{\mu\nu} = 0$ follows (see reference [243]).

An analogous phenomenon occurs for electromagnetism. The free Maxwell equations $\mathcal{F}_{\mu} = \Box\varphi_{\mu} - \partial_{\mu}\partial\cdot\varphi = 0$ also fail to determine the vector potential φ_{μ} completely

7 With a certain exception, as we will see.

due to the differential identity $\partial \cdot \mathcal{F} = 0$. This reduces the number equations to three, leaving one undetermined component of φ_μ corresponding precisely to the gauge freedom. Also for spin 1, we have the identity $\partial_\rho F_{\mu\nu} + \partial_\mu F_{\nu\rho} + \partial_\nu F_{\rho\mu} = 0$ for the field strength tensor $F_{\mu\nu}$.

This suggests two directions of generalization to higher spin: either for the $m = s$ higher spin (and higher derivative) curvatures, or for certain combinations of the $m = 2$ Christoffel symbols. Let us follow this latter direction here.

For spin 1, we have $\partial \cdot \mathcal{F} = 0$, and for 2 we have $\partial \cdot G_\mu = \partial \cdot \mathcal{F}_\mu - \frac{1}{2}\partial_\mu \mathcal{F}' = 0$ in terms of the linearized Einstein tensor (see equation (4.13)). This generalizes immediately to $\partial \cdot \mathcal{F}_{\mu_2\mu_3} - \frac{1}{2}\partial_{(\mu_2}\mathcal{F}'_{\mu_3)} = 0$ for spin 3. However, for spin 4 and higher a new phenomenon appears, and we get, in condensed notation

$$\partial \cdot \mathcal{F}_{(s-1)} - \frac{1}{2}\partial_{(1}\mathcal{F}'_{s-2)} = -\frac{3}{2}\partial_{(1}\partial_2\partial_3\varphi''_{n-4)} \tag{5.39}$$

The term of right was called a *classical anomaly* in [274]. For double traceless fields – where the right-hand side is zero – we have a higher spin *Bianchi identity*.

5.2.2 Lagrangians

A straightforward – after the fact – approach to finding a Lagrangian for an arbitrary spin s field is to follow de Wit and Freedman. In terms of the Fronsdal tensor and its trace, make the ansatz

$$\mathcal{L} = \frac{1}{2}\varphi_{\mu_1\ldots\mu_s}\mathcal{F}^{\mu_1\ldots\mu_s} - a\varphi'_{\mu_3\ldots\mu_s}\mathcal{F}'^{\mu_3\ldots\mu_s} \tag{5.40}$$

where a is a coefficient to determine. The double tracelessness constraint is assumed to hold, which by the way, implies double tracelessness of the Fronsdal tensor. Then perform a gauge transformation on the action $S = \int \mathcal{L} d^4x$. Assuming the gauge parameter to be traceless from spin 3 onwards, we get

$$\delta S = -\int \xi_{\mu_2\ldots\mu_s}\left(\frac{s}{2}\partial \cdot \mathcal{F}^{\mu_2\ldots\mu_s} - \frac{2a}{s-1}\partial^{(\mu_2}\mathcal{F}'^{\mu_3\ldots\mu_s)}\right)dx^4 \tag{5.41}$$

However, from the definition of the Fronsdal tensor, we have the Bianchi identity for double traceless fields (see (5.39))

$$\partial \cdot \mathcal{F}_{\mu_2\ldots\mu_s} - \frac{1}{2}\partial_{(\mu_2}\mathcal{F}'_{\mu_3\ldots\mu_s)} = 0 \tag{5.42}$$

Thus the gauge variation of the action is zero (up to a surface term) precisely with $a = s(s-1)/8$. The Lagrangian can be written very smartly as

$$\mathcal{L} = \frac{1}{2}\varphi_{\mu_1\ldots\mu_s}(\mathcal{F}^{\mu_1\ldots\mu_s} - \eta^{(\mu_1\mu_2}\mathcal{F}'^{\mu_3\ldots\mu_s)}) = \frac{1}{2}\varphi^{(s)} \cdot (\mathcal{F}^{(s)} - \eta^{(12}F'^{s-2)}) \tag{5.43}$$

The Euler–Lagrange equations that follow from varying the action (5.43) are

$$\mathcal{F}_{\mu_1\dots\mu_s} - \frac{1}{2}\eta_{(\mu_1\mu_2}\mathcal{F}'_{\mu_3\dots\mu_s)} = 0 \tag{5.44}$$

Computing the trace of this equation yields $\mathcal{F}'_{\mu_3\dots\mu_s} = 0$, which reinserted yields the Fronsdal equations. In this context – with this spectrum of fields – there is no way to avoid this intermediate step of taking the trace of the Euler–Lagrange equations to arrive at the Fronsdal equations [150] (as will be explained below). The expression in equation (5.44) generalizes the linearized *Einstein tensor* for general relativity

$$G_{\mu_1\dots\mu_s} = \mathcal{F}_{\mu_1\dots\mu_s} - \frac{1}{2}\eta_{(\mu_1\mu_2}\mathcal{F}'_{\mu_3\dots\mu_s)} \tag{5.45}$$

It is divergence free for spin 1 and 2 but not so for higher spin, not even for double traceless fields.[8]

When the Lagrangian in (5.43) is written out explicitly it coincides precisely with the Fronsdal Lagrangian (5.25). However, the Lagrangian clearly needs a deeper analysis, and this is the topic of the next section.

"Varying" practicalities

---

i The following derivation formulas are useful when varying actions:[9]

$$\frac{\partial\phi_{(s')}}{\partial\phi_{(s)}} = \frac{\partial\phi_{\mu_{1'}\dots\mu_{s'}}}{\partial\phi_{\mu_1\dots\mu_s}} = \frac{1}{s!}\sum_{\sigma(s')}\eta_{\mu_1\mu_{1'}}\cdots\eta_{\mu_s\mu_{s'}} = \frac{1}{s!}\sum_{\sigma(s')}\eta^s_{(ss')} \tag{5.46}$$

$$\frac{\partial\phi'_{(s'-2')}}{\partial\phi_{(s)}} = \frac{\partial\phi'_{\mu_{3'}\dots\mu_{s'}}}{\partial\phi_{\mu_1\dots\mu_s}} = \frac{2}{s!}\eta_{(\mu_1\mu_2}\sum_{\sigma(s'-2')}\eta_{\mu_3\mu_{3'}}\cdots\eta_{\mu_s)\mu_{s'}} = \frac{2}{s!}\eta_{(12}\sum_{\sigma(s'-2')}\eta^{s-2}_{s)s'} \tag{5.47}$$

$$\frac{\partial\phi''_{(s'-4')}}{\partial\phi_{(s)}} = \frac{\partial\phi''_{\mu_{5'}\dots\mu_{s'}}}{\partial\phi_{\mu_1\dots\mu_s}} = \frac{8}{s!}\eta_{(\mu_1\mu_2}\eta_{\mu_3\mu_4}\sum_{\sigma(s'-4')}\eta_{\mu_5\mu_{5'}}\cdots\eta_{\mu_s)\mu_{s'}} = \frac{8}{s!}\eta_{(12}\eta_{34}\sum_{\sigma(s'-4')}\eta^{s-4}_{s)s'} \tag{5.48}$$

The permutations are over the primed indices μ' and the symmetrization over the unprimed indices μ. When the first derivative contracts into an object with full symmetry over $\mu_1\dots\mu_s$, a factor $s!$ cancels the $1/s!$. When the second derivative contracts into an object with symmetry over $\mu_3\dots\mu_s$, a factor $(s-2)!$ appears resulting in a factor $\frac{1}{s(s-1)}$. For the third derivative, the corresponding factor becomes $\frac{3}{s(s-1)(s-2)(s-3)}$.

5.2.3 Understanding the Lagrangian

Consider a free field theory with field equations $K\phi = 0$ with kinetic operator K involving two derivatives. We want to derive such field equations as the Euler–Lagrange

---

8 Note the factor of 1/2 in front of the F' term in (5.44) and (5.45) but not in (5.43).

9 We use the symbol ϕ for a generic field, reserving the symbol φ specifically for a higher spin field.

equations of a Lagrangian $\mathcal{L}$. In simple enough examples,[10] it works with $\mathcal{L} = \frac{1}{2}\phi(K\phi)$. However, as we have seen, if the fields are subject to constraints or subsidiary conditions, a more complicated Lagrangian is needed. This is also the case if the kinetic operator involves traces of the fields. We will now analyze this for the case of the Fronsdal field equations.

We start by separating out the two first terms of the Fronsdal tensor (not involving the trace of the field) and put $\mathcal{E}_{\mu_1\dots\mu_s} = \Box\varphi_{\mu_1\dots\mu_s} - \partial_{(\mu_1}\partial\cdot\varphi_{\mu_2\dots\mu_s)}$. We take as the provisional action $\frac{1}{2}\int\varphi^{(s)}\cdot\mathcal{F}_{(s)}$ written

$$S_1 = \frac{1}{2}\int\varphi^{\mu_1\dots\mu_s}\mathcal{E}_{\mu_1\dots\mu_s} + \frac{1}{2}\int\varphi^{\mu_1\dots\mu_s}\left(\partial_{(\mu_1}\partial_{\mu_2}\varphi'_{\mu_3\dots\mu_s)}\right) \tag{5.49}$$

For the time being neglecting the double tracelessness constraint on the field, we get the variation

$$\begin{aligned}\delta S_1 &= \int\mathcal{E}_{\mu_1\dots\mu_s}\delta\varphi^{\mu_1\dots\mu_s} + \frac{1}{2}\int\left(\partial_{(\mu_1}\partial_{\mu_2}\varphi'_{\mu_3\dots\mu_s)} + \eta_{(\mu_1\mu_2}\partial\cdot\partial\cdot\varphi_{\mu_3\dots\mu_s)}\right)\delta\varphi^{\mu_1\dots\mu_s}\\ &= \int\mathcal{F}_{\mu_1\dots\mu_s}\delta\varphi^{\mu_1\dots\mu_s}\\ &\quad - \frac{1}{2}\int\partial_{(\mu_1}\partial_{\mu_2}\varphi'_{\mu_3\dots\mu_s)}\delta\varphi^{\mu_1\dots\mu_s} + \frac{1}{2}\int\eta_{(\mu_1\mu_2}\partial\cdot\partial\cdot\varphi_{\mu_3\dots\mu_s)}\delta\varphi^{\mu_1\dots\mu_s}\end{aligned} \tag{5.50}$$

We see that we cannot get the Fronsdal equations directly, since (as seen in the last line) the variation: (i) lacks one term, and (ii) produces one term to much. Since the problem obviously stems from the trace term in the field equations, one can attempt to remedy it by adding a term – the same as in the de Wit–Freedman action in (5.40) – involving the trace of the Fronsdal tensor where we put the double trace φ'' to zero in $\mathcal{F}'$.

$$\begin{aligned}S_2 &= -a\int\varphi'^{\mu_1\dots\mu_{s-2}}\mathcal{F}'_{\mu_1\dots\mu_{s-2}}\\ &= -a\int\varphi'^{\mu_1\dots\mu_{s-2}}\left(2\Box\varphi'_{\mu_1\dots\mu_{s-2}} - 2\partial\cdot\partial\cdot\varphi_{\mu_1\dots\mu_{s-2}} + \partial_{(\mu_1}\partial\cdot\varphi'_{\mu_2\dots\mu_{s-2})}\right)\end{aligned} \tag{5.51}$$

The variation becomes (still neglecting the double tracelessness of the field)

$$\begin{aligned}\delta S_2 = -a\binom{s}{2}^{-1}\int\delta\varphi^{\mu_1\dots\mu_s}\big(&4\eta_{(\mu_1\mu_2}\Box\varphi'_{\mu_3\dots\mu_s)}\\ &- 2\eta_{(\mu_1\mu_2}\partial\cdot\partial\cdot\varphi_{\mu_3\dots\mu_s)} - 2\partial_{(\mu_1}\partial_{\mu_2}\varphi'_{\mu_3\dots\mu_s)}\\ &- 2\eta_{(\mu_1\mu_2}\partial_{(\mu_3}\partial\cdot\varphi'_{\mu_4\dots\mu_s))}\big)\end{aligned} \tag{5.52}$$

The inverted binomial factor in the variation δS_2 comes from the combinatorics of differentiating the trace of the field. The missing term (corresponding to the term without

10 When K is a second-order differential operator not involving any traces.

η in the formula above) to build up the Fronsdal tensor in the variation of the action can now gotten by choosing $a = \binom{s}{2}/4$, the same as we found before. With this choice, we get the variation of the action

$$\delta(S_1 + S_2) = \int (\mathcal{F}_{\mu_1 \ldots \mu_s} - \frac{1}{2}\eta_{(\mu_1\mu_2}\mathcal{F}'_{\mu_3\ldots\mu_s)})\delta\varphi^{\mu_1\ldots\mu_s}, \tag{5.53}$$

from which follows the field equations (5.44).

So far so good, but it remains to understand the double trace constraint. We have performed the above calculations with free variations $\delta\varphi^{\mu_1\ldots\mu_s}$. This is in principle wrong if the field is constrained by $\varphi''_{\mu_5\ldots\mu_s} = 0$. Let us first note that our calculations are unaffected by this issue for spin 1, 2 and 3.

But in principle one should use projections onto subspaces of variations that obey the double trace condition. This would introduce the need for further terms in the action involving products of double traces of the fields and double traces of the Fronsdal tensor. However, since the double trace of the Fronsdal tensor is zero for double traceless fields, as seen from equation (5.14), one can actually ignore this complication, as argued in [150].

Let us now turn to the gauge variation of the action. Most of the job is already done, since we can use equation (5.53) with $\delta\varphi^{\mu_1\ldots\mu_s}$ a gauge transformation. Performing a partial integration, we get

$$\begin{aligned}
\delta(S_1 + S_2) &= -\int \xi^{(\mu_2\ldots\mu_s}\partial^{\mu_1)}\left(\mathcal{F}_{\mu_1\ldots\mu_s} - \frac{1}{2}\eta_{(\mu_1\mu_2}\mathcal{F}'_{\mu_3\ldots\mu_s)}\right) \\
&= -s\int \xi^{\mu_2\ldots\mu_s}\left(\partial\cdot\mathcal{F}_{\mu_2\ldots\mu_s} - \frac{1}{2}\partial_{(\mu_2}\mathcal{F}'_{\mu_3\ldots\mu_s)} - \frac{1}{2}\eta_{(\mu_2\mu_3}\partial\cdot\mathcal{F}'_{\mu_4\ldots\mu_s)}\right)
\end{aligned} \tag{5.54}$$

In this formula, we recognize the first two terms as the left-hand side of the higher spin Bianchi identity (5.39). The third term is fairly complicated and can be read of from formula (5.16).

It is then clear that in order to have a gauge invariant action, two conditions must be met: (i) the field must be double traceless so that the Bianchi identity holds, and (ii) the gauge parameter must be traceless so that the third term vanishes. Now remember that for the Fronsdal field equations to be gauge invariant, it is enough to have a traceless parameter, while the double tracelessness for the field is needed to get the correct count of degrees of freedom. Now we see that the double tracelessness for the field is also needed for the invariance of the action.[11]

11 This nice discussion is from [274].

5.3 The triplet and minimal approaches

There are various approaches to Minkowski higher spin fields that aim to circumvent the awkward tracelessness and double tracelessness constraint on the gauge parameters and fields, respectively. There is a quite extensive literature on this subject.

5.3.1 Triplet higher spin fields

The name "triplet formulation" was coined in connection to the return to the BRST-approach to free higher spin (see Chapter 2).[12] It can be developed independent of BRST and it was indeed one of the clues – as was string field theory – to my own work on higher spin in the mid 1980s (see comments in Section 2.11). The basic idea is to consider the divergences $\partial \cdot \varphi$ and traces φ' that occur in the Fronsdal Lagrangian and field equations as independent fields, subsequently to be related to the conventional higher spin gauge fields φ through field equations. We will treat the field equations here, the Lagrangian to be reviewed in the section on the BRST approach.

Consider first spin 1 which is almost trivial. The following system of field equations is equivalent to the usual spin 1 equation:

$$\left.\begin{aligned} \Box\varphi_\mu - \partial_\mu H &= 0 \\ H - \partial\cdot\varphi &= 0 \end{aligned}\right\} \Rightarrow \Box\varphi_\mu - \partial_\mu \partial\cdot\varphi = 0 \tag{5.55}$$

The gauge transformations are $\delta\varphi_\mu = \partial_\mu\xi$ and $\delta H = \Box\xi$.

For spin 2, we need a further field C to play the role of the trace of $\varphi_{\mu\nu}$:

$$\left.\begin{aligned} \Box\varphi_{\mu\nu} - \tfrac{1}{2}(\partial_\mu H_\nu + \partial_\nu H_\mu) &= 0 \\ H_\mu - (2\partial\cdot\varphi_\mu + \partial_\mu C) &= 0 \\ C + \varphi' &= 0 \end{aligned}\right\} \Rightarrow \Box\varphi_{\mu\nu} - \partial_{(\mu}\partial\cdot\varphi_{\nu)} + \partial_\mu\partial_\nu\varphi' = 0 \tag{5.56}$$

The gauge transformations are $\delta\varphi_{\mu\nu} = \partial_{(\mu}\xi_{\nu)}$, $\delta H_\mu = 2\Box\xi_\mu$ and $\delta C = -2\partial\cdot\xi$.

For general spin s, we get (in condensed notation)

$$\left.\begin{aligned} \Box\varphi^{(s)} - \tfrac{1}{s}\partial^{(1}H^{s-1)} &= 0 \\ H^{(s-1)} - s\partial\cdot\varphi^{(s-1)} - \tfrac{1}{s-1}\partial^{(1}C^{s-2)} &= 0 \\ C^{(s-2)} + \tbinom{s}{2}\varphi'^{(s-2)} &= 0 \\ C'^{(s-4)} &= 0 \end{aligned}\right\} \Rightarrow \Box\varphi^{(s)} - \partial^{(1}\partial\cdot\varphi^{s-1)} + \partial^{(1}\partial^2\varphi'^{s-2)} = 0 \tag{5.57}$$

Double tracelessness of $\varphi^{(s)}$, again effective from spin 4 onwards, is enforced by the two last equations. It also follows by direct computation on the second equation that

12 The term "triplet" seems to have occurred in print first in [274, 275].

the auxiliary field $H^{(s-1)}$ is traceless. The corresponding gauge transformations work out to

$$\delta\varphi^{(s)} = \partial^{(1}\xi^{s-1)} \tag{5.58}$$

$$\delta H^{(s-1)} = s\Box\xi^{(s-1)} \tag{5.59}$$

$$\delta C^{(s-2)} = -s(s-1)\partial\cdot\xi^{(s-2)} \tag{5.60}$$

where the gauge parameter is traceless from spin 3 onwards. This requirement follows from demanding gauge invariance for the third and fourth equations of (5.57) which define the properties of the C field. The two first equations of (5.57) are actually gauge invariant even without requiring $\xi' = 0$. The two trace equations are not gauge invariant without requiring $\xi' = 0$.

So far we have just reversed engineered the Fronsdal equations. The number of independent field components are unchanged as the fields $H^{(s-1)}$ and $C^{(s-2)}$ are auxiliary and can be solved for, as is indeed done above. The trace constraints are still in force so this formulation should really be designated "constrained triplet formulation".

When we rederive these equations in the BRST approach, we will see that the equation for H comes out exactly as here, but C is an independent field.

5.3.2 Nonlocal minimal approach

The algebraic, trace constraints on gauge parameters and double trace constraints on fields – although it is fairly simple to understand why they appear from counting arguments – have always been considered at least awkward, if not downright mysterious, and there have consequently been several attempts to circumvent them.[13] The constrained triplet system discussed in Section 5.3.1 may be seen as one step toward such a goal. It works by introducing an extra field C which still suffers a trace constraint, although the constraint is "moved", so to speak, from the higher spin field φ itself to the auxiliary C. As we will discuss in connection with the BRST formulation (see Section 5.4.2), it is in fact possible to drop the trace constraint on C, and thus get an *unconstrained* but *nonminimal* model, nonminimal in the sense of introducing for each spin s a support of lower spin fields.

One approach that circumvents the trace constraints while not introducing extra fields is the nonlocal theory of D. Francia and A. Sagnotti [273]. As we saw in Section 5.2, the generalized Christoffel symbols of order $m = s$ are gauge invariant for

13 It could be argued, however, that it is the single trace constraints that are "foreign" to higher spin gauge fields. The double tracelessness constraint is an effect of Fronsdal's decision to work with a double traceless tensor $\varphi^{(s)}$ rather than with the two traceless tensors $\varphi^{(s)}$ and $\varphi^{(s-2)}$ inherited from the massive theory. In the massive theory, trace constraints are natural.

spin s. However, since they are of higher order in derivatives, their traces and divergences do not serve conveniently as components of equations of motion. Instead – in the Fronsdal approach – one generalizes the spin 2 equation of motion, which is based on the trace of the second-order Christoffel symbol $\Gamma_{\rho_1\rho_2;\mu_1\mu_2}$, to all higher spin. The price to pay for this is two-fold: (i) the gauge parameter must be traceless, and (ii) the gauge field itself must be double traceless in order to have a Bianchi identity.

The nonlocal approach generalizes the lower spin cases in another direction. The spin 1 field equation $\partial^{\rho_1} R_{\rho_1;\mu_1} = 0$ is generalized into

$$\frac{1}{\Box^n}\partial \cdot R^{[n]}{}_{;\mu_1\dots\mu_{2n+1}} = 0 \tag{5.61}$$

for odd spin $s = 2n + 1$. The spin 2 field equation $R'{}_{;\mu_1\mu_2} = 0$ is generalized into

$$\frac{1}{\Box^{n-1}} R^{[n]}{}_{;\mu_1\dots\mu_{2n}} = 0 \tag{5.62}$$

for even spin $s = 2n$.

We will study this approach by doing the first two nontrivial cases: spin 3 and spin 4 in the form of examples. For general spin, we refer to the original paper [273] and the review paper [274].

Example 10 (Spin 1 and 2). As a backdrop for the first nontrivial cases, let us record and comment on the formulas for the lower spin fields. Spin 1 is special in its simplicity. The first Christoffel symbol – the field strength – is also the curvature, and we have

$$\Gamma_{\rho_1;\mu_1} = R_{\rho_1;\mu_1} = \partial_{\rho_1}\phi_{\mu_1} - \partial_{\mu_1}\phi_{\rho_1} \tag{5.63}$$

$$\partial \cdot R_{;\mu_1} = \Box\phi_{\mu_1} - \partial_{\mu_1}\partial \cdot \phi = \mathcal{F}_{\mu_1} = 0 \tag{5.64}$$

The Maxwell equation can be written in another way, based on the second-order Christoffel symbol, which is not an entirely natural object for spin 1 but which can nevertheless be defined

$$\Gamma_{\rho_1\rho_2;\mu_1} = \partial_{\rho_1}\partial_{\rho_2}\phi_{\mu_1} - \frac{1}{2}\partial_{\mu_1}\partial_{(\rho_1}\phi_{\rho_2)} \tag{5.65}$$

$$\Gamma'{}_{;\mu_1} = \Box\phi_{\mu_1} - \partial_{\mu_1}\partial \cdot \phi = 0 \tag{5.66}$$

For spin 2, the field equations are naturally written in terms of the second-order Christoffel symbol (the "curvature")

$$\Gamma_{\rho_1\rho_2;\mu_1\mu_2} = \partial_{\rho_1}\partial_{\rho_2}\phi_{\mu_1} - \frac{1}{2}\partial_{(\mu_1}\partial_{(\rho_1}\phi_{\rho_2)\mu_2)} + \partial_{\mu_1}\partial_{\mu_2}\phi_{\rho_1\rho_2} \tag{5.67}$$

$$\Gamma'{}_{;\mu_1\mu_2} = R'{}_{;\mu_1\mu_2} = \Box\phi_{\mu_1\mu_2} - \partial_{(\mu_1}\partial \cdot \phi_{\mu_2)} + \partial_{\mu_1}\partial_{\mu_2}\phi' = 0 \tag{5.68}$$

In these equations, we can see the germs of the two directions of generalization to higher spin: In (5.66) and (5.68), the Fronsdal constrained equations, and in (5.64) and (5.68) the Francia–Sagnotti nonlocal equations. ◂

Example 11 (Spin 3). As we saw in formula (5.7) from spin 3 on, the Fronsdal tensor transforms into an expression involving the trace of the gauge parameter. For spin 3 in particular, $\mathcal{F}_{\mu_1\mu_2\mu_3}$ transforms into $3\partial_{\mu_1}\partial_{\mu_2}\partial_{\mu_3}\xi'$. As Francia and Sagnotti note, there are several nonlocal, higher derivative constructs that transform in the same way

$$\frac{1}{3\Box}\partial_{(\mu_1}\partial_{\mu_2}F'_{\mu_3)} \qquad \frac{1}{3\Box}\partial_{(\mu_1}\partial\cdot F_{\mu_2\mu_3)} \qquad \frac{1}{3\Box^2}\partial_{\mu_1}\partial_{\mu_2}\partial_{\mu_3}\partial\cdot F' \tag{5.69}$$

Of these, the first two terms are equal using the Bianchi identity, and taking traces all three constructs can be turned into each other. Based on these observations, Francia and Sagnotti introduce a second-order Fronsdal tensor

$$\mathcal{F}^{(2)}{}_{\mu_1\mu_2\mu_3} = \mathcal{F}_{\mu_1\mu_2\mu_3} + \frac{1}{6\Box}\partial_{(\mu_1}\partial_{\mu_2}\mathcal{F}'_{\mu_3)} - \frac{1}{2\Box}\partial_{(\mu_1}\partial\cdot\mathcal{F}_{\mu_2\mu_3)} \tag{5.70}$$

which is gauge invariant without imposing any constraint on the gauge parameter. Working it out, we get the field equation

$$\begin{aligned}\mathcal{F}^{(2)}{}_{\mu_1\mu_2\mu_3} &= \Box\phi_{\mu_1\mu_2\mu_3} - \partial_{(\mu_1}\partial\cdot\phi_{\mu_2\mu_3)} + \frac{1}{3}\partial_{(\mu_1}\partial_{\mu_2}\phi'_{\mu_3)}\\ &\quad + \frac{2}{3\Box}\partial_{(\mu_1}\partial_{\mu_2}\partial\cdot\partial\cdot\phi_{\mu_3)} - \frac{1}{\Box}\partial_{\mu_1}\partial_{\mu_2}\partial_{\mu_3}\partial\cdot\phi'\\ &= \frac{1}{\Box}\partial\cdot R'_{\mu_1\mu_2\mu_3} = 0\end{aligned} \tag{5.71}$$

The $\mathcal{F}^{(2)}{}_{\mu_1\mu_2\mu_3}$ tensor satisfies a Bianchi-type identity

$$\partial\cdot\mathcal{F}^{(2)}{}_{\mu_2\mu_3} - \frac{1}{4}\partial_{(\mu_2}\mathcal{F}^{(2)\prime}{}_{\mu_3)} = 0 \tag{5.72}$$

We can now count degrees of freedom. We first do the accounting of equations and field components. There are 20 field equations in (5.71) of which only 10 are differentially independent due to the 10 Bianchi identities (5.72). Thus from the 20 spin 3 field components, 10 are undetermined and this corresponds precisely to the 10 components of the gauge parameter.

On the other hand, the gauge fixing count is a little bit subtler. It is clearly too naive to say that we have 20 field components and 10 gauge parameters, so that $20-2\cdot 10 = 0$ degrees of freedom remains.

Instead, referring back to Section 5.1.1, we find that deDonder gauge fixing condition is still traceless (whereas the gauge parameter has a nonzero trace). We can therefore only gauge fix 9 field components. Furthermore, the gauge variation of the gauge condition is $\delta D_{\mu_2\mu_3} = \Box\xi_{\mu_2\mu_3} - \partial\mu_2\partial_{\mu_3}\xi'$. So in order to be able to regauge field components using a gauge parameter that satisfies the d'Alembertian equation, we must make the further gauge choice $\xi' = 0$. ◂

Example 12 (Spin 4). For spin 4, one again considers the second-order Fronsdal tensor

$$\mathcal{F}^{(2)}{}_{\mu_1\mu_2\mu_3\mu_4} = \mathcal{F}_{\mu_1\mu_2\mu_3\mu_4} + \frac{1}{6\Box}\partial_{(\mu_1}\partial_{\mu_2}\mathcal{F}'_{\mu_3\mu_4)} - \frac{1}{2\Box}\partial_{(\mu_1}\partial\cdot\mathcal{F}_{\mu_2\mu_3\mu_4)} \tag{5.73}$$

This is the same formula as for spin 3. It is gauge invariant without imposing any constraint on the gauge parameter. Working it out, we get the field equation

$$\begin{aligned}\mathcal{F}^{(2)}{}_{\mu_1\mu_2\mu_3\mu_4} &= \Box\phi_{\mu_1\mu_2\mu_3\mu_4} - \partial_{(\mu_1}\partial\cdot\phi_{\mu_2\mu_3\mu_4)} + \frac{1}{3}\partial_{(\mu_1}\partial_{\mu_2}\phi'_{\mu_3\mu_4)}\\ &\quad + \frac{2}{3\Box}\partial_{(\mu_1}\partial_{\mu_2}\partial\cdot\partial\cdot\phi_{\mu_3\mu_4)} - \frac{1}{\Box}\partial_{\mu_1}\partial_{\mu_2}\partial_{\mu_3}\partial\cdot\phi'_{\mu_4} + \partial_{\mu_1}\partial_{\mu_2}\partial_{\mu_3}\partial_{\mu_4}\phi''\\ &= \frac{1}{\Box}R''{}_{\mu_1\mu_2\mu_3\mu_4} = 0\end{aligned} \tag{5.74}$$

The gauge-fixing count for spin 4 and higher is quite complicated. We refer the reader to [274] for the details. ◂

Simple and nice as the formulas (5.61) and (5.62) look, they turn out not to be the entirely correct as was later clarified in [276]. There are ambiguities in the choice of nonlocal terms. A unique form is fixed by requiring the correct coupling $\varphi\cdot J$ to external currents J so that the correct number of degrees of freedom is exchanged. The same result is arrived at through the compensator approach derived from the BRST triplet formulation.

5.3.3 A note on *N*-complexes

The gauge invariance of the higher spin curvatures $R = \Gamma^{(s)}$ can be formalized in terms of so called N-complexes [277] (for reviews and further references, see [278, 279]). The complexes and differentials d introduced in this approach have the property $d^n = 0$ for some positive integer n. They can be used to define the curvatures as $R = d^s\phi = d^{n-1}\phi$. Consequently, under a gauge transformation $\delta\phi = d\xi$, gauge invariance is automatic, since $\delta R = d^{n-1}d\xi = 0$.

Without going into the details of the construction, it is interesting to take a glance at one of its overall features. For spin 1, we have (as usual) the sequence of spaces

$$\Omega_0 \xrightarrow{d} \Omega_1 \xrightarrow{d} \Omega_2 \xrightarrow{d} \Omega_3 \tag{5.75}$$

where Ω_0 is the space of gauge parameters, Ω_1 is the space of gauge fields (a subspace of which are the pure gauge potentials $d\Omega_0$) and Ω_2 is the space of field strengths (curvatures). The identity $d^2 = 0$ ensures that the field strengths do not see the pure gauge potentials. Finally, the space Ω_3 is the space of Bianchi identities.

For spin $s \geq 2$, the corresponding sequence turns out to be

$$\Omega_{s-1} \xrightarrow{d} \Omega_s \xrightarrow{d^s} \Omega_{2s} \xrightarrow{d} \Omega_{2s+1} \tag{5.76}$$

with the analogous (but not identical) interpretation of the spaces Ω_{s-1}, Ω_s, Ω_{2s} and Ω_{2s+1} as gauge parameters, gauge fields, curvatures and Bianchi identities, respectively. It seems that the intriguing point here is the "jumping over" the intermediate Christoffel symbols Γ^m with $2 < m < s$.

5.4 BRST approach to the free theory

In a BRST approach to free higher spin gauge fields, the gauge transformations are generated by the first class constraints of an underlying mechanics model, while the trace conditions can be imposed through second-class constraints. This works well for the free theory and reproduces the Fronsdal theory.

In this section, we will review the approach of [159]. That paper worked with an infinite set of uncoupled harmonic oscillators as appropriate to the tensionless limit of bosonic string theory. The theory therefore contains arbitrary mixed symmetry fields, as is also the case for [158]. Here, we will simplify and just consider one oscillator so that there is just one field of each spin from 0 to ∞.

As told in the historical chapter (see Section 2.11), the BRST method was borrowed from string field theory, but logically it is independent from string theory. Indeed, expanding a spectrum of higher spin over some "internal" variable is a natural thing to do, and the method to express the BRST operator directly in terms of the first-class constraints of some underlying mechanical model, is general.[14] We will see that the BRST approach is in a certain sense the most fundamental; all other formulations can be derived from it, or naturally related to it.

5.4.1 A mechanical model

Underlying string field theory there is a mechanical model, namely the relativistic string.[15] Being a one-dimensional object, the string sweeps out a two-dimensional surface – the world sheet – as it moves in space-time. The string action has reparametrization invariance in the world-sheet coordinates. The string can be viewed as mapping a two-dimensional Lorentzian surface with coordinates (σ, τ) into space-time $x^\mu(\sigma, \tau)$.

14 Many authors have written on the subject of BRST approaches to free massless higher spin theory, both in Minkowski and AdS space-time. In addition to papers cited in Section 2.11, there are [280–285, 241]. For references to massive fields, see the introduction to [282].

15 To be specific, we think of the bosonic string.

The reparametrization invariance is a gauge symmetry and therefore it can be viewed as being generated by first-class constraints of the mechanical string model.[16] These constraints are precisely the Virasoro constraints, obeying the Virasoro first-class constraint algebra. By introducing ghost coordinates corresponding to the Virasoro constraints, the free string field theory can be treated using BRST techniques (for original references, see Section 2.11.1 and the review [181]).

The question now arises: can anything similar be done for higher spin gauge fields? That is, is there any, or perhaps several possible, underlying mechanical models with concomitant reparametrization symmetries and first-class constraints that can be used to set up a field theory? Strangely enough, this very simple question has not been pursued in the literature to any considerable extent. We will return to this topic in more detail in the Volume 2.

Here, we will take as a starting point a very simple model [287, 288]. We start with a classical (or first-quantized) two-particle relativistic mechanical system with centre of motion (x_μ, p_ν) and relative (ξ_μ, π_ν) coordinates and momenta. We do not specify any action, instead working directly from the constraints. For the relative coordinates, we also use holomorphic coordinates classically, or oscillators $(\alpha_\mu, \alpha_\nu^\dagger)$ quantum mechanically. In terms of the relative coordinates and momenta we have

$$\alpha_\mu = \frac{1}{\sqrt{2}}(\xi_\mu + i\pi_\mu) \quad \text{and} \quad \alpha_\mu^\dagger = \frac{1}{\sqrt{2}}(\xi_\mu - i\pi_\mu) \tag{5.77}$$

We take ξ_μ and π_ν to be dimensionless. Classically, we have Poisson brackets

$$\{x_\mu, p_\nu\} = \eta_{\mu\nu} \quad \text{and} \quad \{\xi_\mu, \pi_\nu\} = \eta_{\mu\nu} \tag{5.78}$$

and quantum mechanically

$$[x_\mu, p_\nu] = i\eta_{\mu\nu} \qquad [\xi_\mu, \pi_\nu] = i\eta_{\mu\nu} \qquad [\alpha_\mu, \alpha_\nu^\dagger] = \eta_{\mu\nu} \tag{5.79}$$

Excluding explicit occurrence of the center of motion coordinate x_μ there are six bilinear scalars in terms of these variables

$$G_0 = -p^2 \qquad G_+ = \alpha \cdot p \qquad G_- = \alpha^\dagger \cdot p \tag{5.80}$$

$$T = \frac{1}{2}\alpha \cdot \alpha \qquad T^\dagger = \frac{1}{2}\alpha^\dagger \cdot \alpha^\dagger \qquad N = \frac{1}{2}(\alpha \cdot \alpha^\dagger + \alpha^\dagger \cdot \alpha) = \alpha^\dagger \cdot \alpha + 2 \tag{5.81}$$

From this set, we can choose various linear combinations as first- and second-class constraints by (weakly) equating to zero. Once such a choice is made, ghost coordinates and momenta can be introduced corresponding to the first-class set and the BRST operator Q constructed. Then a free field theory can be set up using BRST techniques.

16 See, for instance, Chapter 2 of [286].

The standard choice is to take the set $\{G_0 = 0, G_+ = 0, G_- = 0\}$ as first class. The algebra of first-class constraints then becomes

$$[G_+, G_-] = -G_0 \qquad [G_+, G_0] = 0 \qquad [G_-, G_0] = 0 \tag{5.82}$$

The operators T, $T^\dagger$ and N span an $\mathfrak{su}(1,1)$ algebra

$$[T, T^\dagger] = N \qquad [N, T^\dagger] = 2T^\dagger \qquad [N, T] = -2T \tag{5.83}$$

The tracelessness constraints (on fields and parameters) are given by $T|\text{state}\rangle = 0$ with the T operator augmented with a ghost contribution. The two operators T and $T^\dagger$ can be regarded as a pair of second-class constraints since we do not require any constraint $N = c$ for some constant c. Doing that would fix the spin to a specific value and we would not have a tower of higher spin fields.

5.4.2 A unified action for integer spin gauge fields

The aim is to collect all the Fronsdal actions for individual higher spin gauge fields into one Lagrangian $\frac{1}{2}\langle\Phi|Q|\Phi\rangle$ with $|\Phi\rangle$ an object that for every spin s contains the triplett fields $\varphi^{(s)}$, $H^{(s-1)}$ and $C^{(s-2)}$ of the reversed engineered Fronsdal theory of Section 5.3.1. Let us review the result of such an endeavor.

The higher spin fields $\varphi^{(s)}$ will be coefficients in an expansion over the Fock space spanned by the creators $\alpha^\dagger_\mu$

$$|\varphi\rangle = |(\varphi_0 + \varphi^\mu \alpha^\dagger_\mu + \varphi^{\mu\nu}\alpha^\dagger_\mu \alpha^\dagger_\nu + \cdots)|\text{vac}\rangle \tag{5.84}$$

acting on a vacuum state $|\text{vac}\rangle$ yet to be specified. The exact coefficients in the expansion must also be fixed in order to get a real nontrivial action. We will do that in Section 5.4.3.

Corresponding to the first-class constraints $G_0 = 0$, $G_+ = 0$ and $G_- = 0$ chosen above, there will a Grassmann ghost variables (c^0, c^+, c^-) with conjugates (b_0, b_+, b_-). The nonzero anticommutators are

$$\{c^0, b_0\} = \{c^+, b_+\} = \{c^-, b_-\} = 1 \tag{5.85}$$

These ghosts satisfy the following Hermitian conjugation properties:

$$(c^-)^\dagger = c^+ \qquad (b_-)^\dagger = b_+ \qquad (c^0)^\dagger = c^0 \qquad (b_0)^\dagger = b_0 \tag{5.86}$$

This ensures that the BRST operator (in momentum space)

$$\begin{aligned} Q &= c^0 G_0 + c^+ G_+ + c^- G_- + c^+ c^- b_0 \\ &= -c^0 p^2 + c^+ \alpha \cdot p + c^- \alpha^\dagger \cdot p + c^+ c^- b_0 \end{aligned} \tag{5.87}$$

is Hermitian. The BRST operator is constructed according to the general algorithm described in Section 3.3.3. The last term in the BRST operator comes from the only nonzero structure constant of the constraint algebra (5.82). The BRST operator is therefore nilpotent with $Q^2 = \frac{1}{2}\{Q, Q\} = 0$.[17]

The self-conjugate "zero-mode" pair (c^0, b_0) require a little bit of care with respect to the vacuum. Let $|-\rangle$ denote a vacuum, defined by $c^0|-\rangle = 0$.[18] Then the conjugate ghost b_0 creates a new state $|+\rangle = b_0|-\rangle$. Then we can just as well think of $|+\rangle$ as a vacuum annihilated by b_0 with respect to which the state $|-\rangle$ is given by $|-\rangle = c^0|+\rangle$ since $c^0|+\rangle = c^0 b_0|-\rangle = \{c^0, b_0\}|-\rangle = |-\rangle$. Since any vacuum state $|0\rangle$ ought to be Hermitian in the sense $(\langle 0|)^\dagger = |0\rangle$, we now see that $\langle +|+\rangle = \langle -|-\rangle = 0$ whereas $\langle +|-\rangle = \langle +|-\rangle = 1$. For the rest of the ghosts, we choose c^+ and b_- to be creators. The following formulas collect the properties of the vacua $|-\rangle$ and $|+\rangle$:

$$c^-|+\rangle = c^-|-\rangle = b_+|+\rangle = b_+|-\rangle = 0 \tag{5.88}$$

$$c^0|-\rangle = b_0|+\rangle = 0 \qquad b_0|-\rangle = |+\rangle \qquad c^0|+\rangle = |-\rangle \tag{5.89}$$

$$\langle +|-\rangle = \langle -|+\rangle = 1 \qquad \langle +|+\rangle = \langle -|-\rangle = 0 \tag{5.90}$$

Then we must assign mechanical *ghost numbers* $\mathrm{gh_m}(\cdot)$ to all objects in the theory. These are collected in Table 5.2 along with data on Grassmann parity $\rho(\cdot)$ and mass dimension $d(\cdot)$. We also collect the structure of the ghost complex in Table 5.3.

Table 5.2: Properties of objects.

Properties/Objects		$\alpha_\mu, \alpha_\mu^\dagger$	p_μ	c^+, c^-	b_+, b_-	c^0	b_0	$\|+\rangle$	$\|-\rangle$
Mechanical ghost number	$\mathrm{gh_m}(\cdot)$	0	0	1	−1	1	−1	−1/2	1/2
Grassman parity	$\rho(\cdot)$	0	0	1	1	1	1	0	1
Dimension	$d(\cdot)$	0	1	1	−1	0	0	0	0

Table 5.3: Ghost complex structure.

$\mathrm{gh_m}(\cdot)$	**3/2**	**1/2**	**−1/2**	**−3/2**
		$\|-\rangle$	$\|+\rangle$	
	$c^+\|-\rangle$	$c^+\|+\rangle$	$c^+b_-\|+\rangle$	$b_-\|+\rangle$
		$c^+b_-\|-\rangle$	$b_-\|-\rangle$	

17 There is an entertaining sign ambiguity in Q. Writing $c^0G_0 + \sigma c^+G_+ + \sigma^* c^- G_- + c^+c^-b_0$ with a complex parameter σ and demanding Q to be formally Hermitian ($Q^\dagger = Q$) one finds that nilpotency only requires $\sigma = \pm 1$ or $\sigma = \pm i$. This reflects the possibility to work in configuration space or momentum space.

18 This is one possible choice. The issue was discussed in [166].

We can expand the higher spin fields and the auxiliary fields over the ghost complex as

$$|\Phi\rangle = (\varphi + Cc^{+}b_{-} + Hb_{-}c^{0})|+\rangle \tag{5.91}$$

where the fields φ, H and C themselves are expansions over the oscillators as in formula (5.84). The gauge parameters are expanded as

$$|\Xi\rangle = \xi b_{-}|+\rangle \tag{5.92}$$

All component fields and gauge parameters are Grassmann even as well as Hermitian.

Finally, we must enhance the second class constraints T and $T^{\dagger}$ with ghost contributions

$$T = \frac{1}{2}\alpha \cdot \alpha + b_{+}c^{-} \qquad T^{\dagger} = \frac{1}{2}\alpha^{\dagger} \cdot \alpha^{\dagger} + c^{+}b_{-} \tag{5.93}$$

so that $[T, Q] = [T^{\dagger}, Q] = 0$.

We now have everything needed to write down the Lagrangian, field equations and gauge transformations under which the Lagrangian, as well as the field equations are invariant.

$$\mathcal{L} = \frac{1}{2}\langle\Phi|Q|\Phi\rangle \tag{5.94}$$

$$Q|\Phi\rangle = 0 \tag{5.95}$$

$$\delta|\Phi\rangle = Q|\Xi\rangle \tag{5.96}$$

The content of the field equations (5.95) can be made more explicit as

$$(G_0\varphi + G_{-}H)|-\rangle = 0 \tag{5.97}$$

$$(G_{+}\varphi - G_{-}C - H)c^{+}|+\rangle = 0 \tag{5.98}$$

$$(G_0 C + G_{+}H)c^{+}b_{-}|-\rangle = 0 \tag{5.99}$$

These equations have become known as the *triplet equations* [274, 289]. The ghosts no longer play any role, except showing that the equations sit in the three levels of the ghost number 1/2 sector of the theory (see Table 5.3). It is however convenient to have this form of the field equations when computing the action. Likewise, the gauge transformations for the component fields can be written as

$$\delta\varphi|+\rangle = G_{-}\xi|+\rangle \tag{5.100}$$

$$\delta Hb_{-}|-\rangle = -G_0\xi b_{-}|-\rangle \tag{5.101}$$

$$\delta Cc^{+}b_{-}|+\rangle = G_{+}\xi c^{+}b_{-}|+\rangle \tag{5.102}$$

Again, the ghosts play no other role here than to show that the gauge transformations sit in the ghost number −1/2 sector. Gauge invariance of the field equations under these transformations is equivalent to the constraint algebra. No trace constraints are needed for the gauge parameters and the component fields need not be double traceless. Thus the BRST formulation is *unconstrained* in this sense.

To make contact with the Fronsdal theory, they should be supplied by the trace constraint

$$T|\Phi\rangle = 0 \Rightarrow \begin{cases} (C - \frac{1}{2}\alpha \cdot \alpha\varphi)|+\rangle = 0 \\ \alpha \cdot \alpha C c^{+} b_{-}|+\rangle = 0 \\ \alpha \cdot \alpha H b_{-}|-\rangle = 0 \end{cases} \tag{5.103}$$

These equations imply that the component auxiliary fields H are traceless, but more importantly that[19]

$$C_{\mu_3 \dots \mu_s} = -\binom{s}{2}\varphi'_{\mu_3 \dots \mu_s} \quad \text{and} \quad \varphi''_{\mu_5 \dots \mu_s} = 0 \tag{5.104}$$

We now turn to the question of extracting the Fronsdal equations. The second field equation (5.98) can be solved algebraically for the auxiliary field H. Doing that, and inserting the result in the first field equation (5.97) also using the trace equation $C = \frac{1}{2}\alpha \cdot \alpha\varphi$ yields

$$\begin{aligned} &\left(G_0\varphi + \frac{1}{2}G_- G_+ \varphi - \frac{1}{4}G_- G_- \alpha \cdot \alpha\varphi\right)|+\rangle \\ &= \frac{1}{2}\left(p^2\varphi + \alpha^\dagger \cdot p\,\alpha \cdot p\varphi - \frac{1}{2}\alpha^\dagger \cdot p\,\alpha^\dagger \cdot p\,\alpha \cdot \alpha\varphi\right)|+\rangle 0 \end{aligned} \tag{5.105}$$

With a bit of imagination, one can already here discern the Fronsdal field equations. Indeed, using the component expansion of φ from formula (5.84) and doing the oscillator algebra, one gets precisely

$$p^2\varphi_{\mu_1 \dots \mu_s} - p_{(\mu_1} p \cdot \varphi_{\mu_2 \dots \mu_s)} + p_{(\mu_1} p_{\mu_2} \varphi'_{\mu_3 \dots \mu_s)} = 0 \tag{5.106}$$

which squares nicely with (5.57). However, as compared to our reverse engineering of the Fronsdal equations, we see that we get more equations in the BRST approach. This can be clarified in two ways. Computing certain traces and divergences of the reversed engineered equations one can see that they actually contain the extra field equation for the C field implicitly. On the other hand, the existence of the extra field equation (5.99) can be understood from the action yielding the equations. So let us turn to this.

5.4.3 Expansion of the action

In formula (5.94), we wrote the Lagrangian as $\frac{1}{2}\langle\Phi|Q|\Phi\rangle$ since we wanted to keep open the question of whether we worked in configuration space or momentum space. It is

19 The minus sign will be explained below. See comment after formula (5.112).

also very common, and convenient, in BRST approaches (as is indeed the cases in its string theory origins) to include the space-time (or momentum-energy) integrals in $\langle | \, | \rangle$. Let us investigate this question as it will prompt us to elaborate on the oscillator expansion of the fields as in formula (5.84).

A note on the formalism and momentum space vs. configuration space

i

It is convenient to write the action and field equations abstractly as in (5.94) and (5.95). It allows us to switch between momentum space and configuration space representations quickly. With $\partial_\mu = ip_\mu$, we have $G_0 = -p^2 = \Box$, $G_+ = \alpha \cdot p = -i\alpha \cdot \partial$ and $G_- = \alpha^\dagger \cdot p = -i\alpha^\dagger \cdot \partial$. The formalism is a hybrid between a first-quantized mechanical model and classical fields, that is, we treat the oscillators and ghosts as q-numbers whereas fields are classical on c-number space-time x^μ or on momentum-energy p_μ. The correspondence $p_\mu = -i\partial_\mu$ is an effect of a Fourier transform, and p_μ is an eigenvalue of the momentum operator. This boils down to the following correspondence:

$$S = -\frac{1}{2}\int d^4x \partial_\mu \varphi(x)\partial^\mu \varphi(x) = \frac{1}{2}\int d^4p \varphi(-p)p_\mu p^\mu \varphi(p) \tag{5.107}$$

and correspondingly for other kinds of kinetic terms in the action.

Now including either a space-time or an energy-momentum integral in the inner product, and using (5.97)–(5.99), we find

$$\begin{aligned} S &= \frac{1}{2}\langle \Phi | Q | \Phi \rangle \\ &= \frac{1}{2}\langle - | (\varphi G_0 \varphi - C G_0 C - H^2 + H G_+ \varphi + \varphi G_- H - H G_- C - C G_+ H) | + \rangle \end{aligned} \tag{5.108}$$

It is clear that the first three terms are diagonal in oscillators, but that for the last four terms there must be an offset by one oscillator in the fields in order to get a nonzero result when the oscillator in G_- and G_+ are taken into account. Take, for instance, the fourth and fifth term

$$\begin{aligned} &\langle - | \int d^4x (H(x)(-i\alpha \cdot \partial)\varphi(x) + \varphi(x)(-i\alpha^\dagger \cdot \partial)H(x)) | + \rangle \\ &= -i \sum_{n=1} \langle - | \int d^4x (H^{(n-1)} \cdot \alpha^{(n-1)} \alpha \cdot \partial \varphi^{(n)} \cdot \alpha^{\dagger(n)} \\ &\qquad + \varphi^{(n)} \cdot \alpha^{(n)} \alpha^\dagger \cdot \partial H^{(n-1)} \cdot \alpha^{\dagger(n-1)}) | + \rangle \end{aligned} \tag{5.109}$$

There are two problems with this expression: it is a total derivative and it is not real. Both problems are solved by a simple device: in the field expansion of formula (5.84) for the ket field $|\varphi\rangle$, we make a field redefinition $\varphi^{(n)} \to i^n \varphi^{(n)}$ and similarly for and $H^{(n)}$ and $F^{(n)}$. Correspondingly, in the bra field $\langle \varphi |$, we get $(\varphi^{(n)})^\dagger \to (-i)^n \varphi^{(n)}$. The effect of this is to leave the diagonal terms in the action (5.108) unchanged, while we get a crucial factor of i in the terms (5.109). The terms become

$$
\begin{aligned}
-i\sum_{n=1}\langle -|\int d^4x((-i)^{n-1}i^nH^{(n-1)}\cdot\alpha^{(n-1)}\alpha\cdot\partial\varphi^{(n)}\cdot\alpha^{\dagger(n)} \\
+(-i)^ni^{n-1}\varphi^{(n)}\cdot\alpha^{(n)}\alpha^{\dagger}\cdot\partial H^{(n-1)}\cdot\alpha^{\dagger(n-1)})|+\rangle \\
=\sum_{n=1}\langle -|\int d^4x(H^{(n-1)}\cdot\alpha^{(n-1)}\alpha\cdot\partial\varphi^{(n)}\cdot\alpha^{\dagger(n)} \\
-\varphi^{(n)}\cdot\alpha^{(n)}\alpha^{\dagger}\cdot\partial H^{(n-1)}\cdot\alpha^{\dagger(n-1)})|+\rangle \\
=2\sum_{n=1}n!\int d^4xH^{(n-1)}\cdot(\partial\cdot\varphi^{(n-1)})
\end{aligned}
\tag{5.110}
$$

The last two terms in the action (5.108) are computed in the same way. All in all, the action becomes

$$
\begin{aligned}
S=\frac{n!}{2}\sum_{n=1}\int d^4x\Big(\varphi^{(n)}\Box\varphi^{(n)}-\frac{1}{n(n-1)}C^{(n-2)}\Box C^{(n-2)}-\frac{1}{n}H^2 \\
+2H^{(n-1)}\cdot(\partial\cdot\varphi^{(n-1)})-\frac{2}{n}C^{(n-2)}\cdot(\partial\cdot H^{(n-2)})\Big)
\end{aligned}
\tag{5.111}
$$

From this action, we can derive the Fronsdal action by first by using the algebraic field equation for H

$$
H^{(n-1)}=n\partial\cdot\varphi^{(n-1)}+\frac{1}{n-1}\partial^{(1}C^{n-2)}
\tag{5.112}
$$

and then the trace condition $C^{(n-2)}=-\binom{n}{2}\varphi'^{(2-n)}$ from equation (5.104). The minus sign comes from the field redefinition $\varphi^{(n)}\to i^n\varphi^{(n)}$ discussed above. We get precisely the Lagrangian of (5.25) with the inessential factor $n!$. This numerical factor can be absorbed by a further trivial field redefinition $\varphi\to\frac{1}{\sqrt{n}}\varphi$.

Unconstrained BRST triplet equations

For the record, and clarification, let us write down the triplet system of field equations that result from working out the component equations (5.97)–(5.99). The field equations are

$$
\Box\varphi^{(n)}-\frac{1}{n}\partial^{(1}H^{n-1)}=0
\tag{5.113}
$$

$$
H^{(n-1)}-n\partial\cdot\varphi^{(n-1)}-\frac{1}{n-1}\partial^{(1}C^{n-2)}=0
\tag{5.114}
$$

$$
\Box C^{(n-2)}+(n-1)\partial\cdot H^{(n-2)}=0
\tag{5.115}
$$

The gauge transformations that follow from (5.100)–(5.102) are

$$
\delta\varphi^{(n)}=-\frac{1}{n}\partial^{(1}\xi^{n-1)}
\tag{5.116}
$$

$$\delta H^{(n-1)} = -\Box\xi^{(n-1)} \tag{5.117}$$

$$\delta C^{(n-2)} = (n-1)\partial\cdot\xi^{(n-2)} \tag{5.118}$$

To get complete agreement with the reversed engineered gauge transformations of equations (5.58)–(5.60), we make a parameter redefinition $\xi^{(n-1)} \to -n\xi^{(n-1)}$.

Now one can readily check gauge invariance. We note that no trace constraints are needed on the gauge parameters. This set of higher spin field equations may thus be termed *unconstrained*.

Note, and this is important, that nothing at all is claimed about traces on fields, field equations or parameters. In particular, C is not the trace of φ. Furthermore, and consequently, H is not traceless. So although the equations may look the same, there are important differences between the reversed engineered equations and the equations derived from the BRST approach. This will be further investigated in the next question box.

Since there are no trace constraints on the fields and parameters, the degree of freedom count is simple. The H field is auxiliary and carry no independent degrees of freedom. The φ and the C fields carry together $\binom{s+3}{s} + \binom{s+1}{3}$ d. o. f.. From this, we subtract gauge and regauge components equal $2\binom{s+2}{3}$ to arrive at $s+1$. Thus we have a spectrum of fields with spins s, $s-2$, $s-3$ down to spin 1 or spin 0. That is, summing $2+2+2+\cdots+1$ or $2+2+2+\cdots+0$ yield $s+1$.

Unconstrained reversed engineered triplet equations?

?

In the light of the above understanding of the unconstrained triplet equations, it may be interesting to ask if one could have arrived at these equations from the reversed engineered Fronsdal equations?

Clearly, one can derive the reversed engineered equations from the BRST equations by imposing the trace constraints $H'^{(n-2)} = 0$, $C^{(n-2)} = -\binom{n}{2}\varphi'^{(n-2)}$, $C'^{(n-4)} = 0$ and $\xi'^{(n-3)} = 0$. The field equation for C then becomes the trace of the field equation for φ, and it can be dropped. We then have the reversed engineered equations of Section 5.3.1.

The inverse problem, finding the unconstrained formulation from the Fronsdal reversed engineered formulation, can be approached in the following way. As a first step, note that the field equations for φ and H are the same in both formulations, as are the gauge transformations. These field equations are therefore gauge invariant without assuming a traceless parameter.

As a second step, drop the trace conditions $C^{(n-2)} = -\binom{n}{2}\varphi'^{(n-2)}$ and $C'^{(n-4)} = 0$. Then tracelessness of H then no longer follows (as it should not).

Now we have no equation for C. As third step, guided by observations done above, one may compute the trace of the field equation for $\varphi^{(n)}$. One then gets an equation that looks like the BRST equation for $C^{(n-2)}$ if one interprets $\varphi'^{(n-2)}$ as $-\binom{n}{2}^{-1}C^{(n-2)}$ and takes $H^{(n-1)}$ traceless. This is clearly a dubious procedure: first dropping the trace conditions on the fields, then using them anyway as hints to the correct unconstrained triplet equations. This illogical procedure is instead a reflection of the consistency of introducing trace constraints in the BRST formulation. While the procedure is plausible as a "method of discovery", it does not make much sense as a "method of justification".

Another approach to the inverse problem, that seems not to have been explored in the literature, is to modify the gauge transformations with terms involving the trace of the gauge parameter.

5.4.4 Zero-tension limit of the Virasoro algebra

The zero-tension limit of the Virasoro algebra can be performed as follows (as done in [159]). Take as the basic mass-shell Virasoro generator

$$L_0 = -\frac{1}{2}\Box + \frac{1}{\alpha'}\sum_{n>0} \alpha_{-n} \cdot \alpha_n - \frac{1}{\alpha'} \tag{5.119}$$

where α' (of mass dimension −2) have been reinserted to make the formula dimensionally correct. Since $[L_m, L_n] \sim L_0$, the Virasoro generator L_m have dimension 1 and we find

$$L_m = i\alpha \cdot \partial + \frac{1}{\sqrt{\alpha'}}\sum_{n>0} \alpha_n \cdot \alpha_{m-n} \tag{5.120}$$

The algebra then reads (without the central extension term)

$$\begin{aligned}
[L_m, L_{-m}] &= 2mL_0 \\
[L_0, L_m] &= -(1/\alpha')mL_m \\
[L_m, L_n] &= (1/\sqrt{\alpha'})(m-n)L_{m-n} \quad \text{with} \quad m, n \neq 0
\end{aligned} \tag{5.121}$$

In the limit $\alpha' \to \infty$ these generators and their algebra turns into a higher spin algebra with an infinite number of oscillators. Truncation to a finite number of oscillators can be done.

5.5 "Minimal" and the other approaches

Since Wigner, and as explicated by Weinberg and others (as noted in the historical chapter) we know what particles – representations of the Poincaré group – there can be in a special relativistic theory. As we have also seen, such particles, in particular higher spin particles, can be represented in many reasonable ways as covariant field theories. We have seen the dichotomies: reducible/irreducible, constrained/unconstrained, minimal/nonminimal and local/nonlocal. These dichotomies are furthermore related to each other, and the ensuing picture may appear quite confusing. It turns out, as we have already alluded to, that they are all derivative of the BRST-approach. In this section, we will analyze this situation in some more detail.

5.5.1 Compensator minimal approach

There are interesting relations between the two unconstrained formulations (nonlocal minimal and BRST-triplet) and a third one, the *compensator minimal approach*, that we will now introduce and study. The original paper is [275] by A. Sagnotti and M. Tsulaia. The theory is further developed in [289] by Francia and Sagnotti. A review can be found in [290].

A starting point is the BRST triplet system. The auxiliary field equation for the H field is used to substitute for it everywhere. What then results is a *doublet system*[20] (φ, C) with field equations, gauge transformations and a Lagrangian that can be computed from the BRST system. The field equations become

$$\Box\varphi^{(s)} - \partial^{(1}\partial\cdot\varphi^{s-1} - \binom{s}{2}^{-1}\partial^{(1}\partial^{2}C^{s-2)} = 0 \tag{5.122}$$

$$\Box C^{(s-2)} + \binom{s}{2}\partial\cdot\partial\cdot\varphi^{(s-2)} + \frac{1}{2}\partial^{(1}\partial\cdot C^{s-3)} = 0 \tag{5.123}$$

This system – still unconstrained – reduces to the Fronsdal system upon imposing the trace conditions (5.104). The first equation becomes the Fronsdal field equation and the second equation is then the trace of the first. One may ask if imposing the trace constraints can be viewed as a gauge choice? This is indeed the case as can be surmised by checking how $C^{(s-2)} + \binom{s}{2}\varphi'^{(s-2)}$ transforms. One then finds

$$\delta(C^{(s-2)} + \binom{s}{2}\varphi'^{(s-2)}) = \binom{s}{2}\partial^{(1}\xi'^{s-3)} \tag{5.124}$$

This suggests using the trace of the gauge parameter to gauge $C^{(s-2)} + \binom{s}{2}\varphi'^{(s-2)}$ to zero. We will find that it is natural to define a new field $\alpha^{(s-3)}$ defined through

$$\partial^{(1}\alpha^{s-3)} = \varphi'^{(s-2)} + \binom{s}{2}^{-1}C^{(s-2)} \tag{5.125}$$

transforming as

$$\delta\alpha^{(s-3)} = \xi'^{(s-3)} \tag{5.126}$$

The field α is called a spin $s-3$ *compensator* in [275]. Equation (5.125) can also be viewed as imposing the trace constraint only up to a pure gauge.

To proceed with the theory, let us return to the field equation (5.122) and write it in terms of the Fronsdal tensor as

$$\mathcal{F}^{(s)} = \partial^{(1}\partial^{2}\varphi'^{s-2)} + \binom{s}{2}^{-1}\partial^{(1}\partial^{2}C^{s-2)} \tag{5.127}$$

This again suggests introducing the field $\alpha^{(s-3)}$ to write

$$\mathcal{F}^{(s)} = 3\partial^{(1}\partial^{2}\partial^{3}\alpha^{s-3)} \tag{5.128}$$

20 Also referred to as the *reduced triplet system*.

This field equation is still unconstrained gauge invariant as both sides transform in the same way due to equations (5.126) and (5.7). Next, we rewrite the field equation (5.123) substituting for $C^{(s-2)}$ using the definition (5.125). This results in a fairly complicated equation

$$\Box\varphi'^{(s-2)} - \partial\cdot\partial\cdot\varphi^{(s-2)} - \frac{1}{2}\partial^{(1}\partial\cdot\varphi'^{s-3)} = \frac{3}{2}\Box\partial^{(1}\alpha^{s-3)} + \partial^{(1}\partial^2\partial\cdot\alpha^{s-4)} \tag{5.129}$$

However, the first three terms can be recognized as being the same as in the formula for the trace of the Fronsdal tensor (5.13). Using this, we get

$$\mathcal{F}'^{(s-3)} - \partial^{(1}\partial^2\varphi''^{s-4)} = 3\Box\partial^{(1}\alpha^{s-3)} + 2\partial^{(1}\partial^2\partial\cdot\alpha^{s-4)} \tag{5.130}$$

The final step of rewriting consists in computing the trace of both sides of the field equation (5.128) and using this to substitute for the combination of terms $\mathcal{F}'^{(s-3)} - 3\Box\partial^{(1}\alpha^{s-3)}$. The result is

$$\partial^{(1}\partial^2\varphi''^{s-4)} = \partial^{(1}\partial^2(4\partial\cdot\alpha^{s-4)} + \partial^{(3}\alpha'^{s-5)}) \tag{5.131}$$

This equation can be satisfied by[21]

$$\varphi''^{(s-4)} = 4\partial\cdot\alpha^{(s-4)} + \partial^{(1}\alpha'^{s-5)} \tag{5.132}$$

The two field equations (5.128) and (5.132) are still gauge invariant under the unconstrained gauge transformations. We see that gauging α to zero not only yields the Fronsdal field equations, but also enforces double tracelessness. The trace of the gauge parameter is used up in the process. The two equations are consistent in that the Bianchi identity (5.39) applied to the first equation (5.128) yields

$$\partial^{(1}\partial^2\partial^3\varphi''^{s-4)} = \partial^{(1}\partial^2\partial^3(4\partial\cdot\alpha^{s-4)} + \partial^{(1}\alpha'^{s-5)})) \tag{5.133}$$

One interpretation of this is to regard the field α as parameter field that parametrizes the difference between the Fronsdal system and the doublet unconstrained system.

Since in the resulting field equations (5.128) and (5.132) there is no reference to the C field, it should be possible to motivate the equations directly from the Fronsdal theory. This can indeed be done. Guided by the unconstrained gauge variation (5.7) of the Fronsdal equations, one postulates fully gauge invariant field equations

$$\mathcal{F}_{(s)} - 3\partial_{(1}\partial_2\partial_3\alpha_{s-3)} = 0 \tag{5.134}$$

with α a new field transforming as in equation (5.126) by definition. Then applying the Bianchi identity (5.39) to the field equation (5.128) again yields (5.133) which we can satisfy by equation (5.132) for the double trace of φ.

---

21 Up to discrete degrees of freedom corresponding to nondynamical integration "functions".

Formal solution of the *C*-equation and an expression for *α*. Case of spin 3

The field equation (5.123) for the C field can be formally inverted to get an expression for C that in its turn can be used to get an explicit, nonlocal, expression for the compensator α. For spin 3, we have

$$\Box C + \frac{1}{2}\partial_\mu \partial \cdot C = l_\mu \tag{5.135}$$

for a left-hand side l_μ which for spin 3 is $-3\partial \cdot \partial \cdot \varphi_\mu$. We make the ansatz

$$C_\mu = \frac{a}{\Box} l_\mu + \frac{b}{\Box^2}\partial_\mu \partial \cdot l \tag{5.136}$$

and find $a = 1$ and $b = -1/3$ so that

$$C_\mu = -\frac{3}{\Box}\partial \cdot \partial \cdot \varphi_\mu + \frac{1}{\Box^2}\partial_\mu \partial \cdot \partial \cdot \partial \cdot \varphi \tag{5.137}$$

Next using equation (5.125), we learn

$$\partial_\mu \alpha = \varphi'_\mu + \frac{1}{3}C_\mu \quad \Rightarrow \quad \alpha = \frac{1}{\Box}\left(\partial \cdot \varphi' + \frac{1}{3}\partial \cdot C\right) \tag{5.138}$$

We then get an expression for α

$$\alpha = \frac{1}{\Box}\partial \cdot \varphi' - \frac{2}{3\Box^2}\partial \cdot \partial \cdot \partial \cdot \varphi = \frac{1}{3\Box^2}\partial \cdot \mathcal{F}' \tag{5.139}$$

where the last equality follows from comparing to (5.16).

5.5.2 An action for the compensator minimal approach

The field equations for the compensator approach can be derived from an action, that apart from the higher spin field φ and the compensator field α, also involves a Lagrange multiplier field β. The Lagrangian was derived, in a rather complicated way, in [289], but a much simplified derivation can be found in [276] that we follow.

Guided by the field equations (5.134), we introduce the unconstrained gauge invariant tensor $\mathcal{A}$ defined by

$$\mathcal{A}_{(s)} = \mathcal{F}_{(s)} - 3\partial_{(1}\partial_2\partial_3\alpha_{s-3)} \tag{5.140}$$

It satisfies a Bianchi-type identity

$$\partial \cdot \mathcal{A}_{(s-1)} - \frac{1}{2}\partial_{(1}\mathcal{A}'_{s-1)} = -\frac{3}{2}\partial_{(1}\partial_2\partial_3(\varphi''_{s-4)} - 4\partial \cdot \alpha_{s-4)} - \partial_{(1}\alpha'_{s-5))}) \tag{5.141}$$

Then consider the following tentative Lagrangian:

$$\mathcal{L}_0 = \frac{1}{2}\varphi_{(s)}\left(\mathcal{A}_{(s)} - \frac{1}{2}\eta_{(12}\mathcal{A}'_{s-2)}\right) \tag{5.142}$$

and compute its unconstrained gauge variation. Due to the gauge invariance of A we only get contributions from the variation of φ. These work out to

$$\begin{aligned}\delta\mathcal{L}_0 &= -\frac{1}{2}s\xi_{(s-1)}\cdot\left(\partial\cdot\mathcal{A}_{(s-1)} - \frac{1}{2}\partial_{(1}\mathcal{A}'_{s-2)}\right) + \frac{3}{4}\binom{s}{3}\xi'_{(s-3)}\cdot\partial\cdot\mathcal{A}'_{(s-3)}\\ &= -3\binom{s}{4}\partial\cdot\partial\cdot\partial\cdot\xi_{(s-4)}(\varphi'' - 4\partial\cdot\alpha - \partial\alpha')_{(s-4)} + \frac{3}{4}\binom{s}{3}\xi'_{(s-3)}\cdot\partial\cdot\mathcal{A}'_{(s-3)}\end{aligned} \tag{5.143}$$

where we have used (5.141). Since $\varphi'' - 4\partial\cdot\alpha - \partial\alpha'$ is an unconstrained gauge invariant quantity, the first term can be compensated for by a Lagrange multiplier contribution to the action

$$\mathcal{L}_1 = 3\binom{s}{4}\beta_{(s-4)}(\varphi'' - 4\partial\cdot\alpha - \partial\alpha')_{(s-4)} \tag{5.144}$$

with the Lagrange multiplier β transforming as

$$\delta\beta_{(s-4)} = \partial\cdot\partial\cdot\partial\cdot\xi_{(s-4)} \tag{5.145}$$

The second term in the variation of $\mathcal{L}_0$ is compensated for by the contribution

$$\mathcal{L}_2 = -\binom{3}{4}\alpha_{(s-3)}\partial\cdot\mathcal{A}'_{(s-3)} \tag{5.146}$$

The full Lagrangian is thus the sum $\mathcal{L}_0 + \mathcal{L}_1 + \mathcal{L}_2$.

5.5.3 Current exchanges in the nonlocal unconstrained formulation

The nonlocal, geometric, free field equations of formulas (5.61) and (5.62) turn out not to be quite correct. They are the simplest of their kind, but they are not unique in that higher divergences and traces of the curvatures – compensated by higher inverse powers of $\Box$ – can be added. The problem arises when the higher spin fields are coupled to external currents J via a source term $\varphi\cdot J$. Then one can, and must, demand that the correct number of physical components are exchanged, and that is not the case for the simplest nonlocal theories.

We will not review the details of this rather elaborate discussion, but refer the reader to the original paper [276]. It turns out that the BRST approach, when reduced to the local compensator form and the further reduced to the nonlocal form, does in fact yield the correct nonlocal theory. Thus it seems safe to say that the BRST unconstrained theory is the more fundamental one.

5.5.4 Maxwell-like equations

The Fronsdal equations for higher spin gauge fields could be considered *Einstein-like* in that the form of the equations are actually the same as the spin 2 field equations with no new terms added. The only new properties are the traceless gauge parameters and the double tracelessness for the fields that sets in at spin 3 and spin 4, respectively.

An alternative would be to consider *Maxwell-like* field equations that resemble the Maxwell spin 1 equations in form, adding no new terms for higher spin. This scheme has been analyzed in the literature in [291–293]. Consider then the Maxwell-like field equations in terms of the *Maxwell tensor* $\mathcal{M}$

$$\mathcal{M}_{\mu_1 \dots \mu_s} = \Box \varphi_{\mu_1 \dots \mu_s} - \partial_{(\mu_1} \partial \cdot \varphi_{\mu_2 \dots \mu_s)} = 0 \tag{5.147}$$

In order for these equations to be invariant under the standard higher spin gauge transformations, one must require the gauge parameter to be divergence-free, that is, $\partial \cdot \xi = 0$. The analysis of this system is actually a bit tricky (see Section 5.1.1 for the method).

First, for spin higher than 1, the Maxwell equations (5.147) lead to a differential constraint on the fields. Computing the divergence of equation (5.147), one gets

$$\partial_{(\mu_4} \partial \cdot \partial \cdot \varphi_{\mu_3 \dots \mu_s} = 0 \quad \Rightarrow \quad \partial \cdot \partial \cdot \varphi_{\mu_3 \dots \mu_s} = 0 \tag{5.148}$$

up to nondynamical degrees of freedom. This constraint affects the count of physical field degrees of freedom. For the field itself, we get the number of components

$$\#\varphi = \binom{s+3}{3} - \binom{s+1}{3} = s^2 + 2s + 1 \tag{5.149}$$

while the divergence free gauge parameter has the following number of components:

$$\#\xi = \binom{s+2}{3} - \binom{s+1}{3} = \frac{1}{2}(s^2 + s) \tag{5.150}$$

Then subtracting gauge and regauge degrees of freedom,[22] we get $s+1$ physical degrees of freedom. This, again, corresponds to a decreasing spectrum of fields with spin s, $s-2$, $s-3$ down to either spin 1 or spin 0.

If one desires an irreducible field, one can demand the field and the gauge parameter to be traceless. In such a case, the corresponding count goes as follows. Taking care not to double-count the trace of the double-divergence, we get for the field

$$\#\varphi = \binom{s+3}{3} - \binom{s+1}{3} - \left(\binom{s+1}{3} - \binom{s-1}{3} \right) = 4s \tag{5.151}$$

22 This is consistent, since just as for the de Donder gauge condition for Fronsdal fields, the gauge parameter contains the same number of components as the gauge condition $\partial \cdot \varphi = 0$ (which implies $\Box\varphi = 0$ so that the fields are massless), and to stay in the gauge it is sufficient to use a gauge parameter that satisfies $\Box\xi = 0$, thus allowing for regauging.

For the gauge parameter, we get

$$\#\xi = \binom{s+2}{3} - \binom{s}{3} - \left(\binom{s+1}{3} - \binom{s-1}{3}\right) = 2s - 1 \tag{5.152}$$

Again subtracting gauge and regauge degrees of freedom, we now get 2 physical degrees of freedom corresponding to a single higher spin field.

A minor worry to address

We computed the divergence of the Maxwell-like equation (5.147) and found a differential constraint (5.148) on the field. One may worry why a similar conclusion does not result from computing the divergence of the Fronsdal equation? However, using equations (5.13) and (5.15) we find that $2\partial \cdot F = \partial F'$.

5.6 Mixed symmetry fields

Mixed symmetry higher spin fields occur in string theory, in that case of massive nature. The free field theory of mixed symmetry fields, massive and massless, has been studied by many authors. One motivation quoted for studying mixed symmetry gauge fields is a possible connection between higher spin gauge theory and string theory. The subject was first approached by T. Curtright in 1980 in [192], by Aulakh et al. in [188] and explored in detail by Labastida and Morris in [193] and by Labastida in [194, 195, 197]. The subject was also treated in [185] using the BRST methods of [158] and [159].

A mixed symmetry field can be written $\varphi_{\mu_1 \dots \mu_{s_1}; \nu_1 \dots \nu_{s_2}; \dots}$ with several separately symmetrized index sets $\{\mu_{s_i}\}$, $\{\nu_{s_j}\}$ etc.. The theory calls for an elaborate formalism. We will not go deeper into this subject here, but refer the reader to the comprehensive work of [189, 190] as well as to the exhaustive review [191]. Further references can be found in these works.

5.7 The frame-like formulation

The *frame-like* formulation of free higher spin field theory – in flat space-time – was set up by M. Vasiliev in a paper from 1980 [294] and elaborated in [295]. It is modeled on the tetrad formulation of general relativity generalizing the vierbein $e_\mu{}^a$ and Lorentz connection $\omega_\mu{}^{ab}$ fields.

Consider first spin 2. We may choose a notation so that $\varphi_\mu{}^a$ stands for the weak field in the sense that the vierbein proper $e_\mu{}^a$ is expanded (around Minkowski spacetime) as

$$e_\mu{}^a = \delta_\mu{}^a + \kappa \varphi_\mu{}^a \tag{5.153}$$

in analogy with $g_{\mu\nu} = \eta_{\mu\nu} + \kappa\varphi_{\mu\nu}$. Then from $g_{\mu\nu} = \eta_{ab}e_\mu{}^a e_\nu{}^b$ (see formula (4.50)), we get to order κ,

$$\varphi_{\mu\nu} = \eta_{ab}(\delta_\mu{}^a\varphi_\nu{}^b + \delta_\nu{}^b\varphi_\mu{}^a) = \varphi_{\nu,\mu} + \varphi_{\mu,\nu} \tag{5.154}$$

Thus, the metric-like spin 2 field φ is the symmetrized frame-like field φ. The comma separating the indices is needed here, as the second index is a lowered frame index. We will however continue to use $e_\mu{}^a$ (and not $\varphi_\mu{}^a$) temporarily. In the vierbein formulation of gravity, one also considers the Lorentz connection $\omega_\mu{}^{ab}$. Normally, the Lorentz connection is expressed in terms of the vierbein, as we studied in Section 4.5.5, but it can also be considered as an independent gauge field, at least until dynamics is introduced.

5.7.1 Vasiliev fields – version 1

We now follow [295]. The frame-like fields, representing spin s, are $e_{\mu,a_1\ldots a_{s-1}}$ and $\omega_{\mu,b,a_1\ldots a_{s-1}}$. The comma again serves to separate indices with different properties. The a_i indices can be thought of as indicating the spin s of the fields. The fields are symmetric in the a_i indices.

These fields generalize the spin 2 fields, written as $e_{\mu,a}$ and $\omega_{\mu,b,a}$. The antisymmetry in a, b for the spin 2 connection $\omega_{\mu,b,a}$ is generalized to requiring that the complete symmetrization in the tangent space indices is zero, that is,

$$\omega_{\mu,(a_1,a_2\ldots a_s)} = 0 \tag{5.155}$$

Furthermore, we have the contraction properties

$$e_{\mu,}{}^c{}_{ca_1\ldots a_{s-3}} = 0 \quad \text{and} \quad \omega_{\mu,b,}{}^c{}_{ca_1\ldots a_{s-3}} = 0 \quad \text{for } s \geq 3 \tag{5.156}$$

The following contraction property follows from (5.155) and (5.156):

$$\omega_{\mu,}{}^c{}_{,ca_1\ldots a_{s-2}} = 0 \quad \text{for } s \geq 2 \tag{5.157}$$

The metric-like field is given by the completely symmetric part of the frame-like field

$$\varphi_{\mu_1\mu_2\ldots\mu_s} = \varphi_{(\mu_1,\mu_2\ldots\mu_s)} \tag{5.158}$$

The tracelessness of the frame fields e together with the definition (5.158) implies the double tracelessness of the metric-like fields ϕ.[23]

23 This fact is sometimes offered as an explanation for the double tracelessness property. What it does is to reduce the double tracelessness of the metric-like fields to the tracelessness in the fiber indices for the frame fields.

An action for a massless spin s field can be written in terms of these fields as

$$S_s(e,\omega) \sim \epsilon^{\mu\nu\rho\sigma}\epsilon^{abc}{}_{\sigma} \int d^4x \Big[\omega_{\rho,b,a}{}^{d(s-2)} \Big(\partial_\mu e_{\nu,d(s-2)c} - \frac{1}{2}\omega_{\mu,\nu,d(s-2)c} \Big) \Big] \tag{5.159}$$

The fields are written in a condensed notation.

Condensed notation for frame-like fields

A variant of condensed notation is employed in the AdS higher spin literature. Upper or lower indices, denoted by the same letter, are considered as symmetrized and instead of writing $\phi_{a_1 \ldots a_s}$, one writes $\phi_{a(s)}$ indicating within the parentheses the number of symmetrized indices. Writing for instance $\phi_{a,a(s)}$ thus means that all $s+1$ a-indices are symmetrized. The symmetrizing weight is taken to 1 in contrast to the early literature [295] which used $1/s!$. The summation convention is the following: summation is carried out over the maximum possible number of upper and lower indices denoted by the same letter. Symmetrization is always carried out before summation. In this notation, the frame-like higher spin fields are written as $e_{\mu,a(s-1)}$ and $\omega_{\mu,b,a(s-1)}$.

The action (5.159) is invariant under local transformations with three independent kinds of parameters $\xi_{a(s-1)}$, $\alpha_{b,a(s-1)}$ and $\beta_{b(2),a(s-2)}$ subject to the following symmetrization and trace conditions, respectively,

$$\alpha_{a,a(s-1)} = 0 \qquad \beta_{ba,a(s-1)} = 0 \tag{5.160}$$

$$\xi^a{}_{a(s-2)} = 0 \qquad \alpha_{b,}{}^a{}_{a(s-2)} = 0 \qquad \beta_{b(2),}{}^a{}_{a(s-2)} = 0 \qquad \text{for } s \geq 3 \tag{5.161}$$

$$\alpha^a{}_{,a(s-1)} = 0 \qquad \beta_b{}^a{}_{,a(s-1)} = 0 \qquad \beta_b{}^b{}_{,a(s-1)} = 0 \qquad \text{for } s \geq 2 \tag{5.162}$$

where the second line of trace conditions are consequences of the first line of trace conditions, using the symmetry conditions (5.160). This, somewhat bewildering, set of conditions can be simplified by translating it into two-component notation. This will be done in Section 5.7.5. The transformation laws are

$$\delta e_{\mu,a(s-1)} = \partial_\mu \xi_{a(s-1)} + \alpha_{\mu,a(s-1)} \tag{5.163}$$

$$\delta\omega_{\mu,b,a(s-1)} = \partial_\mu \alpha_{b,a(s-1)} + \beta_{\mu b,a(s-1)} \tag{5.164}$$

For spin 2, there is no parameter $\beta_{\mu b,a}$, while ξ_a and $\alpha_{\mu,a}$ are linearized local coordinate transformations and Lorentz transformations, respectively.[24]

5.7.2 Counting degrees of freedom

Let us check that the number of propagating field degrees of freedom are correct in the frame-like formulation.

24 The physical interpretation of the transformation parameters for higher spin is just as enigmatic in the frame formulation as in the metric formulation.

The traceless fields $e_{\mu,a(s-1)}$ maintain $4\left(\binom{s+2}{3} - \binom{s}{3}\right) = 4s^2$ field components, whereas the gauge parameters $\xi_{a(s-1)}$, which have the same tangent space properties as the fields, have s^2 components. By the same argument as in Section 5.1.1, this allows for fixing $2s^2$ field components. Next, the parameters $\alpha_{\mu,a(s-1)}$, subject to the symmetry and trace properties (5.160) and (5.161), have $2(s^2 - 1)$ independent components (see box below in Section 5.7.4). Then the transformation with these parameters (see (5.163)) can be used to fix another $2(s^2 - 1)$ field components. The count works out to $4s^2 - 2s^2 - 2(s^2 - 1) = 2$.

5.7.3 Extended frame-like higher spin fields

Referring back to Section 5.7.1, we had the frame-like description of higher spin gauge fields in terms of the pair of fields e and ω with the transformation laws of equations (5.163) and (5.164). We now streamline the notation by renaming the fields $e_{\mu,a(s-1)}$ to $\omega_{\mu,a(s-1)}$ as well as using the symbol ξ also for the parameters α and β since they are anyway distinguished by the index structure. The transformations then read

$$\delta\omega_{\mu,a(s-1)} = \partial_\mu \xi_{a(s-1)} + \xi_{\mu,a(s-1)} \tag{5.165}$$

$$\delta\omega_{\mu,b,a(s-1)} = \partial_\mu \xi_{b,a(s-1)} + \xi_{\mu b,a(s-1)} \tag{5.166}$$

These transformations generalize the familiar results for spin 2. In that case, the parameter $\xi_{\mu b,a}$ is not present. The field $\omega_{\mu,b,a}$ (the Lorentz connection) is auxiliary, as it can be expressed in terms of the vierbein field through the torsion constraint, and the number of propagating field degrees of freedom is 2.

This set of fields and transformations laws in (5.165) and (5.166) is now expanded by adding further auxiliary fields $\omega_{\mu,b(t),a(s-1)}$ with $2 \leq t \leq s-1$. One rationale for this is to think of the parameter $\xi_{\mu b,a(s-1)}$ in (5.166) as the gauge parameter for a new field $\omega_{\mu,b(2),a(s-1)}$ with transformation law $\delta\omega_{\mu,b(2),a(s-1)} = \partial_\mu \xi_{b(2),a(s-1)}$. Then, however, one can contemplate introducing still another parameter $\xi_{\mu b(2),a(s-1)}$ and promoting the transformation law for the new field to

$$\delta\omega_{\mu,b(2),a(s-1)} = \partial_\mu \xi_{b(2),a(s-1)} + \xi_{\mu b(2),a(s-1)} \tag{5.167}$$

Iterating this procedure, we would get

$$\begin{aligned}
\delta\omega_{\mu,b(2),a(s-1)} &= \partial_\mu \xi_{b(2),a(s-1)} + \xi_{\mu b(2),a(s-1)} \\
\delta\omega_{\mu,b(3),a(s-1)} &= \partial_\mu \xi_{b(3),a(s-1)} + \xi_{\mu b(3),a(s-1)} \\
&\vdots \\
\delta\omega_{\mu,b(s-2),a(s-1)} &= \partial_\mu \xi_{b(s-2),a(s-1)} + \xi_{\mu b(s-2),a(s-1)} \\
\delta\omega_{\mu,b(s-1),a(s-1)} &= \partial_\mu \xi_{b(s-1),a(s-1)}
\end{aligned} \tag{5.168}$$

in addition to (5.165) and (5.166). This procedure could have been continued with one more step to $\omega_{\mu,b(s),a(s-1)}$, which would then correspond to the de Wit–Freedman gauge invariant curvature.

Note how the μ and b indices "blend" as $\mu b(i) \to b(i+1)$ in each successive step. This blending is mediated by the background vierbein field. Referring back to the formula (5.153) where we had $e_\mu{}^a = \delta_\mu{}^a + \kappa\varphi_\mu{}^a$ in Minkowski space-time, we can now write in a more general background

$$e_\mu{}^a = h_\mu{}^a + \kappa\varphi_\mu{}^a = h_\mu{}^a + \omega_\mu{}^a \tag{5.169}$$

where $h_\mu{}^a$ is the background vierbein field, and where ω denotes the weak field we will be using. Thus we can write

$$\xi_{\mu b(t),a(s-1)} = h_\mu{}^c \xi_{cb(t),a(s-1)} \tag{5.170}$$

This drastic expansion in the number of fields is not required for the free field theory, but is an inherent feature of the Vasiliev equations. The physical higher spin field components still reside in the coframe-like fields $\omega_{\mu,a(s-1)}$ while the Lorentz-connection-like fields $\omega_{\mu,b,a(s-1)}$ are expressed through zero torsion-like constraints. This is also the case for the extra fields $\omega_{\mu,b(t),a(s-1)}$, starting with spin 3 ($t = 2$).

The extended Vasiliev fields

The extended Vasiliev fields are

$$\omega_{\mu,b(t),a(s-1)} \quad \text{with} \quad 0 \le t \le s-1 \tag{5.171}$$

where we recognize spin 2 for $s = 2$ and $t = 0, 1$. The transformations are given by

$$\delta\omega_{\mu,b(t),a(s-1)} = \partial_\mu \xi_{b(t),a(s-1)} + h_\mu{}^c \xi_{cb(t),a(s-1)} \tag{5.172}$$

where $h_\mu{}^c$ is the background frame field. There is no parameter $\xi_{\mu b(s-1),a(s-1)} \to \xi_{b(s),a(s-1)}$. Fields and parameters are subject to conditions deriving from the conditions of Section 5.7.1. For the fields, we have

$$\omega_{\mu,b(t-1)a,a(s-1)} = 0 \quad \text{for } 1 \le t \le s-1 \tag{5.173}$$

$$\omega_{\mu,b(t),}{}^{c}{}_{ca(s-3)} = 0 \quad \text{for } 0 \le t \le s-1 \text{ and } s \ge 3 \tag{5.174}$$

Note that equation (5.173) states that symmetrizing any b index with all the a indices, yields zero. The index structure of the parameters are the same as for the fields and they satisfy analogous conditions.

5.7.4 Tensor structure and conditions on fields and parameters

Both the fields and the gauge parameters have two types of fiber indices, denoted by a and b. The a indices are related to the spin of the field, whereas the b index is related to the number of derivatives occurring when expressing auxiliary fields in terms of the physical fields.

Let us consider general tensors T with two groups of indices $T_{a(k),b(m)}$, separately symmetric in the a indices and the b indices, and where $m \leq k$. This is precisely the tensor structure of the 1-form gauge fields and the 0-form gauge parameters. It is convenient to interchange the order of the a and b indices since the number of b indices are always less than or equal to the number of a indices. These tensors are subject to the following conditions that generalizes the conditions discussed in Section 5.7.1.

Symmetrizing an a index with the group of b indices on the tensor $T_{a(k),b(m)}$ gives zero. That is,

$$T_{a(k-1),ab(m)} = 0 \tag{5.175}$$

The tensors $T_{a(k),b(m)}$ are traceless in the a indices, that is

$$T_{a(k-2)}{}^{c}{}_{c,b(m)} = 0 \tag{5.176}$$

The two conditions (5.175) and (5.176) taken together imply that the traces over an a and a b index, or over two b indices, are also zero. The tensors $T_{a(k),b(m)}$ constitute mixed symmetry representations of the tangent space symmetry algebra $\mathfrak{so}(d-1,1)$, although the metric signature plays no role for this.

Some facts about 2-row Young tableaux

Young tableaux are a convenient tool for keeping track of index symmetries for tensors and have been extensively used for mixed symmetry fields in higher spin theory in dimensions $d > 4$. They are useful also for the Vasiliev fields since these carry two index sets denoted by a and b.

The general theory of Young tableaux (or Young diagrams) and their use in representation theory can be found in many group theory books. A book dedicated to the subject is [296].[25] Here, we will just cite without proof some basic formulae that will allow us to compute dimensions of tensor representations in a practical way.

The symmetry type of the $T_{a(k),b(m)}$ tensors may be represented by their *Young tableaux*

$$\begin{array}{|l|}\hline \qquad k \qquad \\ \hline \quad m \\ \cline{1-1}\end{array} \tag{5.177}$$

where $\boxed{\qquad k \qquad}$ stands for $\boxed{1}\boxed{2}\boxed{\cdot}\boxed{\cdot}\boxed{\cdot}\boxed{k}$.

25 It seems that treatments of Young diagrams often either provide far too much detail as compared to the particular need one may have at hand, or indeed only discuss one particular application, unfortunately not the needed one! Reference [228] provides more useful details pertaining to the Vasiliev theory. See also the review [297].

The lengths of the successive rows cannot increase. Here it is convenient to switch the order of the a and b indices, as already indicated above. We are using the convention of having symmetry in the rows and antisymmetry in the columns corresponding to fields being 1-forms in the base space index μ. The tracelessness in the row indices are not indicated by the tableau, but must be supplied externally. There are formulas for computing the number of components, but for the purpose of checking the simplest cases it is often more instructive to use the following rules recursively. First, we quote formulas for GL(d).

$$\boxed{\quad k \quad} \otimes \square = \begin{array}{l}\boxed{\quad k \quad}\\ \square\end{array} \oplus \boxed{\quad k+1 \quad} \tag{5.178}$$

$$\begin{array}{l}\boxed{\quad k \quad}\\ \boxed{\; m \;}\end{array} \otimes \square = \begin{array}{l}\boxed{\quad k \quad}\\ \boxed{\; m \;}\\ \square\end{array} \oplus \begin{array}{l}\boxed{\quad k+1 \quad}\\ \boxed{\; m \;}\end{array} \oplus \begin{array}{l}\boxed{\quad k \quad}\\ \boxed{\; m+1 \;}\end{array} \tag{5.179}$$

The principle is to add the new box $\square$ to rows in the given diagram in all ways that result in a new admissible Young tableau. The corresponding formulas for SO(d) are

$$\boxed{\quad k \quad} \otimes \square = \begin{array}{l}\boxed{\quad k \quad}\\ \square\end{array} \oplus \boxed{\quad k+1 \quad} \oplus \boxed{\quad k-1 \quad} \tag{5.180}$$

$$\begin{array}{l}\boxed{\quad k \quad}\\ \boxed{\; m \;}\end{array} \otimes \square = \begin{array}{l}\boxed{\quad k \quad}\\ \boxed{\; m \;}\\ \square\end{array} \oplus \begin{array}{l}\boxed{\quad k+1 \quad}\\ \boxed{\; m \;}\end{array} \oplus \begin{array}{l}\boxed{\quad k \quad}\\ \boxed{\; m+1 \;}\end{array} \oplus \begin{array}{l}\boxed{\quad k-1 \quad}\\ \boxed{\; m \;}\end{array} \oplus \begin{array}{l}\boxed{\quad k \quad}\\ \boxed{\; m-1 \;}\end{array} \tag{5.181}$$

The intuition is that tensoring the given diagram (which have traces removed) with the new vector $\square$ produces the diagrams corresponding to GL(d) (now without traces) but also new diagrams resulting from computing all possible traces of the preceding ones in rows where a box has been added. Since the metric signature plays no role in these combinatorial formulae, they are correct also for Lorentz SO($d-1,1$) tensors.

From these formulae, one can compute the number of components for mixed symmetry tensors occurring in the Vasiliev theory, provided one knows the number of components of the basic fully symmetric and fully antisymmetric tensors. In d dimensions these are

$$\dim \boxed{\quad k \quad} = \binom{k+d-1}{d-1} - \binom{k+d-3}{d-1} \tag{5.182}$$

$$\dim \boxed{\begin{array}{c}\\ k\\ \\ \end{array}} = \binom{d}{k} \quad \text{where } k \leq d \tag{5.183}$$

In $d=4$, we have

$$\dim \boxed{\quad k \quad} = (k+1)^2 \tag{5.184}$$

$$\dim \begin{array}{l}\boxed{\quad k \quad}\\ \boxed{\; m \;}\end{array} = 2(k+m+1)(k-m+1) \quad \text{for } m \geq 1 \tag{5.185}$$

5.7.5 Two-component spinor reformulation

The Vasiliev set of extended higher spin fields (5.171) can be represented in a two-component spinor form. This will bring out a symmetry between the a and b index sets that is not quite obvious. It will also let us represent half-integer spin fields in the same formalism.

That a two-component spinor reformulation should be possible is clear from our general considerations in the historical Chapter 2, and from Sections 3.6.4. A direct transcription using the formula (3.300) may however prove to be cumbersome as we also have to take the two-row tensor structure into account. A more simple counting argument based on the representations of the Lorentz group can instead be tried.

The physical field $\omega_{\mu,a(s-1)}$ has the symmetry of the Young tableau $\boxed{\quad s-1\quad}$. Such a symmetric tensor corresponds to the $D(s-1, s-1)$ representation of the Lorentz group. Its two-component spinor realization is immediately given by a field $\omega_{\mu,\alpha(s-1),\dot{\beta}(s-1)}$. The next tensor with structure $\begin{array}{|c|}\hline \quad s-1 \quad \\ \hline \end{array}\!\!\!\begin{array}{l}\\ \boxed{\,}\end{array}$ has $2(s+1)(s-1)$ components which is precisely twice the number of components in a spinor with index structure $\alpha(s), \dot{\beta}(s-2)$. Adding in a spinor with index structure $\alpha(s-2), \dot{\beta}(s)$, for good measure, we get an exact agreement. Continuing in this way, we can ascertain the following correspondences for $1 \le t \le s-1$,

$$\begin{array}{|l|}\hline \quad s-1 \quad \\ \hline \ \ t \ \ \\ \hline \end{array} \sim D((s-1+t)/2, (s-1-t)/2) \oplus D((s-1-t)/2, (s-1+t)/2) \tag{5.186}$$

$$\omega_{\mu,b(t),a(s-1)} \sim \omega_{\mu,\alpha(s-1+t),\dot{\beta}(s-1-t)} \oplus \omega_{\mu,\alpha(s-1-t),\dot{\beta}(s-1+t)} \tag{5.187}$$

We also have the one already established for the physical field (t=0),

$$\boxed{\quad s-1\quad} \sim D((s-1)/2, (s-1)/2) \tag{5.188}$$

$$\omega_{\mu,a(s-1)} \sim \omega_{\mu,\alpha(s-1),\dot{\beta}(s-1)} \tag{5.189}$$

It should be clear that the counting of components agree. The formula (5.185) yields for the number of tensor components $2(s+t)(s-t)$ and this is precisely the number of components of the two multispinors together. It is convenient to abbreviate the notation further and just write

$$\omega(n,m) = \omega_\mu(n,m)dx^\mu = \omega_{\mu,\alpha(n),\dot{\beta}(m)}dx^\mu \quad \text{with} \quad n+m = 2s-2 \tag{5.190}$$

Thus we will think of $\omega(n,m)$ as a 1-form multi-spinor. Since the transformation parameters have the same fiber index structure as the fields we can write them as 0-form multispinors

$$\xi(n,m) = \xi_{\alpha(n),\dot{\beta}(m)} \quad \text{with} \quad n+m = 2s-2 \tag{5.191}$$

There are a few extra bonuses with the two-component spinor notation. We have not discussed half integer higher spin fields. We now get them for free by just taking s as a half-integer.[26] However, the tensor integer spin fields we have started with are

26 Details are given in [295].

real, and the multispinor fields are complex, so we have to impose reality conditions

$$\omega^{\dagger}_{\mu,\alpha(n),\dot\beta(m)} = \omega_{\mu,\dot\alpha(n),\beta(m)} \tag{5.192}$$

For the half-integer spin fields, this means that the corresponding spinor-tensor fields are Majorana spinors. The same holds for the gauge parameters.

A second bonus with the two-component spinor notation is that the gauge transformation structure becomes particularly clear

$$\delta\omega_{\mu,\alpha(n),\dot\beta(m)} = \partial_\mu\xi_{\alpha(n),\dot\beta(m)} - h_{\mu\alpha}{}^{\dot\sigma}\xi_{\alpha(n-1),\dot\beta(m)\dot\sigma} \quad \text{for } m > n \tag{5.193}$$

$$\delta\omega_{\mu,\alpha(n),\dot\beta(n)} = \partial_\mu\xi_{\alpha(n),\dot\beta(n)} - h_{\mu\alpha}{}^{\dot\sigma}\xi_{\alpha(n-1),\dot\beta(n)\dot\sigma} - h_{\mu\dot\beta}{}^{\sigma}\xi_{\alpha(n)\sigma,\dot\beta(n-1)} \tag{5.194}$$

$$\delta\omega_{\mu,\alpha(n),\dot\beta(m)} = \partial_\mu\xi_{\alpha(n),\dot\beta(m)} - h_{\mu}{}^{\sigma}{}_{\dot\beta}\xi_{\alpha(n)\sigma,\dot\beta(m-1)} \quad \text{for } n > m \tag{5.195}$$

As always, lower or upper indices denoted by the same Greek letter are symmetrized. To get a clear picture of the transformations. Let us write them out explicitly for spin 3 in the shorthand notation.

Example 13 (Spin 3 gauge transformations in two-spinor notation). For spin 3, we have n and m running between 0 and 4. The gauge fields are $\omega(0,4)$, $\omega(1,3)$, $\omega(2,2)$, $\omega(3,1)$ and $\omega(4,0)$. Let us also denote by $h_{1,-1}$ and $h_{-1,1}$ background vierbeins that contract a dotted and inserts an undotted index, or contracts an undotted index and inserts a dotted, respectively. The transformations can then be written in shorthand as follows:

$$\begin{aligned}
\delta\omega(4,0) &= d\xi(4,0)\\
\delta\omega(3,1) &= d\xi(3,1) - h_{-1,1}\xi(4,0)\\
\delta\omega(2,2) &= d\xi(2,2) - h_{1,-1}\xi(1,3) - h_{-1,1}\xi(3,1)\\
\delta\omega(1,3) &= d\xi(1,3) - h_{1,-1}\xi(0,4)\\
\delta\omega(0,4) &= d\xi(4,0) \quad \blacktriangleleft
\end{aligned} \tag{5.196}$$

5.7.6 An intricate problem

This is where we are going to stop, because we are faced with an intricate problem. We have transformation equations, but no field equations, and the first without the second are not so interesting. Of course, from any free field equations in this formalism one must be able to work back to the Fronsdal equations. We know from gravity that the vierbein formulation is more general than the metric formulation, but it is therefore possible to work back the metric formulation. In the higher spin case, we can quote from the Vasiliev paper [295].[27]

27 I have exchanged the original paper references to formulas and literature with the corresponding ones from the present work.

> By the use of the equations of motion $\delta S_s(e,\omega) = 0$, the auxiliary fields $\omega_{\mu,b,a(s-1)}$ can be expressed in terms of derivatives of $e_{\mu,a(s-1)}$ up to the gauge part corresponding to the parameters $\beta_{b(2),a(s-1)}$. Substitution of the corresponding expression for $\omega(e)$ into the action $S_s(e,\omega)$ [(5.159)] gives the action $S_s(e,\omega(e))$ describing the spin s field in terms of $e_{\mu,a(s-1)}$. This action is invariant under the gauge transformations [(5.163)] and is equivalent [294] to the known expressions for the higher spin field actions in terms of symmetric tensors [3, 149, 150].

The equivalence of the vierbein higher spin action (5.159) to the Fronsdal action for symmetric fields can indeed be found in Vasiliev's paper [294]. So far, free vierbein higher spin resembles linearized vierbein gravity.

Let us now return to the transformation formulas for the extended set of fields as given in two-component notation. As already discussed in Section 5.7.2, the gauge transformation formula (5.194) is such that it leaves the physical higher spin field $\omega_{\mu,\alpha(n),\dot\beta(n)}$ with precisely 2 two degrees of freedom. The count is exactly the same: $4(n+1)^2 - 2(n+1)^2 - n(n+2) - n(n+2) = 2$. This, however, means that all freedom in the gauge parameters are used (used to fix the gauge, that is to say). The status of the other transformation equations becomes obscure. They cannot really be gauge transformations. The number of components also does not add up to such an interpretation. This means that all the fields $\omega(m,n)$ with $m \neq n$ must be auxiliary, and completely determined in terms of the physical field $\omega(n,n)$. These are the field equations for the auxiliary fields that we must find. They must be such that the transformations are symmetry transformations.

The situation is analogous to what we have for the generalized Christoffel symbols of de Wit and Freedman. All of them are determined in terms of derivatives on the symmetric tensor higher spin gauge field, and the suffer no independent gauge transformations. Solving this problem is one of the first steps toward the Vasiliev theory.

5.8 Chapter 5 epilogue

I have certainly not dealt with all aspects of free higher spin theory, but hopefully enough to approach the problem of interactions. It should be clear that much of the Minkowski space-time theory can be subsumed under the BRST approach. The frame-like approach provides for another kind of systematics to the theory, offering a possibility to set up the theory in a generally covariant way. This has been exploited in the Vasiliev theory. This is, however, beyond the scope of the present volume. Let us instead turn to the free field theory as it can be developed in the light-front formulation. After that we will have three broad free field theory bases upon which to approach interactions: the metric-like BRST formulation, the covariant frame-like formulation and the light-front formulation.

6 The light-front approach

In this chapter, we will develop the light-front, or light-cone, approach to massless higher spin theory. This was the formalism in which the first cubic self-interaction terms for arbitrary spin were found in 1983 [124, 298]. The structure of the chapter is as follows: In Sections 6.1 and 6.2, we work from the ground up, reviewing the basics of the formulation. Then in Section 6.3 we start afresh, stating our basic conventions (based on previous sections) for free higher helicity fields on the light-front. The interacting theory will be treated in Volume 2 of the present work.

6.1 Origin and overview of the formalism

The light-front formulation of relativistic dynamics was invented by P. A. M. Dirac in 1949 in a paper where he, apart from the "front form", also discussed the more common "instant form" and the not so common "point form" [61]. The light-front form was later rediscovered in 1965 in the context of current algebra as the "infinite momentum frame" by S. Fubini and G. Furlan [299]. For a late 1960s contemporary review, see [300]. The method was extensively applied to problems in strong interaction physics during the 1960s and 1970s and up to the present time. One should therefore be aware of the fact that the light-front approach offers effective methods for phenomenological calculations in quantum field theory. For a history of such applications up to 1980, see [301] which also has a useful bibliography that points to the broad literature on the subject. Here, we are instead interested in using the method for constructing new dynamical systems, as it was originally envisaged by Dirac.

The formalism, as it is applied to the kinds of problems we are here interested in, appears in the higher spin theory, gravity and supergravity literature in a few different guises. The original 1983 papers on higher spin interactions were calculated in configuration space with fields $\phi(x)$ and nonlinear Lorentz (and supersymmetry) transformations $\delta\phi(x)$ written in terms of space-time derivatives on fields.[1] Since the work was in four space-time dimensions, a convenient complex notation could be used where the two $\pm\lambda$ helicity fields could be represented by the complex conjugated pair $\phi, \bar{\phi}$. Likewise, two-dimensional transverse space-time coordinates and derivatives were rewritten into the pairs $x, \bar{x}$ and $\partial, \bar{\partial}$. The remaining directions, related to the "fronts", and denoted by $+$ and $-$, deserve some further attention. The details of all this will be summarized in the box below. This – configuration space formalism – has been developed and extensively used in supergravity research by L. Brink and P. Ramond and collaborators S. Ananth, S. S. Kim and S. Majumdar. The light-cone formalism also

1 See also the author's $N = 1$ supergravity paper [302] for explication of the formalism.

https://doi.org/10.1515/9783110451771-006

played a significant role in the first finiteness proofs for $N = 4$ super-Yang–Mills theory [204, 203] in 1983.

The light-front formalism had previously been used in the early 1980s superstring research by M. B. Green, J. H. Schwarz and L. Brink, indeed going back to the very early bosonic string days.[2] Light-cone superstring theory was formulated in momentum space, and this form was taken over to higher spin theory in a paper from 1987 [205] by I. Bengtsson and N. Linden and myself.

R. Metsaev took up light-front higher spin theory in the late 1980s in a momentum space formalism. It was then employed to investigations into the quartic interactions. This important work went largely unnoticed at the time.[3] It was revived and clarified by M. Ponomarev and E. Skvortsov in [210]. This work and R. Metsaev's work will be discussed in Volume 2 of the present work.

6.1.1 The Dirac forms of relativistic dynamics

In the paper [61], Dirac demands that the physical laws should be invariant under infinitesimal transformations of the coordinates $x_\mu \to a_\mu + \varpi_\mu{}^\nu x_\nu$ and associates "dynamical variables" $F = P^\mu a_\mu + \frac{1}{2} M^{\mu\nu} \varpi_{\mu\nu}$ with such transformations. The Poisson bracket algebra of the P^μ and $M^{\mu\nu}$ is of course the well-known Poincaré algebra which Dirac gives in the paper. Dirac then writes:

> To construct a theory of a dynamical system one must obtain expressions for these ten fundamental quantities [P^μ and $M^{\mu\nu}$] that satisfy these P. b. relations [the Poincaré algebra]. *The problem of finding a new dynamical system reduces to the problem of finding a new solution of these equations.*

Dirac notes that some of the ten fundamental quantities that actually occur in practice in dynamical systems, are simple and others are complicated. The complicated ones he calls *Hamiltonians*. Today we say "kinematical" and "dynamical" generators, respectively. In the example of the *instant form of dynamics* where the dynamical variables are referred to the surface $x^0 = 0$, it turns out that the generators P^1, P^2, P^3 and M^{12}, M^{23}, M^{31} are simple, while the generators P^0 and M^{01}, M^{02}, M^{03} are Hamiltonians. We recognize the latter ones as what we usually call the Hamiltonian and the boost generators.

The last form discussed in the paper is the *front form*. Dirac considers a "[...] three-dimensional surface in space-time formed by a plane wave front advancing with the velocity of light.". Such a surface is called a *front*. He then gives the example $x^0 - x^3 = 0$.

2 See the book [286] for a late 1980s view of string theory. There is also quite a few reviews on bosonic string theory from the 1970s. These references are listed in [303]. For a recent history of string theory, see the book [304].

3 Also by the present author.

For a theory where the dynamical variables refer to such a front, the fundamental quantities associated with transformations that leave the front invariant will be simple. These are P_1, P_2, P_- and $M_{12}, M_{+-}, M_{1-}, M_{2-}$. The remaining ones, P_+, M_{1+}, M_{2+} will be the Hamiltonians.

We now leave the Dirac founding paper and turn to a systematic introduction to light-front dynamics. We note, however, that in the program outlined by Dirac, finding the Hamiltonians is the real difficulty in the construction of a theory of a relativistic dynamical system in the front form. This is precisely what light-front higher spin theory tries to achieve.[4]

6.1.2 Introducing light-fronts

The light-front formulation can be introduced in many ways, but let us imagine a massless particle moving in the x^3 direction with momentum p_3. Since $p_\mu p^\mu = 0$, we can write the momentum four-vector as $p_\mu = (p, 0, 0, p)$. A plane wave describing the particle will then take the form $\exp ip(x^0 + x^3)$. The three-dimensional surface defined by the equation $x^0 + x^3 = t$ for some constant t is a light-front.

For a more systematic introduction, one may follow R. A. Neville and F. Rohrlich [305], and consider two null vectors $n^\mu = \frac{1}{\sqrt{2}}(1, 0, 0, 1)$ and $m^\mu = \frac{1}{\sqrt{2}}(1, 0, 0, -1)$ with the properties $n^2 = m^2 = 0$ and $n \cdot m = -1$. The projections along these vectors define two *light-front*, or *null-plane*, coordinates

$$x^- = -n \cdot x = \frac{1}{\sqrt{2}}(x^0 - x^3) \tag{6.1}$$

$$x^+ = -m \cdot x = \frac{1}{\sqrt{2}}(x^0 + x^3) \tag{6.2}$$

The coordinate x^+ will be interpreted as the light-front time.[5] Then x^- is considered a space coordinate. Raising and lowering + and – indices is done according to $x_+ = -x^-$ and $x_- = -x^+$ as seen from formulas (6.1) and (6.2).

Consequently, the light-front time derivative is $\partial_+ = -\partial^-$ with $\partial_+ x^+ = 1$ and $\partial_- = -\partial^+$ a space derivative with $\partial_- x^- = 1$. The transverse direction may be kept as $x^i = x_i$ for the remaining space indices. In four dimensions it is, however, convenient to combine the transverse directions into complex coordinates and derivatives according to

$$x = \frac{1}{\sqrt{2}}(x^1 + ix^2) \quad \text{and} \quad \bar{x} = \frac{1}{\sqrt{2}}(x^1 - ix^2) \tag{6.3}$$

For easy reference, we collect useful formulas in the box below.

4 This could be designated the *Dirac research program*.

5 Neville and Rohrlich take x^- as time, calling it u. In higher spin theory, x^+ has become an accepted standard.

Summary of light-cone coordinates

Coordinates are give by

$$x^+ = \frac{1}{\sqrt{2}}(x^0 + x^3) \qquad x^- = \frac{1}{\sqrt{2}}(x^0 - x^3) \tag{6.4}$$

$$x = \frac{1}{\sqrt{2}}(x^1 + ix^2) \qquad \bar{x} = \frac{1}{\sqrt{2}}(x^1 - ix^2) \tag{6.5}$$

Derivatives are given by

$$\partial_+ = \frac{\partial}{\partial x^+} = \frac{1}{\sqrt{2}}(\partial_0 + \partial_3) = -\partial^- \qquad \partial_- = \frac{\partial}{\partial x^-} = \frac{1}{\sqrt{2}}(\partial_0 - \partial_3) = -\partial^+ \tag{6.6}$$

$$\partial = \frac{\partial}{\partial \bar{x}} = \frac{1}{\sqrt{2}}(\partial_1 + i\partial_2) \qquad \bar{\partial} = \frac{\partial}{\partial x} = \frac{1}{\sqrt{2}}(\partial_1 - i\partial_2) \tag{6.7}$$

acting on the coordinates according to $\partial_+ x^+ = \partial_- x^- = \partial \bar{x} = \bar{\partial} x = 1$. With a mostly-plus Minkowski metric $- + ++$, the light-front scalar product becomes

$$\begin{aligned} A_\mu B^\mu &= A\bar{B} + \bar{A}B + A_- B^- + A_+ B^+ \\ &= A\bar{B} + \bar{A}B - A^+ B^- - A^- B^+ \end{aligned} \tag{6.8}$$

where the light-front components of the vectors A and B are defined exactly as for the coordinates (6.4) and (6.5). The scalar product can be interpreted such as the light-front metric has signature $+ + --$ in the directions "unbarred", "barred", "plus" and "minus". The light-front d'Alembertian becomes

$$\Box = \partial_\mu \partial^\mu = \partial_i \partial_i - 2\partial_+ \partial_- = 2(-\partial_+ \partial_- + \partial\bar{\partial}) \tag{6.9}$$

Other vectors and tensors can now be written in the light-front basis. For instance, a spin one gauge field has the components $A^+, A^-, A, \bar{A}$. In the light-front gauge, which we will work through below in Section 6.1.4, the space components A and $\bar{A}$ correspond to the physical helicity 1 and –1 fields. As we will see, this generalizes naturally to massless higher spin fields.

It is no coincidence that the light-front coordinates can be written $\frac{1}{\sqrt{2}} x^\mu \sigma_\mu$ in terms of the σ matrices as in formula (3.121). Therefore, the coordinates, or rather the momenta, can be considered as "two-spinor helicity variables" in a certain sense.[6]

6.1.3 The wave equation and the Cauchy problem

In choosing any one coordinate as the time, one is also committed to certain surfaces upon which initial data should be prescribed. In the case of free field equations, it is

6 This can be taken advantage to connect light-front higher spin theory to the spinor helicity formalism of modern amplitude research, as first noted by S. Ananth in [306].

known that one gets a well-posed initial value problem if Cauchy data are provided on a space-like surface. Such a surface has a normal pointing into the forward light-cone. A common and convenient choice is equal time surfaces $x^0 = c$. This is also the quantization surface most often used in quantum field theory. However, the light-front $x^+ = c$ is not a space-like surface. It is a null-plane with its normal n a null vector. In general, it is not sufficient to specify initial data on such a surface. The light-front Cauchy problem is discussed by A. Neville and F. Rohrlich in [305]. These author study conditions under which the massive Klein–Gordon equation in light-front coordinates

$$2\partial_+\partial_-\phi(x) = (\partial_i\partial_i - m^2)\phi(x) \tag{6.10}$$

has a unique solution.

Figure 6.1 shows the x^+-x^- plane in Minkowski space-time with the backward light-cone from the point p. It is intuitively clear that specifying initial data on the null-plane $x^+ = x_0^+$ cannot be sufficient, since waves traveling in the x^- direction are not affected by such conditions.

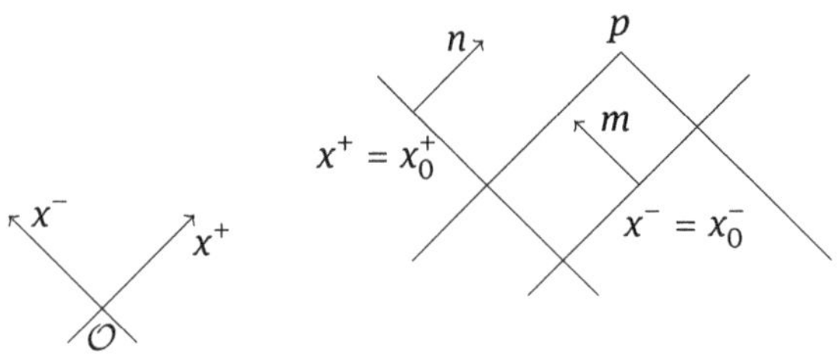

Figure 6.1: Backward light-cone at the point p and null-planes $x^+ = x_0^+$ and $x^- = x_0^-$.

Two theorems are proved, that we quote without providing the solution integrals.

Theorem 1. *The Klein–Gordon equation for $m \geq 0$ is uniquely determined in the region bounded by the wedge formed by null-planes $x^+ = x_0^+$ and $x^- = x_0^-$ if ϕ and the derivatives $\partial_-\phi$ and $\partial_+\phi$ are specified on these planes, respectively. Initial data on the wedge but outside the backward null cone is not necessary.*

Theorem 2. *Given ϕ and $\partial_-\phi$ on the null plane $x^+ = x_0^+$ and the asymptotic condition $\lim_{x^-\to\pm\infty}\phi = 0$, the Klein–Gordon equation has a unique solution in the half-space $x^+ \geq x_0^+$.*

It may seem a bit strange that it is the "space derivative" $\partial_-\phi = -\partial^+\phi$ that should be specified, along with the field ϕ itself, on the null-plane $x^+ = x_0^+$, and not the time derivative $\partial_+\phi$. However, this is as it should be. The Klein–Gordon equation is integrated in two steps. First the "space" integral over x^- is done. Then it is clear that the equation is effectively first order in the light-front time derivative. Therefore, no light-front time derivative of the field must be specified, but rather the light-front space derivative. The reader who is interested in these matters may consult the literature.

For our purposes, it is not necessary to go deeper as we are primarily interested in the "algebraic" problem of constructing interactions on the light-front. We say "algebraic" since our computations will be formal and perturbative in the sense of not worrying about matters of analysis. Solutions to nonlinear wave equations are anyway far beyond our objectives.

Light-cone or light-front?

---

i The designations "light-cone" and "light-front" are often used interchangeably, perhaps even randomly, in the literature to name this particular approach to field theory. Conceptually, however, *light-cone* refers to a three-dimensional surface $x^2 = 0$ in Minkowski space-time, or possibly, to the surface and the interior of the cone. *Light-front*, or equivalently *null-plane*, refers to a surface such as $x^+ = 0$, tangential to the light-cone. With this understanding, we need not be too pedantic about the usage of the words, as long as the meaning is clear from the context.

6.1.4 Light-front gauge fixing for spin 1

It is interesting to contrast covariant gauge fixing with light-front gauge fixing. Here, we will perform it for spin 1. We will use the notation of Section 6.1.2. The light-front gauge is chosen by setting

$$A^+ = 0 \tag{6.11}$$

and then studying the wave equations for the other components. First, note that in this gauge we get

$$\partial \cdot A = \partial_i A^i - \partial^+ A^- \tag{6.12}$$

We start with the wave equation in the "+" direction. It becomes

$$-\partial^+(\partial_i A^i - \partial^+ A^-) = 0 \tag{6.13}$$

Here, we at once encounter two slight subtleties having to do with the light-front x^- direction. The equation tells us that $\partial_i A^i - \partial^+ A^-$ is a constant function of x^-. To move on, we will ignore this subtlety, choosing the constant function to be zero. We can then formally solve equation (6.13) as

$$A^- = \frac{1}{\partial^+}\partial_i A^i \tag{6.14}$$

This is the, related, second subtlety. The operator $1/\partial^+$ is to be thought of as an integral operator. We will return to these questions in Section 6.2.1.

Next, we compute the wave equation in the transverse j directions. The computation gives

$$\Box A^j - \partial^j(\partial_i A^i - \partial^+ A^-) = \Box A^j = 0 \tag{6.15}$$

where we have used (6.14). Thus the two components of A^j describe a massless transverse vector field. Finally, the wave equation in the "–" direction becomes

$$\Box A^- - \partial^-(\partial_i A^i - \partial^+ A^-) = \Box A^- = 0 \tag{6.16}$$

which is formally consistent with (6.14) since

$$\Box A^- = \Box\left(\frac{1}{\partial^+}\partial_i A^i\right) = \frac{1}{\partial^+}\partial_i \Box A^i = 0 \tag{6.17}$$

Barring the x^- subtleties, what we see here is that one component of the vector potential A^+ is set to zero, another A^- is (almost) trivially solved for, while the remaining transverse components A^i describe two dynamical field degrees of freedom. The corresponding gauge fixing can be performed for all integer and half-integer gauge fields.

6.1.5 Spin-Lorentz transformations for spin 1

The spin part of a Lorentz transformation can be parametrized by six infinitesimal parameters $\varpi_{i+}, \varpi_{i-}, \varpi_{+-}$ and $\varpi_{ij} = \varpi$. Consider now such a transformation of the A^+ component

$$\delta_s A^+ = \varpi^+{}_+ A^+ + \varpi^+{}_- A^- + \varpi^+{}_i A^i = \varpi^+{}_i A^i \tag{6.18}$$

in the gauge $A^+ = 0$ and since $\varpi^+{}_- = -\varpi^{++} = 0$. We see that the vector field gets transformed out of the gauge. To stay in the gauge, one has to perform a compensating gauge transformation, or regauge transformation

$$\delta_\xi A^+ = \partial^+ \xi = -\varpi^+{}_i A^i \quad \Rightarrow \quad \xi = -\frac{1}{\partial^+}\varpi^+{}_i A^i \tag{6.19}$$

We can also compute the spin transformation of a physical component

$$\delta_s A^i = \varpi^i{}_j A^j + \varpi^i{}_+ A^+ + \varpi^i{}_- A^- + \partial^i \xi = \varpi^i{}_j A^j - \varpi^{i+}\frac{\partial_j}{\partial^+}A^j - \varpi^{+j}\frac{\partial^i}{\partial^+}A_j \tag{6.20}$$

where we are careful to remember the compensating gauge transformation.

This formula will look much nicer if we pass to complex notation. With the real $\varpi = \varpi^1{}_2 = -\varpi^2{}_1$ and ϖ^{+i} recombined into

$$\varpi^+ = \frac{1}{\sqrt{2}}(\varpi^{1+} + i\varpi^{2+}) \quad \text{and} \quad \bar{\varpi}^+ = \frac{1}{\sqrt{2}}(\varpi^{1+} - i\varpi^{2+}) \tag{6.21}$$

we get

$$\delta_s A = -i\varpi A - \varpi^+ \frac{\bar{\partial}}{\partial^+} A + \bar{\varpi}^+ \frac{\partial}{\partial^+} A \tag{6.22}$$

The transformation for $\bar{A}$ is the complex conjugate of this formula.

The corresponding analysis may be performed for free gauge fields of any spin. It is however easier to take advantage of the fact that for a field ϕ_λ with helicity λ one must have $\delta_\varpi \phi_\lambda = -i\varpi\lambda\phi_\lambda$ and then rely on the closure of the Poincaré algebra to arrive at

$$\delta_s \phi_\lambda = \lambda\left(-i\varpi - \varpi^+ \frac{\bar{\partial}}{\partial^+} + \bar{\varpi}^+ \frac{\partial}{\partial^+}\right)\phi_\lambda \tag{6.23}$$

6.2 Light-front basics in some detail

There are some delicate points when doing classical and quantum field theory on the light-front. They are all, in one way or another, connected to the, so-called, $1/p^+$ problem. The literature on the problem is quite extensive, and the conclusions seem to range from – to put it a bit bluntly – "there is no real problem" to "the method is inconsistent". As is often the case with field theory questions of this nature, the answer depends on the level of rigor applied, but there is also a practical point to it when concrete calculations has to be done. Judging from the large number of light-front and infinite-momentum applications to, for instance, strong interaction physics, one could perhaps risk arguing that, on sociological grounds, the method is basically sound. It is anyway beyond the present work to enter too deep into these questions. For the application, we have in mind we will content with a fairly formal treatment of the $1/p^+$'s. Our level of rigor will be defined in the box below.

6.2.1 Dirac analysis of the free field Lagrangian

For the canonical analysis of the free field theory, we start from the Lagrangian (1.7) for two scalar fields ϕ_1 and ϕ_2 which we combine into ϕ and $\bar{\phi}$ according to

$$\phi = \frac{1}{\sqrt{2}}(\phi_1 + i\phi_2) \quad \text{and} \quad \bar{\phi} = \frac{1}{\sqrt{2}}(\phi_1 - i\phi_2) \tag{6.24}$$

The interpretation – to be eventually done – is that ϕ_1 and ϕ_2 are the remaining two physical components of a higher spin massless tensor or tensor-spinor field after light-front gauge fixing in four dimensions. The Lagrangian, written with light-front derivatives, is

$$L = \int d^3x\mathcal{L} = -\int d^3x(\phi\partial^-\partial^+\bar{\phi} - \phi\partial\bar{\partial}\bar{\phi}) \tag{6.25}$$

The canonical conjugate momenta are defined by[7]

$$\pi(x) = \frac{\delta L}{\delta(\partial^-\bar{\phi}(x))} = \frac{1}{2}\partial^+\phi(x) \quad \text{and} \quad \bar{\pi} = \frac{\delta\mathcal{L}}{\delta(\partial^-\phi)} = \frac{1}{2}\partial^+\bar{\phi} \tag{6.26}$$

Space dependence will be suppressed in formulas whenever not needed for clarity, as in the equation for $\bar{\pi}$. Since ∂^+ is a space derivative, we see that the canonical momenta do not depend on the time derivative of the fields, signaling the presence of constraints.[8] We therefore have two primary constraints

$$\chi = \pi - \frac{1}{2}\partial^+\phi \approx 0 \quad \text{and} \quad \bar{\chi} = \bar{\pi} - \frac{1}{2}\partial^+\bar{\phi} \approx 0 \tag{6.27}$$

where $\approx$ means "weakly zero" in the Dirac sense. To get the Dirac procedure started, we impose the naive Poisson brackets

$$\{\phi(x), \bar{\pi}(y)\} = \{\bar{\phi}(x), \pi(y)\} = \delta(x^- - y^-)\delta^2(x_i - y_i) \equiv \delta^3(x - y) \tag{6.28}$$

We work with Poisson brackets here, eventually to be turned into equal x^+ commutators.[9] All other brackets between phase space variables, except those in (6.28) are naively zero.

We start by computing the Hamiltonian

$$H = \int d^3x(\pi\partial^-\bar{\phi} + \bar{\pi}\partial^-\phi) - L = \int d^3x\phi\partial\bar{\partial}\bar{\phi} \tag{6.29}$$

The Hamiltonian is not unique due to the presence of constraints, so we must add them to H with multiplier fields $u(x)$ and $\bar{u}(x)$ to get the effective Hamiltonian $\tilde{H}$

$$\tilde{H} = H + \int d^3x(\bar{u}(x)\chi(x) + u(x)\bar{\chi}(x)) \tag{6.30}$$

Next, we look for secondary constraints by requiring $\partial^-\chi = \partial^-\bar{\chi} = 0$. The results are[10]

$$\partial^-\chi(x) = \partial\bar{\partial}\phi(x) - \partial^+u(x) = 0 \tag{6.31}$$

---

7 The choice of which canonical momenta to be barred or unbarred is conventional.

8 See Section 3.2.2 for the Dirac analysis of constrained systems.

9 In bracket expressions such as in (6.28), it is to be understood that $x, y, \ldots$ in the right-hand side generalized functions stand for spatial coordinates x^-, x_i with $i = 1, 2$ or $x^+, x, \bar{x}$.

10 In this computation, and others of a similar nature in this section, we freely allow ourselves to use $\partial_y^+\delta(x^- - y^-) = -\partial_x^+\delta(x^- - y^-)$ and partial integrations with respect to x^- discarding surface terms. Furthermore, equalities involving constraints are really weak equalities $\approx$.

$$\partial^- \bar{\chi}(x) = \partial\bar{\partial}\bar{\phi}(x) - \partial^+ \bar{u}(x) = 0 \tag{6.32}$$

Thus no new constraints appear, instead we get equations for the multiplier fields $u(x)$ and $\bar{u}(x)$. Next, we compute the matrix $C(x,y)$ of constraints

$$C(x,y) = \begin{pmatrix} \{\chi(x),\chi(y)\} & \{\chi(x),\bar{\chi}(y)\} \\ \{\bar{\chi}(x),\chi(y)\} & \{\bar{\chi}(x),\bar{\chi}(y)\} \end{pmatrix} = -\begin{pmatrix} 0 & 1 \\ 1 & 0 \end{pmatrix} \partial_x^+ \delta(x^- - y^-)\delta^2(x_i - y_i) \tag{6.33}$$

This matrix can be formally inverted in the sense that

$$\int d^3z C(x,z)C^{-1}(z,y) = \delta^3(x-y) \tag{6.34}$$

with the result

$$C^{-1}(x,y) = \begin{pmatrix} 0 & 1 \\ 1 & 0 \end{pmatrix} \epsilon(x^- - y^-)\delta^2(x_i - y_i) \tag{6.35}$$

where the step function (half-sign function) ϵ satisfies (see box below)

$$\partial_x^+ \epsilon(x^- - y^-) = -\delta(x^- - y^-) \tag{6.36}$$

One could perhaps suspect that this cavalier treatment is not without its subtleties, but this is the way it is conventionally done, and we will conform to that tradition. For a deeper discussion, see [307].

It is now possible to compute the Dirac brackets (see Section 3.2.5). The non-zero matrix components of C^{-1} are $C^{-1}_{\chi\bar{\chi}}$ and $C^{-1}_{\bar{\chi}\chi}$. The Dirac bracket can then be computed from the formula

$$\begin{aligned} \{A(x), B(y)\}_D = \{A(x), B(y)\} & \\ - \int d^3z d^3w \{A(x), \chi(z)\} C^{-1}_{\chi\bar{\chi}}(z,w) \{\bar{\chi}(w), B(y)\} & \\ - \int d^3z d^3w \{A(x), \chi(\bar{z})\} C^{-1}_{\bar{\chi}\chi}(z,w) \{\chi(w), B(y)\} & \end{aligned} \tag{6.37}$$

The nonzero Dirac brackets become[11]

$$\{\phi(x), \bar{\phi}(y)\}_D = \epsilon(x^- - y^-)\delta^2(x_i - y_i) \tag{6.38a}$$

$$\{\phi(x), \bar{\pi}(y)\}_D = \frac{1}{2}\delta(x^- - y^-)\delta^2(x_i - y_i) \tag{6.38b}$$

$$\{\pi(x), \bar{\pi}(y)\}_D = \frac{1}{4}\partial_x^+ \delta(x^- - y^-)\delta^2(x_i - y_i) \tag{6.38c}$$

---

11 One must remember to compute all brackets between all canonical variables.

All other brackets between canonical variables compute to zero. However, since in light-front field theory, one seldom works with the canonical momenta, we also quote

$$\{\phi(x), \partial_y^+ \bar{\phi}(y)\}_D = \delta(x^- - y^-)\delta^2(x_i - y_i) \tag{6.39}$$

This bracket is consistent with the bracket (6.38b) upon using the second of the constraints in (6.27) which can now be taken as a strong equation.[12]

Some authors have a factor of 1/2 in the bracket (6.39), but that is conventional. The relevant point is that this Dirac bracket is nonzero whereas the corresponding Poisson bracket that the analysis started from is zero.[13] In any case, the bracket (6.39) will actually be our basic bracket. Formally, we could write it as

$$\{\phi(x), \bar{\phi}(y)\}_D = \frac{1}{\partial_y^+}\delta(x^- - y^-)\delta^2(x_i - y_i) \tag{6.40}$$

We will however never use it in this form, but rather the corresponding quantum commutator transformed to momentum space. It will be introduced in Section 6.3.1.

Interpretation of $1/\partial^+$ and $1/p^+$

Let us focus on the x^- direction only. From the equation,

$$\partial^+ g(x_+) = f(x_+) \tag{6.41}$$

it is clear that $g(x_+)$ is a primitive function to $f(x_+)$ and as such involves an undetermined constant. From this simple observation, we can devise a fairly rigorous interpretation of the operator $1/\partial^+$ when we formally write

$$g(x_+) = \frac{1}{\partial^+} f(x_+) \tag{6.42}$$

For that, integrate both sides of equation (6.41) from L to x_+,

$$\int_L^{x_+} \partial'^+ g(x'_+)dx'_+ = \int_L^{x_+} f(x'_+)dx'_+ \Rightarrow g(x_+) - g(L) = \int_L^{x_+} f(x'_+)dx'_+ \tag{6.43}$$

Then do the same with L replaced by L' with $L < x_+ < L'$. Adding the two resulting equations yields

$$2g(x_+) = g(L) + g(L') + \int_L^{x_+} f(x'_+)dx'_+ - \int_{x_+}^{L'} f(x'_+)dx'_+ \tag{6.44}$$

12 Note that $\partial^+\epsilon(x^-) = -\delta(x^-)$ with the minus sign due to the light-front metric.

13 Starting from the Lagrangian (6.25) multiplied by a constant k, the Dirac bracket (6.39) would come out with a factor $1/k$ on the right-hand side.

This we write as

$$g(x_+) = \frac{g(L) + g(L')}{2} + \int_L^{L'} \epsilon(x_+ - x'_+) f(x'_+) dx'_+ \tag{6.45}$$

where

$$\epsilon(x_+) = \begin{cases} +\frac{1}{2} & \text{for } x_+ > 0 \\ -\frac{1}{2} & \text{for } x_+ < 0 \end{cases} \quad \text{with} \quad \partial^+ \epsilon(x_+) = \frac{\partial}{\partial x_+} \epsilon(x_+) = \delta(x_+) \tag{6.46}$$

The arbitrary integration constant in $g(x_+)$ is represented by $\frac{1}{2}(g(L) + g(L'))$ in (6.45). The arbitrariness can be removed by imposing boundary conditions such as

$$\lim_{L \to -\infty} g(L) = \lim_{L' \to \infty} g(L') = 0 \tag{6.47}$$

Compare to the Neville–Rohrlich theorems of Section 6.1.3.

6.3 Free higher helicity fields

With a background of the above considerations, we now retreat a little to set up a formalism for approaching the problem of interactions for massless higher spin fields on the light-front. In four space-time dimensions dimension, all gauge fields have 2 physical degrees of freedom. This simplifies the mathematics since a complexified notation can be used throughout. Thus we work with the field ϕ_λ and its complex conjugate $\bar{\phi}_\lambda$ of helicities λ and $-\lambda$, respectively. These will be called *helicity fields*. The wave equations are

$$\partial^- \phi_\lambda = \frac{\partial \bar{\partial}}{\partial^+} \phi_\lambda \quad \text{and} \quad \partial^- \bar{\phi}_\lambda = \frac{\partial \bar{\partial}}{\partial^+} \bar{\phi}_\lambda \tag{6.48}$$

irrespective of helicity. Indeed, gauge fixing the four-dimensional Fronsdal [3] equations for arbitrary spin, yield these equations, and helicity data is encoded in the Lorentz transformations as shown in Section 6.1.5.

6.3.1 Fields, Fourier transforms and commutators

We will think of the higher helicity fields as quantum fields so that the Dirac brackets of Section 6.2.1 will be turned into equal light-front time commutators according to the scheme of (1.1). That is, we take as our quantization rule

$$\text{If} \quad \{A(x), B(y)\}_D = C \quad \text{then} \quad [A(x), B(y)]_{x^+ = y^+} = iC \tag{6.49}$$

Let us start with two real fields $\phi_k(x)$ such as the transverse components of a light-front higher helicity field. The basic equal x^+ commutator is thus taken as

$$[\phi_k(x), \partial_y^+ \phi_l(y)]_{x^+=y^+} = i\delta^3(x-y)\delta_{kl} \tag{6.50}$$

For Fourier transform pairs, we choose

$$\begin{aligned} \phi_k(x) &= \frac{1}{(2\pi)^{3/2}} \int d^3p \phi_k(p) e^{ip\cdot x} \\ \phi_k(p) &= \frac{1}{(2\pi)^{3/2}} \int d^3x \phi_k(x) e^{-ip\cdot x} \end{aligned} \tag{6.51}$$

The complex field is introduced by defining

$$\phi = \frac{1}{\sqrt{2}}(\phi_1 + i\phi_2) \quad \bar{\phi} = \frac{1}{\sqrt{2}}(\phi_1 - i\phi_2) \tag{6.52}$$

both for x-space and p - space fields. In terms of the complex field, the nonzero equal time commutator becomes

$$[\phi(x), \partial_y^+ \bar{\phi}(y)]_{x^+=y^+} = i\delta^3(x-y) \tag{6.53}$$

Fourier transforming, we get the momentum space equal time commutator, using $q^+ e^{-iq\cdot y} = i\partial_y^+ e^{-iq\cdot y}$ and partially integrating

$$\begin{aligned} [\phi_k(p), q^+ \phi_l(q)] &= \frac{1}{(2\pi)^3} \int d^3x d^3y [\phi_k(x), \phi_l(y)] i\partial_y^+ e^{-i(p\cdot x + q\cdot y)} \\ &= \frac{1}{(2\pi)^3} \int d^3x d^3y \delta_{kl} \delta^3(x-y) e^{-i(p\cdot x+q\cdot y)} \\ &= \delta_{kl}\delta^3(p+q) \end{aligned} \tag{6.54}$$

For complex fields, this yields the convenient form

$$[\phi(p), \bar{\phi}(q)]_{x^+=y^+} = \frac{\delta^3(p+q)}{q^+} \tag{6.55}$$

To maintain fields of all helicities, we introduce a two-dimensional complex internal Fock space spanned by oscillator pairs $(\beta, \bar{\alpha})$ and $(\bar{\beta}, \alpha)$ where

$$[\beta, \bar{\alpha}] = [\bar{\beta}, \alpha] = 1 \tag{6.56}$$

thus taking α and $\bar{\alpha}$ as creation operators and β and $\bar{\beta}$ as annihilation operators.[14]

---

14 This is a change of notation in reference to the notation of my papers, where the double use of † as designating a creation operator and the notation for Hermitian conjugation, caused a potential conflict with the complex notation.

Using this, we collect all helicities in a Fock space field,

$$|\Phi(p)\rangle = \sum_{\lambda=0}^{\infty} \frac{1}{\sqrt{\lambda!}} \left(\phi_\lambda(p)\bar{\alpha}^\lambda + \bar{\phi}_\lambda(p)\alpha^\lambda \right) |0\rangle \tag{6.57}$$

In formulas like this, p is short for $p, \bar{p}$ and $\gamma = p^+$. The conjugated Fock space field is given by

$$|\Phi(p)\rangle^\dagger = \langle \Phi(-p)| = \sum_{\lambda=0}^{\infty} \frac{1}{\sqrt{\lambda!}} \langle 0| \left(\bar{\phi}_\lambda(-p)\beta^\lambda + \phi_\lambda(-p)\bar{\beta}^\lambda \right) \tag{6.58}$$

The fields ϕ and $\bar{\phi}$ being functions of $x^+, p^+, p, \bar{p}$, carry mass dimension −2. This follows from the Fourier transform (6.51) since $\phi(x)$ has dimension 1 as usual. The vacuum and the oscillators are dimensionless, hence the Fock space field $|\Phi(p)\rangle$ also carry dimension −2.

The light-front internal Fock-space

---

i The complex pairs of oscillators $(\beta, \bar{\alpha})$ and $(\bar{\beta}, \alpha)$ can be introduced in the following way from a pair of oscillators $(\alpha_i, \alpha_i^\dagger)$, $i = 1, 2$, with the usual commutators $[\alpha_i, \alpha_j^\dagger] = \delta_{ij}$:

$$\beta = \frac{1}{\sqrt{2}}(\alpha_1 + i\alpha_2) \qquad \bar{\beta} = \frac{1}{\sqrt{2}}(\alpha_1 - i\alpha_2) \tag{6.59}$$

$$\alpha = \frac{1}{\sqrt{2}}(\alpha_1^\dagger + i\alpha_2^\dagger) \qquad \bar{\alpha} = \frac{1}{\sqrt{2}}(\alpha_1^\dagger - i\alpha_2^\dagger) \tag{6.60}$$

The commutation relations of (6.56) are satisfied. The operators α and $\bar{\alpha}$ are creation operators. One can compute, for instance, $\bar{\alpha} = \beta^\dagger$ and $\alpha = \bar{\beta}^\dagger$, but the use of † for designating creation operators is confusing together with the complex notation, and will not be used.

6.3.2 The light-front Poincaré algebra

We start from the four-dimensional Poincaré algebra as given in formulas (3.98)–(3.100) and rewrite it in terms of light-front coordinates and momenta. The algebra consists of 45 commutators, 22 of which are nonzero.

In light-front dynamics, the Poincaré generators split into a set $\mathcal{K}$ of *kinematic generators* and a set $\mathcal{D}$ of *dynamic generators*. The kinematic generators are those that leave the light-front $x^+ = 0$ invariant, while the dynamic generator transform out of the front. We get[15]

$$\mathcal{K} = \{p^+, p, \bar{p}\} \cup \{j, j^{+-}, j^+, \bar{j}^+\} \tag{6.61}$$

$$\mathcal{D} = \{p^-\} \cup \{j^-, \bar{j}^-\} \tag{6.62}$$

---

15 As there is an obvious symmetry between + and – directions, it should be noted that this particular split corresponds to taking x^+ as the light-front time. Also note that, although we refer to $\mathcal{K}$ and $\mathcal{D}$ as sets here, they are actually Lie algebras, as will become clear.

Transformations off the null plane

Physically, the Hamiltonian h transforms the null plane $x^+ = 0$ into another null plane $x^+ = \tau$ infinitesimally near. The dynamic Lorentz generators – which can be thought of as boosts – generate infinitesimal rotations of the plane $x^+ = 0$ around the surface of the light-cone $x^2 \leq 0$, thus also transforming off the plane.

In the free field theory, all the generators are linearly realized on the fields. The dynamical generators – the Hamiltonians in Dirac's terminology – will be nonlinearly realized in an interacting theory. The idea behind the light-front approach to interacting higher spin field theory is to try to construct interactions by an ansatz-verification method.

The light-front momentum operators $p^-, p^+, p, \bar{p}$ are defined analogous to other light-front vectors. The angular momentum generators need a little more work, but the idea is that every tensor index μ is rewritten in terms of the directions "plus", "minus", "unbarred" and "barred". Doing this, we get the following short calculations:

$$j = j_{12} \tag{6.63}$$

$$j^{+-} = \frac{1}{\sqrt{2}}(j^{+0} - j^{+3}) = \frac{1}{2}(j^{00} + j^{30} - j^{03} - j^{33}) = j^{30} \tag{6.64}$$

$$j^{+} = \frac{1}{\sqrt{2}}(j^{1+} + ij^{2+}) = \frac{1}{2}(j^{10} + j^{13} + ij^{20} + ij^{23}) \tag{6.65}$$

$$\bar{j}^{+} = \frac{1}{\sqrt{2}}(j^{1+} - ij^{2+}) = \frac{1}{2}(j^{10} + j^{13} - ij^{20} - ij^{23}) \tag{6.66}$$

$$j^{-} = \frac{1}{\sqrt{2}}(j^{1-} + ij^{2-}) = \frac{1}{2}(j^{10} - j^{13} + ij^{20} - ij^{23}) \tag{6.67}$$

$$\bar{j}^{-} = \frac{1}{\sqrt{2}}(j^{1-} - ij^{2-}) = \frac{1}{2}(j^{10} - j^{13} - ij^{20} + ij^{23}) \tag{6.68}$$

The algebra of these generators can then be worked out. It will be convenient to organize the resulting commutators in three groups, each containing three subtypes. This will facilitate the task of controlling them all when constructing nonlinear interaction representations.

$\mathcal{K} - \mathcal{K}$ commutators

The 21 commutators of this type have the following algebraic structure:

$$[\mathcal{K}, \mathcal{K}] \subset \mathcal{K} \qquad \#9$$

$$[\mathcal{K}, \mathcal{K}] \subset \mathcal{D} \qquad \#0$$

$$[\mathcal{K}, \mathcal{K}] = \emptyset \qquad \#12$$

The three types of right-hand sides of the commutators will be called *linear*, *non-linear* and *zero*, respectively. The nonzero commutators of this type are

$$[j, p] = ip \qquad [j, \bar{p}] = -i\bar{p} \tag{$\mathcal{KK}$.1}$$

$$[j^+, \bar{p}] = ip^+ \qquad [\bar{j}^+, p] = ip^+ \tag{$\mathcal{KK}$.2}$$

$$[j^{+-}, p^+] = ip^+ \tag{$\mathcal{KK}$.3}$$

$$[j, j^+] = j^+ \qquad [j, \bar{j}^+] = -\bar{j}^+ \tag{$\mathcal{KK}$.4}$$

$$[j^{+-}, j^+] = ij^+ \qquad [j^{+-}, \bar{j}^+] = i\bar{j}^+ \tag{$\mathcal{KK}$.5}$$

Commuting two kinematic generators can never give a dynamic generator. This part of the algebra, being satisfied by the free theory by construction therefore has no further consequences for the interactions.

$\mathcal{K} - \mathcal{D}$ commutators

The 21 commutators of this type have the following algebraic structure:

$$\begin{array}{ll} [\mathcal{K}, \mathcal{D}] \subset \mathcal{K} & \#6 \\ [\mathcal{K}, \mathcal{D}] \subset \mathcal{D} & \#7 \\ [\mathcal{K}, \mathcal{D}] = \emptyset & \#8 \end{array}$$

The nonzero commutators of the first subtype are

$$[p^+, j^-] = -ip \qquad [p^+, \bar{j}^-] = -i\bar{p} \tag{$\mathcal{KD}$.1}$$

$$[j^+, p^-] = ip \qquad [\bar{j}^+, p^-] = i\bar{p} \tag{$\mathcal{KD}$.2}$$

$$[j^+, \bar{j}^-] = ij^{+-} - j \qquad [\bar{j}^+, j^-] = ij^{+-} + j \tag{$\mathcal{KD}$.3}$$

These commutators tell us that the kinematic transformations commute with the nonlinear part of the dynamic transformations. In practice, therefore, since the free part is satisfied by construction, they form a set of zero commutators together with the third subtype. These are

$$[p, p^-] = 0 \qquad [\bar{p}, p^-] = 0 \qquad [p^+, p^-] = 0 \tag{$\mathcal{KD}$.4}$$

$$[p, j^-] = 0 \qquad [\bar{p}, \bar{j}^-] = 0 \tag{$\mathcal{KD}$.5}$$

$$[j^+, j^-] = 0 \qquad [\bar{j}^+, \bar{j}^-] = 0 \tag{$\mathcal{KD}$.6}$$

$$[j, p^-] = 0 \tag{$\mathcal{KD}$.7}$$

Together these will fix some of the structure of the interaction terms. The nonzero commutators of the second subtype are

$$[j^{+-}, p^-] = -ip^- \tag{$\mathcal{KD}$.8}$$

$$[\bar{p}, j^-] = -ip^- \qquad [p, \bar{j}^-] = -ip^- \tag{$\mathcal{KD}$.9}$$

$$[j^{+-}, j^{-}] = -ij^{-} \qquad [j^{+-}, \bar{j}^{-}] = -i\bar{j}^{-} \qquad (\mathcal{KD}.10)$$

$$[j, j^{-}] = j^{-} \qquad [j, \bar{j}^{-}] = -\bar{j}^{-} \qquad (\mathcal{KD}.11)$$

These work order by order in the interaction and fix still more of the structure. There is no mixing of interaction terms of different order in these commutators. Taken together, the $\mathcal{K} - \mathcal{D}$ commutators determine the general form of the interactions up to p^+-structure.

$\mathcal{D} - \mathcal{D}$ commutators

The 3 commutators of this type have the following algebraic structure:

$$[\mathcal{D}, \mathcal{D}] \subset \mathcal{K} \qquad \#0$$

$$[\mathcal{D}, \mathcal{D}] \subset \mathcal{D} \qquad \#0$$

$$[\mathcal{D}, \mathcal{D}] = \emptyset \qquad \#3$$

With much of the structure already determined, the $\mathcal{D}-\mathcal{D}$ commutators yield recursive differential equations in the $\gamma_r = p_r^+$ for the higher spin fields Φ_r entering an interaction term. Note that in super-Poincaré algebras there are $[\mathcal{D}, \mathcal{D}] = \mathcal{D}$ commutators. In the case that is discussed here, without supersymmetry, there are only the zero commutators

$$[h, j^{-}] = 0 \quad [h, \bar{j}^{-}] = 0 \quad [j^{-}, \bar{j}^{-}] = 0 \qquad (6.69)$$

The third commutator contain all information, as performing it requires the first two.

In the generic case, once cubic interaction terms are found, they will lead to higher order terms. Commuting two transformations on the cubic level will in general not be zero, but lead to quartic level transformations.[16]

The p^+ structure in interacting theory

We will indeed see that it is the detailed p^+ structure that is left undetermined by kinematical part of the algebra. Instead it is determined by the dynamical commutators. This may be a significant point: the light-front buys us quite a few simplifications – if we are prepared to work in a non-covariant formalism – but the price to pay is that the really hard difficulties are concentrated to the p^+ structure. ↯

A tentative connection to BRST theory

Given that all interaction data will be carried by the deformations of j^- and $\bar{j}^-$, the equation $[j^-, \bar{j}^-] = 0$, although being a commutator, resembles the $\{Q, Q\} = 0$ equation of deformed BRST-theory. Such de- ?

16 There is, however, a special case – discovered by R. R. Metsaev – where the cubic theory closes the algebra by itself [208, 209]. The theory is however not unitary.

formations result in strongly homotopy, or L_∞ algebras [308]. Support for this connection comes from Siegel's and Zwiebach's work in the 1980s deriving the BRST gauge fixed string theory from light-front string theory by introducing ghost coordinates and constructing the BRST and anti-BRST operators out of the dynamical Lorentz operators [164, 309].

6.3.3 The free theory light-front Poincaré generators

The next step is to write down an explicit realization of the light-front Poincaré algebra. We start in a first quantized setting with the covariant generators given by

$$P_\mu = p_\mu \quad \text{and} \quad J_{\mu\nu} = x_\mu p_\nu - x_\nu p_\mu + M_{\mu\nu} \tag{6.70}$$

where we have $[x_\mu, p_\nu] = i\eta_{\mu\nu}$. The spin Lorentz generators $M_{\mu\nu}$ will be specified in terms of the two-dimensional oscillators (6.56). In complex notation, a linear realization can be expressed as

$$j = i(x\bar{p} - \bar{x}p) + M \qquad j^{+-} = i\frac{\partial}{\partial y}y = iy\frac{\partial}{\partial y} + i \tag{6.71}$$

$$j^+ = xy \qquad \bar{j}^+ = \bar{x}y \tag{6.72}$$

$$j^- = xh + ip\frac{\partial}{\partial y} - \frac{i}{y}Mp \qquad \bar{j}^- = \bar{x}h + i\bar{p}\frac{\partial}{\partial y} + \frac{i}{y}M\bar{p} \tag{6.73}$$

where the momenta p^+ is denoted by y and the Hamiltonian p^- is denoted by h and given by

$$h = \frac{p\bar{p}}{y} \tag{6.74}$$

The helicity contributions to j, j^- and $\bar{j}^-$ is carried by

$$M = \alpha\bar{\beta} - \bar{\alpha}\beta \tag{6.75}$$

So far, this is a hybrid notation where we think of x and $\bar{p}$ as first quantized with

$$[x, \bar{p}] = [\bar{x}, p] = i \tag{6.76}$$

but we treat x^- and p^+ explicitly in terms of

$$p^+ = y \quad \text{and} \quad x^- = -i\frac{\partial}{\partial y} \tag{6.77}$$

This turns out to be convenient [310].[17] Further convenience will be gained by representing x and $\bar{x}$ as

$$x = i\frac{\partial}{\partial \bar{p}} \quad \text{and} \quad \bar{x} = i\frac{\partial}{\partial p} \tag{6.78}$$

17 We differ from [205] which used $\beta = 2p^+$.

Lightfrontiana: α, β, γ or η

The motivation for introducing a new symbol for the momenta p^+ is readability and aesthetics, apart from the fact that p^+ plays quite a prominent role in light-front physics. In the light-cone superstring papers of the 1980s, p^+ was denoted by α. As also the oscillators of string theory were denoted by α's and we had the α', too, the present author, in connection to working on the paper [205], decided to use β instead. This was taken up by Metsaev and most subsequent authors on light-front higher spin theory. Later on, I found β a bit typographically unaesthetic, so changed over to γ (that looks great both in handwriting and print). One alternative would be to return to η that was used in the 1960s research on the infinite momentum frame; see, for instance, [311–313]. I am now using β and $\bar{\beta}$ for transverse annihilators. Notation, notation, notation, always notation!

The generator j measures helicity. All basic variables are eigenvectors of j with eigenvalues equal to the helicity of the variable. Consider $[j, v] = \lambda v$. Explicitly, we get the following helicity assignments:

$$\lambda = 1 \qquad x, p, \alpha, \beta \tag{6.79}$$

$$\lambda = 0 \qquad x^-, p^+ \tag{6.80}$$

$$\lambda = -1 \qquad \bar{x}, \bar{p}, \bar{\alpha}, \bar{\beta} \tag{6.81}$$

Satisfying the Poincaré algebra

The orbital part of the generator j^- is, to start with, given by $j^- = xp^- - x^-p$. Then p^- is replaced by the free Hamiltonian (6.74) and x^- is replaced according to (6.77). The very same replacements are made for $\bar{j}^-$. Furthermore, j^{+-} is given by $j^{+-} = x^+p^- - x^-p^+$. Then we set $x^+ = 0$ and again x^- according to (6.77). The Poincaré algebra is still satisfied after these substitutions are made, as can be explicitly checked.

The precise connection between these generators and field theory operators can be worked out as follows. The basic equal time light-front field commutator is (as shown in Section 6.3.1)

$$[\phi(x), \partial_y^+ \bar{\phi}(y)]_{x^+=y^+} = i\delta^3(x-y) \tag{6.82}$$

It translates into the corresponding momentum space equal time commutator

$$[\phi(p), q^+ \bar{\phi}(q)]_{x^+=y^+} = \delta^3(p+q) \tag{6.83}$$

or

$$[\phi(p), \bar{\phi}(q)]_{x^+=y^+} = \frac{\delta^3(p+q)}{q^+} \tag{6.84}$$

Now, taking the Hamiltonian as a template, we want to generate transformations

$$\delta_H \phi(p) = h\phi(p) \quad \text{with} \quad h = \frac{p\bar{p}}{p^+} \tag{6.85}$$

Then try the field theory operator

$$H = \int q^+ dq^+ d^2q \bar{\phi}(-q) \frac{q\bar{q}}{q^+} \phi(q) \tag{6.86}$$

and compute

$$\begin{aligned}
\delta_H \phi(p) &= [\phi(p), H] \\
&= \int q^+ dq^+ d^2q ([\phi(p), \bar{\phi}(-q)]_{x^+=y^+}) \frac{q\bar{q}}{q^+} \phi(q) \\
&= -\int q^+ dq^+ d^2q ([\phi(p), \bar{\phi}(q)]_{x^+=y^+}) \frac{q\bar{q}}{q^+} \phi(-q) \\
&= -\int dq^+ d^2q \delta^3(p+q) \frac{q\bar{q}}{q^+} \phi(-q) = \frac{p\bar{p}}{p^+} p^+ \phi(p)
\end{aligned} \tag{6.87}$$

where in second equality we have performed the change of variables $q \to -q$. For any one Poincaré generator g, the analogous computation can be performed and we define the field theory generator G corresponding to g,

$$\delta_G \phi(p) = [\phi(p), G] \quad \text{with} \quad G = \int q^+ dq^+ d^2q \bar{\phi}(-q) g \phi(q) \tag{6.88}$$

The generalization to Fock fields is given by

$$G = \frac{1}{2} \int y dy dp d\bar{p} \langle \Phi | g | \Phi \rangle \tag{6.89}$$

In particular, for the Hamiltonian, we write

$$H = \frac{1}{2} \int y dy dp d\bar{p} \langle \Phi | h | \Phi \rangle \tag{6.90}$$

In an interacting field theory, the dynamical transformations become nonlinear and the dynamical generators acquire contributions cubic in fields.

6.3.4 Configuration space vs. momentum space representation

The momentum space form of the Hamiltonian operator of equation (6.86) is consistent with the configuration space form

$$H = -i \int d^3x \partial^+ \bar{\phi}(x) \frac{\partial\bar{\partial}}{\partial^+} \phi(x) \tag{6.91}$$

as may be shown by Fourier transforming between momentum space and configuration space using the transform pairs of (6.51). Using the equal time commutator (6.82), it follows that

$$\delta_H \phi(x) = [\phi(x), H] = \frac{\partial\bar{\partial}}{\partial^+} \phi(x) \tag{6.92}$$

For any of the free theory generators, we have

$$G = -i \int d^3x \partial^+ \bar{\phi}(x) g \phi(x) \tag{6.93}$$

with the generalization to Fock fields

$$G = -\frac{i}{2} \int d^3x \partial^+ \langle \Phi(x) | g | \Phi(x) \rangle \tag{6.94}$$

The one point to remember here is to use the appropriate representation for the generators g: Momenta represented by derivatives in configuration space, and vice versa in momentum space. As we will be using the momentum space representation, this means working with coordinates represented as in formulas (6.77) and (6.78). Having this clear, we can now move on to discuss the Hermiticity properties of the generators.

6.3.5 Hermiticity properties of light-front Poincaré generators

In Section 3.3.1, we discussed the property of Hermiticity for operators in abstract quantum mechanics. We now have to apply this to a study of the properties of the light-front Poincaré generators. It may seem that this is a trivial issue as the covariant generators $P_\mu = p_\mu$ and $J_{\mu\nu} = x_\mu p_\nu - x_\nu p_\mu$, are obviously Hermitian. However, there are some complications related to light-front peculiarities such as the complexified notation, setting $x^+ = 0$, taking $p^- = p\bar{p}/p^+$ and the nontrivial integration measure $p^+ dp^+$, that make Hermiticity well worth a study.

For instance, naive Hermiticity checking of the generators (6.71)–(6.72), along the lines $p_\mu^\dagger = p_\mu$ and $J_{\mu\nu}^\dagger = (x_\mu p_\nu - x_\nu p_\mu)^\dagger = p_\nu x_\mu - p_\mu x_\nu = x_\mu p_\nu - x_\nu p_\mu$, does not work by itself. It works for j^+ and $\bar{j}^+$. It works for the orbital part of j as the following short calculation shows

$$j^\dagger = (i(x\bar{p} - \bar{x}p))^\dagger = -i(p\bar{x} - \bar{p}x) = i(x\bar{p} - \bar{x}p) \tag{6.95}$$

where we use the commutators (6.76) in the last equality. However, for j^{+-} this does not work. The noncommutativity of $p^+ = y$ and $x^- = -i\partial/\partial y$ is not compensated by the noncommutativity x^+ and p^- since we have set $x^+ = 0$. The remedy is to consider $p^+ j^{+-}$ where, as we will see, the factor p^+ comes from the integration measure $p^+ dp^+ d^2p$. Then we have

$$(p^+ j^{+-})^\dagger = \left(ip^+ \frac{\partial}{\partial y} y \right)^\dagger = -iy \frac{\overleftarrow{\partial}}{\partial y} p^+ = iy \frac{\partial}{\partial y} p^+ = p^+ j^{+-} \tag{6.96}$$

In the second and third equalities, we have employed the usual pragmatic algorithm for Hermiticity checking from elementary quantum mechanics (see box below).

Similarly, for the conjugated pair j^- and $\bar{j}^-$, we must consider p^+j^- and $p^+\bar{j}^-$ as the following calculation shows

$$(p^+j^-)^\dagger = \left(p^+\left(xh + ip\frac{\partial}{\partial y}\right)\right)^\dagger = h\bar{x}p^+ - i\frac{\overleftarrow{\partial}}{\partial y}\bar{p}p^+ = p^+\bar{x}h + p^+[h,\bar{x}] + i\frac{\partial}{\partial y}\bar{p}p^+$$
$$= p^+\bar{x}h - i\bar{p} + ip^+\bar{p}\frac{\partial}{\partial y} + i\bar{p} = p^+\left(\bar{x}h + i\bar{p}\frac{\partial}{\partial y}\right) = p^+\bar{j}^- \tag{6.97}$$

Here, we understand that, while x and p^- commutes, x and $h = p\bar{p}/y$ does not. This cannot be compensated for by commuting p and $\partial/\partial y$. However, the p^+ factor restores Hermiticity in the sense that j^- and $\bar{j}^-$ is a conjugated pair. With this understanding, we now have for the Poincaré generators

$$(p^+g)^\dagger = p^+g \quad \text{for } g = \{p^+, h, j, j^{+-}\} \tag{6.98}$$
$$(p^+g)^\dagger = p^+\bar{g} \quad \text{for } g = \{p, \bar{p}, j^+, \bar{j}^+, j^-, \bar{j}^-\} \tag{6.99}$$

Hermiticity checking algorithm in elementary quantum mechanics

! In elementary one-dimensional quantum mechanics, we take, perhaps without further thinking, the operators x and p as Hermitian in the sense $x^\dagger = x$ and $p^\dagger = p$. For x, this is indeed trivial, but for p, represented as $p = -id/dx$ we need to remember that the operator act on a wave function. To be completely clear how the Hermiticity checking algorithm works, we introduce the *operations of complex conjugation and transposition*. Let G denote an operator of some sort, and Φ and Ψ wave functions. If $G\Phi = \Psi$, then the complex conjugate G^* of G is defined by

$$G^*\Phi^* = \Psi^* \tag{6.100}$$

The transposition G^T of G is defined by

$$\int \Psi(G^T\Phi)dx = \int \Phi(G\Psi)dx \tag{6.101}$$

For the momentum operator $P = -id/dx$ in particular, we have

$$P^* = id/dx \quad \text{for complex conjugation} \tag{6.102}$$
$$P^T = -i\frac{\overleftarrow{d}}{dx} \quad \text{for transposition} \tag{6.103}$$

The switch of direction of action for the derivative is a direct consequence of the definition (6.101). So is the rule $(GF)^T = F^TG^T$. Consider now the expectation value $\langle G\rangle = \int \Psi^*G\Psi dx$ and compute its complex conjugate $\langle G\rangle^*$ and require it to be real

$$\langle G\rangle^* = \left(\int \Psi^*G\Psi dx\right)^* = \int (G\Psi)^*\Psi dx \stackrel{?}{=} \int \Psi^*G\Psi dx \tag{6.104}$$

The first equality is the definition of the expectation value complex conjugated, the second equality is the norm property, and the last equality is the reality requirement (marked by a '?' since it is what needs to be checked). To show reality, we must transfer the action of the operator from $(G\Psi)^*$ to $G\Psi$. This is done in two steps, first using the definition of the complex conjugate of the operator (6.100), second using the definition of the transpose of the operator (6.101). Performing these two steps yield

$$\int (G\Psi)^* \Psi dx = \int G^* \Psi^* \Psi dx = \int \Psi^* (G^*)^T \Psi dx \tag{6.105}$$

The reality requirement then implies $(G^*)^T = G$, which we recognize as Hermiticity $G^\dagger = G$ with $(G^*)^T = G^\dagger$. Taking momentum as an example, the calculation runs as follows:

$$\langle P \rangle^* = \left(\int \Psi^* \left(-i\frac{d}{dx}\Psi \right) \right)^* dx = \int \left(-i\frac{d}{dx}\Psi \right)^* \Psi dx = \tag{6.106}$$

$$= \int i\frac{d}{dx}\Psi^* \Psi dx = \int \Psi^* (i\frac{\overleftarrow{d}}{dx}\Psi) dx = \int \Psi^* \left(-i\frac{d}{dx}\Psi \right) dx \tag{6.107}$$

where the equalities are effected by, in turn: first definition, norm property, complex conjugation, transposition, and finally partial integration. Pragmatically, the operations of complex conjugation and transposition, combined into Hermitian conjugation followed by partial integration, are packaged into the algorithm

$$\left(-i\frac{d}{dx} \right)^\dagger = i\frac{\overleftarrow{d}}{dx} = -i\frac{d}{dx} \tag{6.108}$$

supplied with the additional rule for products of operators $(GF)^T = F^T G^T$.

As we have seen here, questions about Hermiticity, and unitarity, of operators are ultimately dependent on the Hilbert space of states that they "live in".

With this groundwork done, we can now check the Hermiticity properties of the field theory Poincaré operators. We want to show that either $G^\dagger = G$ or $G^\dagger = \bar{G}$. In order to do that, we customize the Hermiticity checking algorithm according to

$$G^\dagger = \int dq^+ d^2q (q^+ g\phi(q))^\dagger \phi(q) = \int dq^+ d^2\bar{\phi}(-q)\overleftarrow{g}^* q^+ \phi(q) \tag{6.109}$$

where $\phi(q)^\dagger = \bar{\phi}(-q)$ and $\bar{\phi}(-q)^\dagger = \phi(q)$. By writing $\overleftarrow{g}^*$ we denote the effect of computing $g^\dagger$. This involves complex conjugation, transposition of products operators including reversing the direction of action of the y derivatives. Next, using the commutators $[x, \bar{p}] = [\bar{x}, p] = i$ and partial integration with respect to y, we should arrive back at G or $\bar{G}$ as appropriate. That this is indeed the case, follows from the discussion above leading up to equations (6.98) and (6.99). Note that it works just as well representing x and $\bar{x}$ as derivatives according to (6.78).

6.4 Chapter 6 epilogue

A book must stop somewhere, and let us stop with a little appetizer. The following observation has never, to the best of my knowledge, been exploited in attempts to construct higher order interactions for massless higher spin fields on the light-front.

The commutators between j^{+-} and the rest of the Poincaré generators offer a way to split the algebra into three subalgebras. Any operator A that satisfies

$$[j^{+-}, A] = igA \tag{6.110}$$

is said to have *goodness* g. Then referring back to the list of commutators in Section 6.3.3, we can read of the following goodness classification[18] of the Poincaré generators:

$$\mathcal{G}_+ = \{\gamma, j^+, \bar{j}^+\} \quad \text{with } g = +1 \tag{6.111}$$

$$\mathcal{G}_0 = \{p, \bar{p}, j, j^{+-}\} \quad \text{with } g = 0 \tag{6.112}$$

$$\mathcal{G}_- = \{h, j^-, \bar{j}^-\} \quad \text{with } g = -1 \tag{6.113}$$

These are three subalgebras (although not invariant) of the Poincaré algebra, and $\mathcal{G}_+$ and $\mathcal{G}_-$ are Abelian, while $\mathcal{G}_0$ is non-Abelian. We also recognize the set of kinematical generators as $\mathcal{K} = \mathcal{G}_+ \cup \mathcal{G}_0$. Again looking back at the commutators of Section 6.3.3, it is clear that the algebra of kinematical generators is a semidirect product of the algebra of goodness 0 generators with those of goodness +1, that is,

$$\mathcal{K} = \mathcal{G}_+ \rtimes \mathcal{G}_0 \tag{6.114}$$

Indeed, since $[\mathcal{G}_0, \mathcal{G}_+] \subset \mathcal{G}_+$, we see that $\mathcal{G}_+$ is an invariant subalgebra of $\mathcal{K}$ (but not of the full Poincaré algebra).[19] By symmetry, the semidirect product $\mathcal{G}_- \rtimes \mathcal{G}_0$ also forms a 7-parameter subalgebra.

What is perhaps more intriguing is the fact that the goodness split is actually also a triangular decomposition of the light-front Poincaré algebra. Therefore, we have

$$[\mathcal{G}_0, \mathcal{G}_\pm] \subseteq \mathcal{G}_\pm \qquad [\mathcal{G}_+, \mathcal{G}_-] \subseteq \mathcal{G}_0 \tag{6.115}$$

One cannot escape thinking of this is terms of raising and lowering operators. The question is: can this be exploited in some way for nonlinear representations of the light-front Poincaré algebra? Especially since the algebra is noncompact and the unitary representations are infinite-dimensional. It turns out, actually, that there is a nonunitary "finite-dimensional" purely cubic theory on the light-front. It is in principle the theory found in 1983 and 1987, elaborated by Metsaev in [208, 209] and clarified by D. Ponomarev and E. Skvortsov in [210].

18 The terminology of "goodness" derives from the infinite momentum approach to quantum field theory. The three cases have been referred to as "good", "bad" and "terrible" respectively.

19 Compare how the translations form an invariant Abelian subalgebra of the full Poincaré algebra, leading to the semidirect product structure of translations with Lorentz transformations.

A Epilogue

It is time to finish this manuscript. I have to apologize to the reader for having written so many pages and still only treated the free field theory in Minkowski space-time, and having to ask for patience until the second volume on the interacting theory will hopefully arrive in a couple of years. However, upon looking back at the text, I do think that I have managed to stay true to the vision outlined in the preface and in the introductory chapter: the "rethinking" vision, for short.

Furthermore, the text is not all about free field theory, there are actually quite a lot on the interacting theories of spin 1 and spin 2. What we know and understand about these theories – certainly much more than I have managed to capture – together with the free field theory of higher spin, provide the bases from which all attempts toward interactimg higher spin theories have sprung. This is so for the *Vasiliev theory*, the *light-front approach* and the various *covariant Minkowski approaches* that has been studied.

The plan for the second volume of the present work is to review these main different approaches to interacting higher spin gauge theory. I will refrain from here passing any judgment on their relative merits and shortcomings. It is still early times. However, I would like to end with the following thought.

Comparing spin 1 Yang–Mills theory and spin 2 gravitational theory, they share properties, but they differ in certain respects. They can be understood "geometrically", provided one has a general enough concept of geometry. They can be understood "gauge theoretically", provided on has a general enough concept of gauge theory. They can be understood "deformation theoretically", provided one has a general enough concept of deformation theory. With general enough concepts, all three aspects seem to merge.

It seems to me that one must ask, if these conceptual schemes are enough to really understand the nature of higher spin theory? Or if they must be transcended in one way or another? I, for one, do not know.

https://doi.org/10.1515/9783110451771-007

Bibliography

[1] P. A. M. Dirac. Relativistic wave equations. *Proc. R. Soc. A*, 155:447–459, 1936.
[2] E. Majorana. Teoria relativistica di particelle con momento intrinseco arbitrario. *Nuovo Cimento*, 9:335–344, 1932.
[3] C. Fronsdal. Massless fields with integer spin. *Phys. Rev. D*, 18:3624–3629, 1978.
[4] E. P. Wigner. On unitary representations of the inhomogeneous Lorentz group. *Ann. Math.*, 40:149–204, 1939.
[5] A. Pais. *Inward Bound: Of Matter and Forces in the Physical World*. Oxford University Press, 1986.
[6] J. Fang and C. Fronsdal. Massless fields with half-integer spin. *Phys. Rev. D*, 18:3630–3633, 1978.
[7] S. Weinberg. Photons and gravitons in S-matrix theory: Derivation of charge conservation and equality of gravitational and inertial mass. *Phys. Rev.*, 135:B1049–B1056, 1964.
[8] J. Fang and C. Fronsdal. Deformations of gauge groups: Gravitation. *J. Math. Phys.*, 20(11):2264–2271, 1979.
[9] C. Fronsdal. Singletons and massless, integral-spin fields on de Sitter space. *Phys. Rev. D*, 20:848–856, 1978.
[10] E. Sezgin and P. Sundell. Higher spin N=8 supergravity. *J. High Energy Phys.*, 9811:016, 1998. arXiv:hep-th/9805125.
[11] Morris Kline. *Why the Professor Can't Teach*. St. Martins Press, 1977.
[12] U. Eco. *The Search for the Perfect Language*. Fontana Press, 1997.
[13] P. A. M. Dirac. A new notation for quantum mechanics. *Math. Proc. Camb. Philos. Soc.*, 35:416–418, 1939.
[14] S. Weinberg. Towards the final laws of physics. In J. C. Taylor, editor, *The 1986 Dirac Memorial Lectures*, pages 61–110. Cambridge University Press, 1987.
[15] E. Segré. *From X-rays to Quarks*. Dover Publications, 1980.
[16] T. Y. Cao. *The Conceptual Development of 20th Century Field Theories*. Cambridge University Press, 1997.
[17] P. A. M. Dirac. *The Principles of Quantum Mechanics*, 4 edition. Oxford University Press, 1958.
[18] S. Weinberg. *The Quantum Theory of Fields*, volume 1, Foundations. Cambridge University Press, 1993.
[19] E. Segré. *From Falling Bodies to Radio Waves*. Dover Publications, 1984.
[20] S. Esposito. Searching for an equation: Dirac, Majorana and the others. *Ann. Phys.*, 327:1617–1644, 2012.
[21] S. J. Gould. *Time's Arrow, Time's Cycle. Myth and Metaphor in the Discovery of Geological Time*. Penguin Books, 1988.
[22] S. S. Schweber. Some chapters for a history of quantum field theory: 1938–1952. In B. S. DeWitt and R. Stora, editors, *Les Houches, Session XL, 1983*, pages 37–220. Elsevier Science Publishers B.V., 1984.
[23] E. M. Corson. *Introduction to Tensors, Spinors, and Relativistic Wave-Equations*. Blackie & Son Limited, 1953.
[24] P. A. M. Dirac. *The Development of Quantum Mechanics*. Gordon and Breach Science Publishers, 1971.
[25] L-V de Broglie. Recherches sur la théorie des quanta. *Ann. Phys.*, 10:t III, 1925. Translated 2004 as 'Researches on the quantum theory' by A. F. Kracklauer.
[26] L. I. Schiff. *Quantum Mechanics*, 3:rd ed. International Student Edition. McGraw-Hill Kogakusha, Ltd., 1968.
[27] J. D. Bjorken and S. D. Drell. *Relativistic Quantum Mechanics*. McGraw-Hill Book Company,

https://doi.org/10.1515/9783110451771-008

1964.
[28] S. Weinberg. The search for unity: Notes for a history of quantum field theory. *Daedaluss*, 106:17–35, 1977.
[29] S. Esposito, E. Recami, A. van der Merwe, and R. Battiston, editors, *Ettore Majorana: Unpublished Research Notes on Theoretical Physics*. Springer, 2008.
[30] G. E. Uhlenbeck and S. Goudsmit. Ersetzung der Hypothese vom unmechanischen Zwang durch eine Forderung bezüglich des inneren Verhaltens jedes einzelnen Elektrons. *Naturwissenschaften*, 47:953–954, 1925.
[31] G. E. Uhlenbeck and S. Goudsmit. Spinning electrons and the structure of spectra. *Nature*, 117:264–265, 1926.
[32] L. H. Thomas. The motion of a spinning electron. *Nature*, 107:514, 1926.
[33] E. Merzbacher. *Quantum Mechanics*, 3:rd ed. John Wiley & Sons, Inc., 1998.
[34] M. Born, W. Heisenberg, and P. Jordan. Zur Quantenmechanik. II. *Z. Phys.*, 35:557, 1926.
[35] W. Pauli. Zur Quantenmechanik des magnetischen Elektrons. *Z. Phys.*, 43:601–623, 1927.
[36] C. G. Darwin. The electron as a vector wave. *Proc. R. Soc. A*, 116:227–253, 1927.
[37] P. A. M. Dirac. The quantum theory of the electron. *Proc. R. Soc. A*, 117:610–624, 1928.
[38] S. S. Schweber. *QED and the Men Who Made it: Dyson, Feynman, Schwinger and Tomonaga*. Princeton University Press, 1994.
[39] H. Kragh. The genesis of Dirac's relativistic theory of electrons. *Arch. Hist. Exact Sci.*, 24:31–67, 1981.
[40] P. A. M. Dirac. The physical interpretation of the quantum dynamics. *Proc. R. Soc. A*, 113:621–641, 1927.
[41] P. Jordan. Über eine neue Begründung der Quantenmechanik. *Z. Phys.*, 40:809–838, 1927.
[42] J. von Neumann. *Matematische Grundlagen der Quantenmechanik*, 2 edition. Springer, 1932.
[43] B. L. van der Waerden. Spinoranalyse. *Nachr. Ges. Wiss. Göttingen Math. Phys.*, page 100–109, 1929. English translation by G. Pasa, see arXiv:1703.09761.
[44] O. Laporte and G. E. Uhlenbeck. Application of spinor analysis to the Maxwell and Dirac equations. *Phys. Rev.*, pages 1380–1397, 1931.
[45] C. G. Darwin. The wave equations of the electron. *Proc. R. Soc. A*, 118:654–679, 1928.
[46] E. Cartan. *The Theory of Spinors*. Dover Publications, 1981 (first publ. in French in 1937).
[47] M. Fierz. Über die relativistische Theorie kräftefreier Teilchen mit beliebigem Spin. *Helv. Phys. Acta*, 12:3–37, 1939.
[48] M. R. de Traubenberg, X. Bekaert and M. Valenzuela. An infinite supermultiplet of massive higher-spin fields. *J. High Energy Phys.*, 0905:118, 2009. arXiv:0904.2533.
[49] R. Casalbuoni. Majorana and the infinite component wave equations. *PoS*, EMC2006:004, 2006. arXiv:hep-th/0610252.
[50] D. M. Fradkin. Comments on a paper by Majorana concerning elementary particles. *Am. J. Phys.*, 34:314–318, 1966.
[51] D. Tz. Stoyanov and I. T. Todorov. Majorana representations of the Lorentz group and infinite-component fields. *J. Math. Phys.*, 9:2146–2167, 1968.
[52] Progress in meson theory in Japan (Introduction by S. Tomonaga), Prog. Theor. Phys. Suppl. 1 (1955).
[53] A. Proca. Sur la théorie ondulatoire des électrons positifs et négatifs. *J. Phys. Radium*, 7:347–353, 1936.
[54] W. Pauli. Relativistic field theories of elementary particles. *Rev. Mod. Phys.*, 13:203–232, 1941.
[55] M. H. L. Pryce. On the neutrino theory of light. *Proc. R. Soc.*, 165(921):247–271, 1938.
[56] R. A. Krajcik and M. M. Nieto. Historical development of the Bhabha first-order relativistic wave equations for arbitrary spin. *Am. J. Phys.*, 45, No. 9:818–822, 1977.

[57] G. Petiau. Contribution à la théorie des equations d'ondes corpusculaires. *Acad. R. Belg. Cl. Sci. Mém. Collect.*, 8(2):1–116, 1936.
[58] N. Kemmer. Quantum theory of Einstein-Bose particles and nuclear interaction. *Proc. R. Soc. A*, 166:127–153, 1938.
[59] N. Kemmer. The particle aspect of meson theory. *Proc. R. Soc. A*, 173:91–116, 1939.
[60] R. J. Duffin. On the characteristic matrices of covariant systems. *Phys. Rev.*, 54:1114, 1938.
[61] P. A. M. Dirac. Forms of relativistic dynamics. *Rev. Mod. Phys.*, 21:392–399, 1949.
[62] A. I. Miller. *Early Quantum Electrodynamics*. Cambridge University Press, 1994.
[63] E. P. Wigner. Relativistische Wellengleichungen. *Z. Phys.*, 124:665–684, 1947.
[64] V. Bargmann and E. P. Wigner. Group theoretical discussion of relativistic wave equations. *Proc. Natl. Acad. Sci.*, 34:211–223, 1948.
[65] H. A. Kramers, F. J. Belinfante, and J. K. Lubanski. Über freie Teilchen mit nichtverschwindender Masse und beliebiger Spinquantenzahl. *Physica*, VIII, no 7:597–627, 1941.
[66] F. J. Belinfante. Undor calculus and charge-conjugation. *Physica*, VI, no 9:849–869, 1939.
[67] O. Klein. Eine Verallgemeinerung der Dirachsen relativistichen Wellengleichungen. *Ark. Mat. Astron. Fys.*, 25A, No 15:1–19, 1936.
[68] E. P. Wigner. Unitary representations of the inhomogeneous lorentz group including reflections. In F. Gursey, editor, *Group Theoretical Concepts in Elementary Particle Physics*, pages 37–80. Gordon and Breach, New York, 1964.
[69] W. Rarita and J. Schwinger. On the theory of particles with half-integral spin. *Phys. Rev.*, 60:61, 1941.
[70] G. Velo and D. Zwanziger. Propagation and quantization of Rarita-Schwinger waves in an external electromagnetic potential. *Phys. Rev.*, 186:1337–1341, 1969.
[71] S. J. Gupta. Fierz-Pauli theory of particles of spin 3/2. *Phys. Rev.*, 95:1334–1341, 1954.
[72] P. A. Moldauer and K. M. Case. Properties of half-integer spin Dirac-Fierz-Pauli particles. *Phys. Rev.*, 102:279–285, 1956.
[73] R. Howe. *Harish-Chandra 1923–1983, A Bigraphical Memoir*. National Academy of Sciences, Washington, 2011.
[74] Harish-Chandra. Infinite irreducible representations of the Lorentz group. *Proc. R. Soc. A*, 189:372–401, 1947.
[75] V. Bargmann. Irreducible unitary representations of the Lorentz group. *Ann. Math.*, 48:568–640, 1947.
[76] I. M. Gelfand, R. A. Minlos, and Z. Ya. Shapiro. *Representations of the Rotation and Lorentz Groups and their Applications*. Dover, 2018.
[77] M. Taketani and S. Sakata. On the wave equation of meson. *Prog. Theor. Phys. (1955)*, Suppl. 1:84–97, 1955. Reprint from Proc. Phys. Math. Soc. Japan 22 (1940 757).
[78] H. J. Bhabha. The theory of the elementary particles. *Rep. Prog. Phys.*, 10:253–271, 1944.
[79] H. J. Bhabha. Relativistic wave equations for the elementary particles. *Rev. Mod. Phys.*, 17:200–216, 1945.
[80] H. J. Bhabha. Relativistic wave equations for the proton. *Proc. Indian Acad. Sci. A* 21:241–264, 1945.
[81] C. Itzykson and J-B. Zuber. *Quantum Field Theory*. McGraw-Hill International Book Company, 1980.
[82] R. H. Good. Properties of the Dirac matrices. *Rev. Mod. Phys.*, 27:187–211, 1955.
[83] H. Umezawa. *Quantum Field Theory*. North-Holland Publ. Comp., 1956.
[84] Harish-Chandra. On relativistic wave equations. *Phys. Rev.*, 71:793–805, 1947.
[85] H. J. Bhabha. On the postulational basis of the theory of the elementary particles. *Rev. Mod. Phys.*, 21:200–216, 1949.

[86] A. H. Taub. Spinor equations for the meson and their solution when no field is present. *Phys. Rev.*, 56:799–810, 1949.
[87] G. Gentile. Sulle equazioni d'onda relativistische di Dirac per particelle con momento intrinsico qualsiasi. *Nuovo Cimento*, 17:5–12, 1940.
[88] E. Wald. On first order wave equations for elementary particles without subsidiary conditions. *Proc. R. Soc. A* 191:253–268, 1947.
[89] K. J. Le Couteur. The structure of linear relativistic wave equations. I. *Proc. R. Soc. Lond. A*, 202:284–300, 1950.
[90] K. J. Le Couteur. The structure of linear relativistic wave equations. II. *Proc. R. Soc. Lond. A*, 202:394–407, 1950.
[91] W. A. Hepner. A canonical transformation in the theory of particles of arbitrary spin. *Phys. Rev.*, 84:744–749, 1951.
[92] F. J. Belinfante. A new form of the Baryteron equation and some related questions. *Nature*, 143:201, 1939.
[93] C. Fronsdal. On the theory of higher spin fields. *Nuovo Cim. Suppl.*, 9:416–443, 1958.
[94] S-J. Chang. Lagrange formulation for systems with higher spin. *Phys. Rev.*, 173(5):1308–1315, 1967.
[95] S. Weinberg. Feynman rules for any spin. *Phys. Rev.*, 133:B1318–B1332, 1963.
[96] S. Weinberg. Feynman rules for any spin. II. Massless particles. *Phys. Rev.*, 134:B882–B896, 1964.
[97] D. L. Pursey. General theory of covariant particle equations. *Ann. Phys.*, 32:157–190, 1965.
[98] W. K. Tung. Relativistic wave equations and Lagrangian field theory for arbitrary spin. *Phys. Rev. Lett.*, 16:763–766, 1966.
[99] W. K. Tung. Relativistic wave equations and field theory for arbitrary spin. *Phys. Rev.*, 156:1385–1398, 1967.
[100] L. P. S. Singh and C. R. Hagen. Lagrangian formulation for arbitrary spin. I. The boson case. *Phys. Rev. D*, 9(4):898–909, 1974.
[101] M. Fierz and W. Pauli. On relativistic wave equations for particles of arbitrary spin in an electromagnetic field. *Proc. R. Soc. A*, 173:211–232, 1939.
[102] S. Weinberg. Feynman rules for any spin. III. *Phys. Rev.*, 181:1893–1899, 1969.
[103] S. N. Gupta. Quantization of Einstein's gravitational field: Linear approximation. *Proc. Phys. Soc.*, 65:161–169, 1952.
[104] K. Bleuler. Eine neue Methode zur Behandlung de longitudinalen und skalaren Photonen. *Helv. Phys. Acta*, 23:567–586, 1950.
[105] B. S. DeWitt. Quantum theory of gravity. I. The canonical theory. *Phys. Rev.*, 160:1113–1148, 1967.
[106] B. S. DeWitt. Quantum theory of gravity. II. The manifestly covariant theory. *Phys. Rev.*, 162:1195–1239, 1967.
[107] B. S. DeWitt. Quantum theory of gravity. III. Applications of the covariant theory. *Phys. Rev.*, 162:1239–1262, 1967.
[108] C. Kiefer. *Quantum Gravity*. Oxford University Press, 2012.
[109] C. Rovelli. *Quantum Gravity*. Cambridge University Press, 2004.
[110] C. J. Isham, R. Penrose, and Sciama D. W., editors, *Quantum Gravity: An Oxford Symposium*. Clarendon Press, Oxford, 1975. Proceedings from the 1974 conference.
[111] C. J. Isham, R. Penrose, and Sciama D. W., editors, *Quantum Gravity: A Second Oxford Symposium*. Clarendon Press, Oxford, 1975. Proceedings from the 1980 conference.
[112] S. N. Gupta. Gravitation and electromagnetism. *Phys. Rev.*, 96:1683–1685, 1954.
[113] N. Rosen. General relativity and flat space. I. *Phys. Rev.*, 57:147–150, 1940.
[114] C. N. Yang and R. L. Mills. Conservation of isotopic spin and isotopic gauge invariance. *Phys.*

Rev., 96:191–195, 1954.
[115] R. Utiyama. Invariant theoretical interpretation of interaction. *Phys. Rev.*, 101:1597–1607, 1956.
[116] R. H. Kraichnan. Special-relativistic derivation of generally covariant gravitation theory. *Phys. Rev.*, 98:1118–1122, 1955.
[117] W. E. Thirring. An alternative approach to the theory of gravitation. *Ann. Phys.*, 16:96–117, 1961.
[118] R. P. Feynman. *Feynman Lectures on Gravitation*. Westview Press 2002.
[119] W. Wyss. Zur unizität der Gravitationstheorie. *Helv. Phys. Acta*, 38:469–480, 1965.
[120] S. Deser. Self-interaction and gauge invariance. *Gen. Relativ. Gravit.*, 1:9–18, 1970.
[121] D. G. Boulware and S. Deser. Classical general relativity derived from quantum gravity. *Ann. Phys.*, 89:193–240, 1975.
[122] F. A. Berends, G. J. H. Burgers, and H. van Dam. On spin three self interactions. *Z. Phys. C*, 24:247–254, 1984.
[123] F. A. Berends, G. J. H. Burgers, and H. van Dam. On the theoretical problems in constructing interactions involving higher-spin massless particles. *Nucl. Phys. B*, 260:295–322, 1985.
[124] A. K. H. Bengtsson, I. Bengtsson, and L. Brink. Cubic interaction terms for arbitrary spin. *Nucl. Phys. B*, 227:31–40, 1983.
[125] A. K. H. Bengtsson. Gauge invariance for spin-3 fields. *Phys. Rev. D*, 32:2031–2036, 1985.
[126] A. K. H. Bengtsson and I. Bengtsson. Massless higher-spin fields revisited. *Class. Quantum Gravity*, 3:927–936, 1986.
[127] T. Ortin. *Gravity and Strings*. Cambridge University Press, 2004.
[128] L. O'Raifeartaigh. *The Dawning of Gauge Theory*. Princeton University Press, 1997.
[129] D. W. Sciama. The analogy between charge and spin in general relativity. In *Recent Developments in General Relativity, Festschrift for Infeld*, pages 415–439. Pergamon Press, Oxford, 1962.
[130] D. W. Sciama. The physical structure of general relativity. *Rev. Mod. Phys.*, 36:463–469 and 1103 (Erratum), 1964.
[131] T. W. B. Kibble. Lorentz invariance and the gravitational field. *J. Math. Phys.*, 2:212–221, 1961.
[132] D. Ivanenko and G. Sardanashvily. The gauge treatment of gravity. *Phys. Rep.*, 94(1), 1983.
[133] G. D. Kerlick, F. W. Hehl, P. von der Heyde and J. M. Nester. General relativity with spin and torsion: Foundations and prospects. *Rev. Mod. Phys.*, 48:393–416, 1976.
[134] T. W. B. Kibble and K. S. Stelle. Gauge theories of gravity and supergravity. In H. Ezawa and S. Kamefuchi, editors, *Progress in Quantum Field Theory: in honour of professor H. Umewaza*, pages 57–81. North-Holland, Amsterdam, 1986.
[135] M. Blagojević and F. W. Hehl, editors, *Gauge Theories of Gravitation, A Reader with Commentaries*. Imperial College Press, 2013.
[136] V. I. Ogievetski and I. V. Polubarinov. Interacting fields of spin 1 and symmetry properties. *Ann. Phys.*, 25:358–386, 1963.
[137] S. Coleman and J. Mandula. All possible symmetries of the S matrix. *Phys. Rev.*, 159:1251–1256, 1967.
[138] L. O'Raifeartaigh. Lorentz invariance and internal symmetry. *Phys. Rev.*, 139:B1052–B1062, 1965.
[139] S. Weinberg. *The Quantum Theory of Fields*, volume 2, Modern Applications. Cambridge University Press, 1996.
[140] D. Z. Freedman, P. van Nieuwenhuizen, and S. Ferrara. Progress towards a theory of supergravity. *Phys. Rev. D*, 13:3214–3218, 1976.
[141] D. Z. Freedman and P. van Nieuwenhuizen. Properties of supergravity theory. *Phys. Rev. D*, 14:912–916, 1976.

[142] S. Deser and B. Zumino. Consistent supergravity. *Phys. Lett.*, 62B:335–337, 1976.
[143] C. Fronsdal. Elementary particles in a curved space. *Rev. Mod. Phys.*, 37:221–224, 1968.
[144] C. Fronsdal. Elementary particles in a curved space. II. *Phys. Rev. D*, 10:589–598, 1973.
[145] C. Fronsdal and R. B. Haugen. Elementary particles in a curved space. III. *Phys. Rev. D*, 12:3810–3818, 1975.
[146] C. Fronsdal. Elementary particles in a curved space. IV. Massless particles. *Phys. Rev. D*, 12:3819–3830, 1975.
[147] C. Fronsdal and J. Fang. Elementary particles in a curved space. V. Massive and massless spin-2 fields. *Lett. Math. Phys.*, 2:391–397, 1978.
[148] C. Fronsdal and M. Flato. Elementary particles in a curved space. VI. One massless particle equals two Dirac singletons. *Lett. Math. Phys.*, 2:421–426, 1978.
[149] T. Curtright. Massless field supermultiplets with arbitrary spin. *Phys. Lett. B*, 85:219–224, 1979.
[150] B. de Wit and D. Z. Freedman. Systematics of higher-spin fields. *Phys. Rev. D*, 21:358–367, 1979.
[151] T. L. Curtright. High spin fields. *AIP Conf. Proc.*, 68:985–988, 1980.
[152] T. Damour and S. Deser. Geometry of spin 3 gauge theories. *Ann. Inst. Henri Poincaré*, 47:277–307, 1987.
[153] F. A. Berends, J. W. van Holten, P. van Nieuwenhuizen, and B. de Wit. On field theory for massive and massless spin-5/2 particles. *Nucl. Phys. B*, 154:261–282, 1979.
[154] F. A. Berends, J. W. van Holten, B. de Wit, and P. van Nieuwenhuizen. On spin-5/2 gauge fields. *J. Phys. A, Math. Gen.*, 13:1643–1649, 1980.
[155] C. Aragone and S. Deser. Consistency problems of hypergravity. *Phys. Lett. B*, 86:161–163, 1979.
[156] C. Aragone and S. Deser. Higher spin vierbein gauge fermions and hypergravities. *Nucl. Phys. B*, 170:329–352, 1980.
[157] N. H. Barth and S. M. Christensen. Arbitrary spin field equations on curved manifolds with torsion. *J. Phys. A, Math. Gen.*, 16:543–563, 1983.
[158] S. Ouvry and J. Stern. Gauge fields of any spin and symmetry. *Phys. Lett. B*, 177:335–340, 1986.
[159] A. K. H. Bengtsson. A unified action for higher spin gauge bosons from covariant string theory. *Phys. Lett. B*, 182:321–325, 1986.
[160] P. Ramond. A pedestrian approach to covariant string theory. *Prog. Theor. Phys. Suppl.*, 86:126–134, 1985. Florida preprint UFTP-85-18.
[161] G. B. West. The construction of gauge invariant actions for arbitrary spin and bosonic string field theories. *Nucl. Phys. B*, 277:125, 1986.
[162] D. Pfeffer, P. Ramond, and V. G. J. Rodgers. Gauge invariant field theory of free strings. *Nucl. Phys. B*, 276:131–172, 1986. Florida preprint UFTP-85-19.
[163] W. Siegel. Covariantly second quantized string. *Phys. Lett. B*, 142:276–280, 1984.
[164] W. Siegel and B. Zwiebach. Gauge string fields from the light-cone. *Nucl. Phys. B*, 282:125–141, 1987.
[165] E. Witten. Noncommutative geometry and string field theory. *Nucl. Phys. B*, 268:253–294, 1986.
[166] M. Kato and K. Ogawa. Covariant quantization of string based on BRS invariance. *Nucl. Phys. B*, 212:443–460, 1982.
[167] S. Hwang. Covariant quantization of the string in dimensions $d = 26$ using a Becchi-Rouet-Stora formulation. *Phys. Rev. D*, 25:2614–2620, 1983.
[168] T. Kugo and I. Ojima. Local covariant operator formalism non-Abelian gauge theories and quark confinement problem. *Prog. Theor. Phys. Suppl.*, 66:1–130, 1979.
[169] L. Brink, P. Di Vecchia, and P. Howe. A locally supersymmetric and reparametrization invariant

action for the spinning string. *Phys. Lett.*, 65B:471–474, 1976.
[170] S. Deser and B. Zumino. A complete action for the spinning string. *Phys. Lett.*, 65B:369–373, 1976.
[171] A. M. Polyakov. Quantum geometry of bosonic string. *Phys. Lett.*, 103B:207–213, 1981.
[172] W. Siegel. Covariantly second quantized string II. *Phys. Lett. B*, 149:157–161, 1984. Correct typeset version: Phys. Lett. B 151 (1984) 391–395.
[173] W. Siegel. Covariantly second quantized string III. *Phys. Lett. B*, 149:162–166, 1984. Correct typeset version: Phys. Lett. B 151 (1984) 396–400.
[174] W. Siegel and B. Zwiebach. Gauge string fields. *Nucl. Phys. B*, 263:105–128, 1986.
[175] D. Friedan. String field theory. *Nucl. Phys. B*, 271:540–560, 1986.
[176] T. Banks and M. E. Peskin. Gauge invariance of string fields. *Nucl. Phys. B*, 264:513–547, 1986.
[177] T. Banks, M. E. Peskin, C. R. Preitschopf, D. Friedan, and E. Martinec. All free string theories are theories of forms. *Nucl. Phys. B*, 274:71–92, 1986.
[178] A. Neveu and P. C. West. Gauge covariant local formulation of bosonic strings. *Nucl. Phys. B*, 268:125–150, 1986.
[179] A. Neveu, H. Nicolai, and P. C. West. New symmetries and ghost structure of covariant string theories. *Phys. Lett. B*, 167:307–314, 1986.
[180] K. Itoh, T. Kugo, H. Kunimoto, and H. Ooguri. Gauge invariant local action of string field from BRS formalism. *Prog. Theor. Phys.*, 75:162–174, 1986.
[181] C. B. Thorn. String field theory. *Phys. Rep.*, 175:1–101, 1989.
[182] W. Siegel. Classical superstring mechanics. *Nucl. Phys. B*, 263:93–104, 1986.
[183] V. Bargmann and I. T. Todorov. Spaces of analytic functions on a complex cone as carriers for the symmetric tensor representation of SO(n). *J. Math. Phys.*, 18:1141–1148, 1977.
[184] Y. Meurice. From points to gauge fields. *Phys. Lett. B*, 186:189–194, 1987.
[185] F. Hussain, G. Thompson, and P. D. Jarvis. Massive and massless gauge fields of any spin and symmetry. *Phys. Lett. B*, 216:139–144, 1989.
[186] M. Bellon and S. Ouvry. D=4 Supersymmetry for gauge fields of any spin. *Phys. Lett. B*, 187:93–96, 1987.
[187] M. Henneaux and C. Teitelboim. First and second quantized point particles of any spin. In C. Teitelboim and J. Zanelli, editors, *Quantum Mechanics of Fundamental Systems 2*, Series of the Centro de Estudios Científicos de Santiago. Plenum Press, N York, 1989.
[188] C. S. Aulakh, I. G. Koh, and S. Ouvry. Higher spin fields with mixed symmetry. *Phys. Lett. B*, 173:284–288, 1986.
[189] J. Mourad A. Campoleoni, D. Francia and A. Sagnotti. Unconstrained higher spins of mixed symmetry. I. Bose fields. *Nucl. Phys. B*, 815:289–367, 2009. arXiv:0810.4350.
[190] J. Mourad A. Campoleoni, D. Francia and A. Sagnotti. Unconstrained higher spins of mixed symmetry. II. Fermi fields. *Nucl. Phys. B*, 828:405–514, 2010. arXiv:0904.4447.
[191] A. Campoleoni. Metric-like Lagrangian formulations for higher-spin fields of mixed symmetry. *Riv. Nuovo Cimento*, 2010: 3-4:123–253, 2010. arXiv:0910.3155.
[192] T. Curtright. Generalized gauge fields. *Phys. Lett. B*, 165:304–308, 1980.
[193] J. M. F. Labastida and T. R. Morris. Massless mixed-symmetry bosonic free fields. *Phys. Lett. B*, 180:101–106, 1986.
[194] J. M. F. Labastida. Massless bosonic free fields. *Phys. Rev. Lett.*, 58:531–534, 1987.
[195] J. M. F. Labastida. Massless fermionic free fields. *Phys. Lett. B*, 186:365–369, 1987.
[196] J. M. F. Labastida and M. Pernici. BRST quantization in the Siegel gauge. *Phys. Lett. B*, 194:511–517, 1987.
[197] J. M. F. Labastida. Massless particles in arbitrary representations of the Lorentz group. *Nucl. Phys. B*, 322:185–209, 1989.

[198] A. Pashnev and M. M. Tsulaia. On the BRST approach to the description of a Regge trajectory. *Unpublished*, 1996. arXiv:hep-th/9611022.

[199] A. Pashnev and M. M. Tsulaia. Dimensional reduction and,BRST approach to the description of a Regge trajectory. *Mod. Phys. Lett. A*, 12:861–870, 1997. arXiv:hep-th/9703010.

[200] A. Pashnev and M. Tsulaia. Description of the higher massless irreducible integer spins in the BRST approach. *Mod. Phys. Lett. A*, 13:1853–1864, 1998. arXiv:hep-th/9803207.

[201] I. L. Buchbinder V.A. Krykhtin. Gauge invariant Lagrangian construction for massive bosonic higher spin fields in Dimensions. *Nucl. Phys. B*, 727:537–563, 2005. arXiv:hep-th/0505092.

[202] A. Fotopoulos and M. Tsulaia. Gauge invariant Lagrangians for free and interacting higher spin fields. A review of the BRST formulation. *Int. J. Mod. Phys. A*, 24:1–60, 2009. arXiv:0805.1346.

[203] L. Brink, O. Lindgren, and B. E. W. Nilsson. The ultra-violet finiteness of the N=4 Yang-Mills theory. *Phys. Lett.*, 123B:323–328, 1983.

[204] S. Mandelstam. Light-cone superspace and the ultraviolet finiteness of the N=4 model. *Nucl. Phys. B*, 213:149–168, 1983.

[205] A. K. H. Bengtsson, I. Bengtsson, and N. Linden. Interacting higher-spin gauge fields on the light front. *Class. Quantum Gravity*, 4:1333–1345, 1987.

[206] E. S. Fradkin and R. R. Metsaev. A cubic interaction of totally symmetric massless representations of the Lorentz group in arbitrary dimensions. *Class. Quantum Gravity*, 8:L89–L94, 1991.

[207] R. R. Metsaev. Generating function for cubic interaction vertices of higher spin fields in any dimension. *Mod. Phys. Lett. A*, 8(25):2413–2426, 1993.

[208] R. R. Metsaev. Poincaré invariant dynamics of massless higher spins: Fourth order analysis on mass shell. *Mod. Phys. Lett. A*, 6:359–367, 1991.

[209] R. R. Metsaev. S-matrix approach to massless higher spins theory: II. the case of internal symmetry. *Mod. Phys. Lett. A*, 6:2411–2421, 1991.

[210] D. Ponomarev and E. Skvortsov. Light-front higher-spin theories in flat space. *J. Phys. A, Math. Theor.*, 50:095401, 09 2016. arXiv:1609.04655.

[211] D. Ponomarev. Off-shell spinor-helicity amplitudes from light-cone deformation procedure. J. High Energy Phys. 12(2016)117, 2016, arXiv:1611.00361.

[212] G. J. H. Burgers. *On the Construction Interactions of Field Theories for Higher Spin Massless Particles*. PhD thesis, Rijksuniversteit, Leiden, 1985.

[213] C. Fronsdal. Some open problems with higher spins. In P. van Nieuwenhuizen and D. Z. Freedman, editors, *Supergravity*, pages 245–249. North-Holland Publishing Company, 1979.

[214] X. Bekaert, N. Boulanger, and S. Cnockaert. Spin three gauge theory revisited. *J. High Energy Phys.*, 0601:052, 2006. arXiv:hep-th/0508048.

[215] X. Bekaert, N. Boulanger, and S. Leclercq. Strong obstruction of the Berends-Burgers-van Dam spin-3 vertex. *J. Phys. A, Math. Theor.*, 43(18):185401, 2010. arXiv:1002.0289.

[216] I. G. Koh and S. Ouvry. Interacting gauge fields of any spin and symmetry. *Phys. Lett. B*, 179:115–118, 1986.

[217] S. Ouvry, L. Cappiello, M. Knecht and J. Stern. BRST construction of interacting gauge theories of higher spin gauge fields. *Ann. Phys.*, 193:10–39, 1989.

[218] M. Knecht F. Fougére and J. Stern. Algebraic construction of higher spin interaction vertices. Preprint IPNO/Th 99-44,LAPP-TH-338-91, 1991.

[219] D. J. Gross and A. Jevicki. Operator formulation of interacting string field theory (I). *Nucl. Phys. B*, 283:1–49, 1987.

[220] D. J. Gross and A. Jevicki. Operator formulation of interacting string field theory (II). *Nucl. Phys. B*, 287:225–250, 1987.

[221] D. J. Gross and A. Jevicki. Operator formulation of interacting string field theory (III). *Nucl.*

Phys. B, 293:29–82, 1987.
[222] A. K. H. Bengtsson. BRST approach to interacting higher-spin gauge fields. *Class. Quantum Gravity*, 5:437–451, 1988.
[223] S. W. MacDowell and F. Mansouri. Unified geometric theory of gravity and supergravity. *Phys. Rev. Lett.*, 38:739–742, 1977.
[224] K. S. Stelle and P. C. West. Spontaneously broken de Sitter symmetry and the gravitational holonomy group. *Phys. Rev. D*, 21:1466–1488, 1980.
[225] E. S. Fradkin and M. A. Vasiliev. On the gravitational interaction of massless higher spin fields. *Phys. Lett. B*, 189:89–95, 1987.
[226] E. S. Fradkin and M. A. Vasiliev. Cubic interaction in extended theories of massless higher spin fields. *Nucl. Phys. B*, 291:141–171, 1987.
[227] C. Iazeolla, X. Bekaert, S. Cnockaert and M. A. Vasiliev. Nonlinear higher spin theories in various dimensions. In G. Bonelli, R. Argurio, G. Barnich and M. Grigoriev, editors, *First Solvay Workshop on Higher-Spin Gauge Theories*, pages 132–197. Université Libre de Bruxelles, International Solvay Institutes for Physics and Chemistry, 2004. arXiv:hep-th/0503128v2.
[228] V. E. Didenko and E. D. Skvortsov. Elements of Vasiliev theory. 2014. arXiv:1401.2975.
[229] M. Henneaux and C. Teitelboim. *Quantization of Gauge Systems*. Princeton University Press, 1992.
[230] P. A. M. Dirac. Generalized Hamiltonian dynamics. *Can. J. Math.*, 2:129–148, 1950.
[231] P. A. M. Dirac. Generalized Hamiltonian dynamics. *Proc. R. Soc. A*, 246:326–332, 1958.
[232] P. A. M. Dirac. *Lectures on Quantum Mechanics*. Belfer Graduate School of Science, Yeshiva University, New York, 1964.
[233] D. Salisbury and K. Sundermeyer. Léon Rosenfeld's general theory of constrained Hamiltonian dynamics. *Eur. Phys. J. H*, 42(1):23–61, 2017.
[234] A. J. Hanson, T. Regge, and C. Teitelboim. *Constrained Hamiltonian Systems*. Accademia Nazionale dei Lincei, 1976.
[235] G. Marmo, N. Mukunda, and J. Samuel. Dynamics and symmetry for constrained dynamics: a geometrical analysis. *Riv. Nuovo Cimento*, 6. N. 2:1–62, 1983.
[236] N. Mukunda and E. C. G. Sudarshan. *Classical Dynamics: A Modern Perspective*. World Scientific, 2015 (first published 1974).
[237] W. M. Seiler and R. W. Tucker. Involution and constrained dynamics I: The Dirac approach. *J. Phys. A, Math. Gen.*, 28:4431–4451, 1995.
[238] R. P. Feynman. Space-time approach to non-relativistic quantum mechanics. *Rev. Mod. Phys.*, 20:367–387, 1948.
[239] P. A. M. Dirac. The Lagrangian in quantum mechanics. *Phys. Z. Sowjetunion*, Band 3, Heft 1:64–72, 1933.
[240] M. Srednicki. *Quantum Field Theory*. Cambridge University Press, 2007.
[241] A. Semikhatov G. Barnich, M. Grigoriev and I. Tipunin. Parent field theory and unfolding in BRST first-quantized terms. *Commun. Math. Phys.*, 260:147–181, 2005. arXiv:hep-th/0406192v3.
[242] G. Barnich and M. Grigoriev. Hamiltonian BRSTand Batalin-Vilkovisky formalism for second quatization of gauge theories. *Commun. Math. Phys.*, 254:581–601, 2005. arXiv:hep-th/0310083.
[243] S. Weinberg. *Gravitation and Cosmology*. John Wiley & Sons, 1972.
[244] R. F. Streater and A. S. Wightman. *PCT, Spin and Statistics and All That*. Benjamin/Cummings, 1964.
[245] W-K. Tung. *Group Theory in Physics*. World Scientific, 1985.
[246] A. O. Barut and R. Rączka. *Theory of Group Representations and Applications*. World Scientific, 1986.

[247] E. P. Wigner. Invariant quantum mechanical equations of motion. *International Atomic Energy Agency (Vienna)*, pages 59–82, 1963.

[248] G. J. Iverson and G. Mack. Quantum fields and interactions of massless particles – the continuous spin case. *Ann. Phys.*, 64:211–253, 1971.

[249] H. Weyl. *The Theory of Groups and Quantum Mechanics*. Dover Publications, 1950. Reprint of the English 1931 edition.

[250] R. Penrose and W. Rindler. *Spinors and Space-Time, Volume 1*. Cambridge Univ. Press, 1984.

[251] Y. Choquet-Bruhat, C. DeWitt-Morette, and M. Dillard-Bleick. *Analysis, Manifolds and Physics*. North-Holland Publishing Company, 1982.

[252] C. Isham. *Lectures on Groups and Vector Spaces*. World Scientific, 1989.

[253] G. F. Simmons. *Introduction to Topology and Modern Analysis*. McGraw-Hill Book Company, 1963.

[254] B. Hartley and T. O. Hawkes. *Rings, Modules and Linear Algebra*. CHapman and Hall, 1970.

[255] T. Gowers (ed.), J. Barrow-Green, and I. Leader (ass. eds.), *The Princeton Companion to Mathematics*. Princeton Univ. Press, 2008.

[256] S. Mac Lane. *Mathematics: Form and Function*. Springer-Verlag, 1986.

[257] R. Bott and L. W. Tu. *Differential Forms in Algebraic Topology*. Springer-Verlag, 1982.

[258] R. W. Sharpe. *Differential Geometry, Cartan's Generalization of Klein's Erlangen Program*. Springer, 1996.

[259] N. J. Hicks. *Notes on Differential Geometry*. D. Van Nostrand Company, 1965.

[260] C. Isham. *Modern Differential Geometry for Physicists*. World Scientific, 1989.

[261] M. Nakahara. *Geometry, Topology and Physics*. Institute of Physics Publishing, 2002.

[262] P. J. Olver. *Applications of Lie Groups to Differential Equations*. Springer, 1993.

[263] J. Fuchs and C. Schweigert. *Symmetries, Lie Algebras and Representations*. Cambridge Monographs on Mathematical Physics. Cambridge University Press, 1997.

[264] H. Georgi. *Lie Algebras in Particle Physics*. Benjamin Cummings Publ. Co., 1982.

[265] A. O. Barut. Unified algebraic construction of representations of compact and non-compact Lie algebras and Lie groups. *Lect. Theor. Phys.*, 9A:125–172, 1967.

[266] D. J. Saunders. *The Geometry of Jet Bundles*. Cambridge University Press, 1989.

[267] V. I. Ogievetski and I. V. Polubarinov. On the meaining of gauge invariance. *Nouvo Cimento*, 23:173–180, 1962.

[268] M. F. Atiyah. *Geometry of Yang-Mills fields*. Acad. Naz. dei Lincei, Lezioni Fermiane, 1979.

[269] T. Eguchi, P. B. Gilkey, and A. J. Hanson. Gravitation, gauge theories and differential geometry. *Phys. Rep.*, 66(6):213–393, 1980.

[270] E. Schrödinger. *Space-Time Structure*. Cambridge University Press, 1950.

[271] R. M. Wald. *General Relativity*. University of Chicago Press, 1984.

[272] M. Ferraris, M. Franciaviglia, and C. Reina. Variational formulation of general relativity from 1915 to 1925 "Palatini's method" discovered by Einstein in 1925. *Gen. Relativ. Gravit.*, 14:243–253, 1982.

[273] D. Francia and A. Sagnotti. Free geometric equations for higher spins. *Phys. Lett. B*, 543:303–310, 2002. arXiv:hep-th/0207002.

[274] D. Francia and A. Sagnotti. On the geometry of higher-spin gauge fields. *Class. Quantum Gravity*, 20:473–486, 2003. arXiv:hep-th/0212185.

[275] A. Sagnotti and M. Tsulaia. On higher spins and the tensionless limit of string theory. *Nucl. Phys. B*, 682:83–116, 2004. arXiv:hep-th/0311257.

[276] J. Mourad, D. Francia and A. Sagnotti. Current exchanges and unconstrained higher spins. *Nucl. Phys. B*, 773:203–237, 2007. arXiv:hep-th/0701163.

[277] M. Dubois-Violette. Generalized differential spaces with $d^n = 0$ and the q-differential calculus. *Czechoslov. J. Phys.*, 46:1227–1233, 1979.

[278] M. Henneaux. n-complexes and higher spin gauge fields. *Int. J. Geom. Methods Mod. Phys.*, 5:1255–1263, 2008. arXiv:0808.1975.
[279] M. Dubois-Violette. Lectures on differentials, generalized differentials and some examples related to theoretical physics. *LPT-ORSAY 00/31*, 1979. arXiv:mat.qa/0005256.
[280] G. Bonelli. On the tensionless limit of bosonic strings, infinite symmetries and higher spins. *Nucl. Phys. B*, 669:159–172, 2003. arXiv:hep-th/0305155.
[281] C. Burdík, A. Pashnev, and M. M. Tsulaia. The Lagrangian description of representations of the Poincaré group. *Nucl. Phys. B, Proc. Suppl.*, 102&103:285–292, 2001.
[282] A. V. Galajinsky, I. L. Buchbinder and V. A. Krykhtin. Quartet unconstrained formulation for massless higher spin fields. *Nucl. Phys. B*, 779:155–177, 2007. arXiv:hep-th/0702161.
[283] I. L. Buchbinder, A. Pashnev, and M. Tsulaia. Lagrangian formulation of the massless higher integer spin fields in the AdS background. *Phys. Lett. B*, 523:338–346, 2001. arXiv:hep-th/0109067.
[284] X. Bekaert, I. L. Buchbinder, A. Pashnev, and M. Tsulaia. On higher spin theory: strings, BRST, dimensional reductions. *Class. Quantum Gravity*, 21:S1457–S1463, 2004.
[285] I. L. Buchbinder, V. A. Krykhtin, and A. Pashnev. BRST approach to Lagrangian construction for fermionic massless higher spin fields. *Nucl. Phys. B*, 711:367–391, 2005.
[286] M. B. Green, J. H. Schwarz, and E. Witten. *Superstring Theory, Volume 1.* Cambridge University Press, 1987.
[287] A. K. H. Bengtsson. Mechanical models for higher spin gauge fields. *Fortschr. Phys.*, 57:499–504, 2009. arXiv:0902.3915.
[288] A. K. H. Bengtsson. BRST theory for continuous spin. *J. High Energy Phys.*, October 2013:108(10), 2013. arXiv:1303.3799.
[289] D. Francia and A. Sagnotti. Minimal local Lagrangians for higher-spin geometry. *Phys. Lett. B*, 624:93–104, 2005. arXiv:hep-th/0507144.
[290] D. Francia and A. Sagnotti. Higher-spin geometry and string theory. *J. Phys. Conf. Ser.*, 33:57–72, 2006. arXiv:hep-th/0601199.
[291] A. Campoleoni and D. Francia. Maxwell-like lagranfians for higher spins. *J. High Energy Phys.*, 1303:168, 2013. arXiv:1206.5877.
[292] E. D. Skvortsov and M. A. Vasiliev. Transverse invariant higher-spin fields. *Phys. Lett. B*, 664:301–306, 2008. arXiv:hep-th/0701278.
[293] D. Francia, S. L. Lyakhovich, and A. A. Sharapov. On the gauge symmetries of Maxwell-like higher-spin Lagrangians. *Nucl. Phys. B*, 881:248–268, 2014. arXiv:1310.8589.
[294] M. A. Vasiliev. Gauge form of description of massless fields with arbitrary spin. *Sov. J. Nucl. Phys.*, 32(3):439–442, 1980.
[295] M. A. Vasiliev. Free massless fields of arbitrary spin in the de Sitter space and initial data for a higher spin superalgebra. *Fortschr. Phys.*, 35:741–770, 1987.
[296] W. Fulton. *Young Tableaux: With Applications to Representation Theory and Geometry.* Cambridge University Press, 1996.
[297] X. Bekaert and N. Boulanger. The unitary representations of the Poincaré group in any spacetime dimension. In *2nd Modave Summer School in Theoretical Physics Modave, Belgium, August 6–12, 2006*, 2006. arXiv:hep-th/0611263.
[298] A. K. H. Bengtsson, I. Bengtsson, and L. Brink. Cubic interaction terms for arbitrarily extended supermultiplets. *Nucl. Phys. B*, 227:41–49, 1983.
[299] S. Fubini and G. Furlan. Renormalization effects for partially conserved currents. *Physics*, 1:229–247, 1965.
[300] H. Leutwyler. Current algebra and lightlike charges. *Springer Tracts Mod. Phys.*, 50:29–41, 1969.
[301] A. Harindranath. An introduction to light front dynamics for pedestrians. In *International*

School on Light-Front Quantization and Non-Perturbative QCD, Ames, Iowa, May 6–June 2, 1996, 1996. arXiv:hep-ph/9612244.

[302] A. K. H. Bengtsson. N=1 supergravity in the light-cone gauge. *Nucl. Phys. B*, 228:190–204, 1983.

[303] J. H. Schwarz. Superstring theory. *Phys. Rep.*, 89:223–322, 1982.

[304] D. Rickles. *A Brief History of String Theory*. Springer, 2014.

[305] R. A. Neville and F. Rohrlich. Quantum field theory off null planes. *Il Nouvo Cimento A*, 1:625–644, 1971.

[306] S. Ananth. Spinor helicity structures in higher spin theories. *J. High Energy Phys.*, 1211:089, 2012. arXiv:1209.4960.

[307] P. J. Steinhardt. Problems of quantization in the infinite momentum frame. *Ann. Phys.*, 128:425–447, 1995.

[308] R. Fulp, T. Lada, and J. Stasheff. Sh-Lie algebras induced by gauge transformations. *Commun. Math. Phys.*, 231:25–43, 2002. math.QA/0012106.

[309] A. K. H. Bengtsson and Noah Linden. Interacting covariant open bosonic strings from the light-cone Ji-. *Phys. Lett. B*, 187:289–294, 1987.

[310] N. Linden. Lorentz generators in light-cone gauge superstring field theory. *Nucl. Phys. B*, 286:429–454, 1986.

[311] K. Bardakci and M. B. Halpern. Theories at infinite momentum. *Phys. Rev.*, 176:1686–1699, 1968.

[312] S-J. Chang and S-K. Ma. Feynman rules and quantum electrodynamics at infinite momentum. *Phys. Rev.*, 180:1506–1513, 1969.

[313] J. B. Kogut and D. E. Soper. Quantum electrodynamics in the infinite-momentum frame. *Phys. Rev. D*, 1:2901–2914, 1970.

Index

Also of Interest

Statistical Physics
Michael V. Sadovskii, 2019
ISBN 978-3-11-064510-1, e-ISBN (PDF) 978-3-11-064848-5

Quantum Field Theory
Michael V. Sadovskii, 2019
ISBN 978-3-11-064515-6, e-ISBN (PDF) 978-3-11-064852-2

Quantum Systems, Channels, Information
A Mathematical Introduction
Alexander S. Holevo, 2019
ISBN 978-3-11-064224-7, e-ISBN 978-3-11-064249-0

Coherent Quantum Physics
A Reinterpretation of the Tradition
Arnold Neumaier, 2019
ISBN 978-3-11-066729-5, e-ISBN: 978-3-11-066738-7

Metrology and Theory of Measurement
Valery A. Slaev. Anna G. Chunovkina, Leonid A. Mironovsky 2020
ISBN 978-3-11-065094-5, e-ISBN: 978-3-11-065250-5

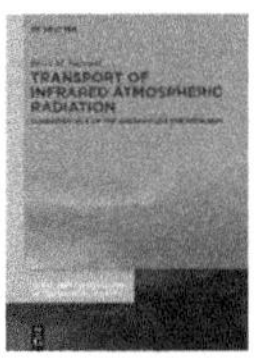

Transport of Infrared Atmospheric Radiation
Boris M. Smirnov, 2020
ISBN 978-3-11-062765-7, e-ISBN: 978-3-11-062875-3

www.ingramcontent.com/pod-product-compliance
Ingram Content Group UK Ltd.
Pitfield, Milton Keynes, MK11 3LW, UK
UKHW050943290926
15470UKWH00005B/1612

9 783110 450538